ENVIRONMENTAL SCIENCE, ENGINEERING AND TECHNOLOGY

IN THE WILDS OF SOUTH AMERICA

ENVIRONMENTAL SCIENCE, ENGINEERING AND TECHNOLOGY

ENVIRONMENTAL SCIENCE, ENGINEERING AND TECHNOLOGY

IN THE WILDS OF SOUTH AMERICA

LEO E. MILLER

snova
New York

Library of Congress Cataloging-in-Publication Data

ISBN: 978-1-53616-074-1

Published by Nova Science Publishers, Inc. † New York

CONTENTS

PREFACE[*]

I have frequently wondered how many of the large number of people who visit natural-history museums have any conception of the appearance and actions, in their wilderness homes, of the creatures they see, and of the experiences of the field-naturalists who visit the little-known places of the earth in search of them.

My experience as a field-naturalist consists of nearly six years of almost continuous exploration in South America, and embraces practically all of the republics of that continent.

The purpose of this narrative is to follow the course of these explorations into the tropical jungles of the Amazon, Paraguay, Orinoco, and others of South America's master rivers, and to the frigid heights of the snow-crowned Andes.

In these jungles one hears the hoarse cough of the jaguar and the scream of long-tailed, multicolored macaws as they fly two by two overhead; the extraordinary chorus of frogs and insects may lull the weary senses to sleep at nightfall, but the dismal roar of howling monkeys is sure to awaken one at dawn. To start at the sudden, long-drawn hiss of a boa or the lightning-like thrust of the terrible bushmaster, the largest of poisonous snakes, and a creature so deadly that a man may die within ten minutes after the fatal stroke, and to shudder as the wild, insane cackle of the wood-rails shatters the brooding silence of the forest, are merely incidents of the explorer's every-day life; and so, too, are visits to deep lagoons teeming with crocodiles, cannibal fishes, and myriads of water-fowl; lengthy sojourns in gloomy forests where orchids droop from moss-draped branches, brilliant butterflies shimmer in the subdued light, and curious animals live in the eternal shadows; and ascents of the stupendous mountain ranges where condors soar majestically above the ruins of Incan greatness. In short, the expeditions recorded in the book lead through remote wilderness where savage peoples and little-known animals spend their lives in stealth and vigilance, all oblivious of the existence of an outer world.

* This is an edited, augmented and reformatted version of "In the Wilds of South America" by Leo E. Miller, originally published by Charles Scribner's Sons, dated 1919. The views, opinions, and nomenclature expressed in this book are those of the authors and do not reflect the views of Nova Science Publishers, Inc.

The explorations here recounted were undertaken by me as a member or leader of the following expeditions, all of which were undertaken under the auspices of the American Museum of Natural History, New York City: Colombia—March, 1911, to September, 1912; Colombia—November, 1914, to April, 1915; Venezuela—November, 1912, to June, 1913; British Guiana—July to October, 1913; Roosevelt-Rondon South American Expedition, mostly in Brazil but covering a part of Paraguay, with stops in Uruguay and Argentina—October, 1913, to June, 1914; Bolivia—May, 1915, to January, 1916, touching at Panama, Ecuador, and Peru en route; Argentina—January to September, 1916. The purpose of these expeditions was to collect birds and mammals; also to study the fauna in general and to make all possible observations regarding the flora, topography, climate, and human inhabitants of the regions visited. The personnel of each expedition is given in the proper place in the text.

I wish to acknowledge my indebtedness to Doctor Frank M. Chapman and to Colonel Theodore Roosevelt for suggesting and encouraging the production of this book, also to Mrs. Alice K. Fraser for the great amount of time and work devoted to typewriting the manuscript.

Leo E. Miller.

PART I. COLOMBIA

Chapter 1

BUENAVENTURA TO CALI, AND THE CAUCA VALLEY

The voyage from Panama to Buenaventura, the more northern of Colombia's two Pacific seaports, requires but two days' time. Owing to numerous reefs and rocks that render navigation perilous along the coast of northwestern South America, it is necessary for ships to sail far out into the Pacific. Banks of low-hanging fog, encountered at frequent intervals, add further to the skipper's difficulties.

The captain of the *Quito* followed a simple plan for finding port. It was his custom to steam in a southerly direction about forty-eight hours, and then head toward the coast. Once in sight of land, there was little difficulty in getting his bearings, although it frequently meant steaming back a distance of ten or fifteen miles.

At noon on the second day out we entered what might be called the belt of perpetual rain, and for three hours water fell in such torrents that it seemed a solid wall. When the deluge had ceased and the last wisps of blue-gray vapor melted into oblivion, the shore-line, dim and distant, could be discerned. The faint outline of a rugged coast became gradually sharper; jagged rocks, frowning precipices, and dark, gloomy forests slowly unfolded themselves to the vision. The magnitude of it all was most impressive.

Then followed a ten-mile sail through the placid water of Buenaventura Bay. Numberless brown pelicans fished in the shallows while others, in long files, alternately sailed and flapped through the air on their way to some isolated nook among the mangroves. The dark, hazy shore-line at the head of the bay gradually dissolved itself into lines of graceful cocoanut-palms and low, thatched huts flanked by a seemingly endless mantle of green. Huge dugout canoes made from logs of great size swarmed out from the water's edge, their dusky paddlers vying with one another in their efforts to be the first to reach the steamer; then the men quarrelled violently among themselves, and also shouted to the persons on the deck, soliciting luggage to take ashore. Before long, trunks were being lowered into some of these wallowing craft while passengers embarked in others, and the paddle of a mile to shore began.

Unfortunately the tide was ebbing, leaving extensive mud-flats exposed along the water-front. As there was no pier it was necessary for the canoemen to carry on their backs the human freight as well as trunks and other luggage through a wide belt of mud and sand.

Our party consisted of Doctor Frank M. Chapman, curator of birds, of the American Museum, Louis Agassiz Fuertes, and myself. At Buenaventura we were joined by William Richardson, who had spent many years as a field-naturalist in Central and South American countries. We were starting on a zoological expedition—a quest for birds and mammals, and also to study the country, life-zones, problems of distribution and many other things inseparable from a biological survey such as we proposed to make. The original plans of the expedition called for a rather short stay; but for me, at least, the experience was destined to cover a period of eighteen months and take me to some of the most remote and wildest portions of the country.

Viewed from the water, Buenaventura appears most unattractive. The row of squat, makeshift huts, built on tall poles, extends far beyond the line of high water; as the tide rises the water swishes and gurgles underneath the houses and the occupants travel about in canoes. Farther from the shore the ground is high and the town is more interesting, though not inviting.

Figure 1. Buenaventura.

The place bears an unenviable reputation. On account of the superabundant rainfall and hot climate, fevers and other life-sapping diseases are rife and few foreigners can withstand the ordeal of a lengthy residence there. This notoriety had reached our ears long before we embarked on the journey; it was, therefore, with a feeling of relief that we learned of the departure of a train for the interior early the next morning. For a distance of twenty-five miles, after leaving Buenaventura, nothing was visible but swamps filled with mangrove thickets. Then the foot-hills of the Andes appeared, the steady climb began and the character of the vegetation changed. Instead of the low, matted growth of shrubbery, there grew trees and palms of goodly size. Stops for wood and water were made frequently; the train usually halted near a collection of native huts, the occupants of which earned their living chopping wood for the railroad company. Each habitation was surrounded by a small clearing in which broad-leaved banana, plantain, and papaya trees grew in wonderful luxuriance. Jungles of tall bamboo bordered the plantations and grew beside the track. Plantains and bamboo seem to be the staples of the people. The former they eat, and of the latter their houses are built. The flimsy structures were ramshackle affairs with ragged, thatched roofs, and fitted well into their surroundings. Frequently we had a fleeting view of the almost nude occupants of the huts, lolling about in the darkened interior.

The first town of any importance was Cisneros. We were delayed an hour at this station because the train from the opposite direction had met with an accident that blocked the track, and, as the people were celebrating one of their numerous *fiestas*, it was impossible to get men to clear away the wreckage without great loss of time.

The railroad continued up the slope, following the winding canyon of the Dagua. It has been said that the cost of constructing it was a million dollars a mile. Tunnels, deep cuts through spurs and ridges, trestles and high bridges followed one another in quick succession. The perpendicular sides of the excavations were covered with long moss and drooping ferns that waved plume-like overhead. Mountain torrents poured their crystal streams from openings in overgrown crevices and were dashed to spray on the rocks below. Hundreds of feet lower down, the Dagua raged within the narrow confines of a rock-bound gorge. Thick jungles, dark and impenetrable, cover the slopes. We were conscious of the perfume of flowers concealed amid the forbidding masses of deepest green. An iguana, fully four feet long and of a bright green color dashed across the track a few feet ahead of the puffing engine; a moment later and the beautiful creature would have been crushed to death. Overhead, flocks of parrots screamed defiance at the lowly, wheezing thing that laboriously made its way farther and farther into their time-hallowed abode; and toucans, clattering their long bills and yelping, performed queer acrobatics in a lofty tree-top. A violent lunge recalled us to earth; the train had stopped for more fuel so the passengers got out and amused themselves touching the sensitive-plants that grew abundantly along the road-bed.

Not long afterward we emerged suddenly into a peculiar region. There was an abrupt end to the gloomy forest, and in its place grew straggling clumps of giant cacti. The dividing-line is as sharp as if cut with a knife. The fauna also is different; instead of brilliantly hued tanagers, trogons and toucans, there are wrens, finches, and other birds of sombre color. This desert-like belt continued for a distance of some miles, and then forest again appeared, on the top only of the ridges, at first, but gradually extending downward until the slopes were entirely covered.

Caldas, the terminus of the railroad, was reached at noon and, after a good deal of bargaining, the proprietor of the Hotel del Valle provided us with a room containing four bare, wooden beds; but fortunately our blanket-bags had come with us, so we rather rejoiced that no bedding was provided by the innkeeper. The buildings comprising the town are scattered here and there in small groups, making it difficult to get a comprehensive idea of their number. The first impression suggests that there is a population of a few hundred only, when it is really several thousand. At this time (April, 1911) Caldas was an attractive spot, as its elevation is two thousand feet, and the country immediately surrounding it is open; but in recent years sufferers from malaria, yellow fever, and other diseases have gone there from Buenaventura to recuperate, and have left the several maladies firmly implanted in the entire region, making it most unhealthful.

A small tent-show was playing at Caldas, and as this was a most unusual occurrence it created a certain amount of furor among the people. It rained heavily the greater part of the afternoon, but darkness had scarcely crept up from the lowland when troops of people, each one carrying a chair or box to sit on, came tramping from all directions, their bare feet making swishing and gurgling sounds as they plodded through mud and water. The elite—even Caldas boasts of a high-class social set—arrived later and stood during practically the entire performance in order to be the better seen and admired by the "common" people.

So far, Richardson had acted as cashier for the party, and it was rather startling to see entries in his journal such as "lunch, $200.00; railroad-tickets, $2,000.00; oranges, $15.00." The Colombian dollar, or *peso*, had depreciated in value until it was worth exactly one cent in United States currency. Practically all the money in circulation was in bills of from one to one hundred pesos, the former predominating. If one had only a hundred one-peso notes, equalling an American dollar, they made quite a bulky parcel; for this reason all the men carry large leather pocketbooks attached to a strap slung across the shoulder, and quite incidentally these containers also hold cigars, matches, and various other little articles dear to the hearts of their owners.

Richardson had arranged for *arrieros* and a caravan of pack-mules to meet us early the following morning, but it was almost noon when they appeared. We were in the land of *mañana*, but had not as yet learned to curb our impatience at the hundred and one exasperating things that were constantly cropping out to impede our progress or upset our plans. One of the first things the visitor to Latin-America must learn is to take things good-naturedly and as easily as possible. If one employs servants regularly it is possible to correct many of their customs that are so annoying to the North American; but the countries, as a whole, cannot be reformed by any one in a single day, and the person who takes things too seriously either lacks a sense of humor or conveys the impression that he is very foolish.

Some of the mules were saddled for riding, while others were equipped with thick pack-saddles made of burlap stuffed with straw. Bags and trunks were brought out, sorted as to weight, and then loaded on the pack-mules, being held in place one on either side of the animal with cowhide thongs. Each mule carried about two hundred and fifty pounds. While adjusting cargoes, the *arrieros*, or drivers, place their poncho over the mules' eyes; otherwise they would not stand for the rather rough treatment to which they are subjected.

The road was fairly wide and good. It followed along the gorge of the Dagua, now a small stream. Within a few hours the village of El Carmen was reached and we dismounted to await the pack-train and incidentally to have lunch at the posada, and to see a cock-fight, for the fiesta of yesterday was still in progress in the rural districts.

We climbed slowly and steadily upward. At fifty-five hundred feet the zone of clouds and vapor appeared; trees, rocks, in fact everything seemed unreal and ghost-like, enveloped in the thick, blue-gray haze that penetrated clothing and sent a piercing chill to the very marrow. Darkness was fast approaching, so we stopped at a wayside hut called El Tigre for the night. The house was damp and cold, as might have been expected, and its occupants were practically without food. A profusion of vegetation grew in the yard; there were roses, geraniums, hibiscus, and9 hydrangeas growing everywhere; monstrous ferns with lace-like leaves formed a thick, velvety background for the brilliant, many-colored blooms. In the garden, blackberries, strawberries, cabbages, coffee, and an edible tuber called *aracacha* grew; there were also a few stunted banana and plantain stalks, but on account of the cold climate it requires two years for them to mature, and the fruit is small and of poor quality.

Thanks to an early start on the following morning, we reached the summit of the range, or the Cordillera Occidental, as it is better known, by ten o'clock. The whole slopes are covered with the densest of subtropical jungles. A steady downpour had fallen the entire morning, against which *ponchos* availed little. A halt of two hours was therefore called at a rather cheerless inn just beyond the pass, named San Antonio; the *señora* who conducted the establishment was glad to see us, for Richardson had apprised her of our coming; she soon had plantains roasting on the embers, and her shop provided sardines for lunch.

The descent of the eastern slope now began. The trail narrowed down and was rough; in places the decline was 45°. On both sides rose the living walls of impenetrable, gloomy jungle. One thing could not fail to impress us, and that was the great, breathless silence of the forest. Where we had expected to find multitudes of gorgeous birds, a babble of animal voices and brilliant flowers, there was only the sombre, silent mass of unvaried green. Within two hours we had left the regions of cold and penetrating mists. For the first time we beheld the beautiful valley of the Cauca far below, spread before our vision like a velvet carpet of softest green that reached the very foot-hills of the Central Range not less than forty miles distant.

The steady, rhythmic skuff of bare or sandal-shod feet, mingled with the louder tramp of mules and discordant cries of the *arrieros*, now reached our ears at frequent intervals, to be followed shortly by the appearance of pack-trains heavily laden with coffee and hides as they swung around a bend in the narrow mountain trail, and we knew that the end of our journey, at least for the present, was near.

Downward we rode, always downward, with the valley still several hundreds of feet below, and the mountains towering thousands of feet in the rear.

Here and there a bit of humanity flashed into view near one of the lonely *haciendas* snugly nestling in some seemingly inaccessible niche in the mountainside. To our right, a solitary monastery perched upon a barren peak, with its separate narrow trail leading from the dizzy height and winding its tortuous course along the jutting precipice until lost in the filmy haze.

Ahead, a black mass that dissolved itself into one immense flock of vultures appeared on the landscape. This was their season of harvest and the quarrelsome scavengers were reluctant to leave their repast—an unfortunate burro that had been abandoned on the trail.

With a feeling of repugnance, we spurred our horses on to greater effort, and at last our anticipations were realized as, rounding an abrupt point, we beheld Cali directly at our feet. A half-hour later we had clattered through a green arch formed by four magnificent ceibas that stood like sentinels guarding the approach to the city, crossed the bridge spanning the Rio Cali, wended our way up the stone-paved streets, and drawn rein in the *patio* of the Hotel Central.

Cali is a typical Colombian city. At first the uniformly low, whitewashed buildings with barred windows, thick adobe walls, and pretty *patios*, or inner courts, thrust themselves forcibly upon the attention, on account of the sharp contrast to the style of architecture to which the American is accustomed; but later one accepts them as a matter of course quite in harmony with the monotonous and easy-going life of most Latin-American cities.

There is nothing particularly modern about Cali; but the city is interesting, perhaps for that very reason. I saw not a single chimney, nor was there a pane of glass anywhere except in the huge cathedral facing the verdure-laden plaza. Churches are numerous, of massive construction, and built in Spanish style. The bells, of which there are many, are suspended in open niches in the towers, covered with verdigris, and keep up an almost continuous clanging.

The streets are narrow and crooked. A stream of water flows through the centre of some of them; this serves both purposes—as a kind of sewage system and also to supply water for various needs, although there is a system of piping in some of the houses, and fountains on a few street corners supply drinking-water to those who care to fetch it. I have seen, on several occasions, children attempting to bathe in the little stream; a short distance below, ducks were swimming in the water; then a person stepping from one of the doorways threw a pailful of garbage into it; finally, someone stepped out and unconcernedly dipped up a pitcherful of the water and took it indoors.

It is quite unusual to see any of the women of the upper class on the streets during the daytime, except on special occasions, or while they are on their way to and from church. They remain secluded in their homes, safe from the gaze of vulgar eyes. Embroidering and music are the chief diversions, and a large number of them are really very accomplished in both lines. It was remarkable to notice how many pianos there were, when we consider that each instrument had to be brought over the Andes slung on poles and carried by mules.

Practically all work is performed by people of the lower class. They toil day and night and, in most instances, for very little remuneration. One may see them engaged in various occupations at all hours of the day; but during the early hours of the morning, long files wend their way down the streets with the public market-place as the point of focus. The huge brick structure is a busy place. It reminds one of an ants' nest with its incoming and departing swarms. Inside the building are rows and heaps of fruit, vegetables, meat, bread, and many other articles. A motley crowd of women fills the place to overflowing; each carries a basket, or wooden tray, on her head into which the purchases are placed, when, after an indefinite amount of bargaining and haggling, they have been consummated. Invariably each receptacle contains a curious collection; a number of green and ripe plantains; a slice of pumpkin; a pepper, garlic, and a tomato; a chunk of meat, and a papaya. Perhaps there may also be a bunch of *yerba buena* and some achiote seeds with which to give a spicy flavor and yellow color to the soup; but these condiments are, unfortunately, used in such quantities that a goodly supply is usually kept on hand even when there is no other food in the house.

The nights are delightful in Cali. A refreshing wind springs up soon after sundown; the military band plays in the plaza, lights twinkle and the breeze sighs through the royal palms and orange-trees scattering broadcast snowy petals and heavy perfume. Only the *gente* are admitted into this little fairy-land. Gayly dressed and highly-rouged women, clothed in the extreme of fashion, parade along the winding walks; but it is considered in bad taste for them to appear without an escort. The poorer class, ragged and barefooted, gathers outside the iron fence and peers through the bars; the children run and play noisily on the neighboring streets. At last the bells in the cathedral boom the hour of ten; the band plays the national anthem, when everyone stands, the men with uncovered heads. Then the crowd disperses quietly and orderly. Soon the town is wrapped in slumber with only the sighing wind and the occasional shrill blasts of police whistles to disturb the drowsy solitude.

It was said, that Cali had a population of forty thousand, but that figure doubtless included the populace of the suburban districts for a considerable distance. The city is bound to grow, however, on account of its favorable location in the fertile Cauca Valley, which is one of the garden spots of all South America.

The Cauca River is about four miles distant from the city, and the settlement of Guanchito is located on the river-bank. A little toy-like train makes frequent trips back and forth between the two points because the *puerto*, as Guanchito is commonly called, is of real importance. Steamers and launches from Cartago take on and discharge passengers and freight, and many rafts laden with green plantains and produce arrive daily. The village presents a scene of great activity during the morning hours; clusters of ragged little booths, like mushrooms, have sprung up during the hours of darkness where women, squatting under the shambling shelters, cook *sancocho* over charcoal braziers; files of *peons* hurry back and forth as they transfer the cargoes from rafts and canoes to the waiting freight-cars; and there is a great deal of good-natured raillery between the slovenly *mozos* who liberally patronize the eating and drinking places, and the stand-keepers who feign an air of coyness withal. Gradually, as the sun mounts higher the crowds grow thinner. Their morning's work over, the people either depart via the waterway they had come, or take the train back to Cali.

An interesting ferry service is maintained at Guanchito. A stout steel cable has been strung across the river, and to a pulley running along it, two chains are fastened, their other ends being tied to either end of the boat. The latter is a huge, flat-bottomed affair, capable of holding many people and horses. Before starting across, the up-stream chain is shortened, so that the side of the boat presents a sharp angle to the current, and the craft is speedily pushed to the other side of the river.

Extensive marshes border the Cauca, a short distance above Guanchito. During the rainy season the water spreads over many miles of land, and is very deep; but in the dry season it recedes rapidly leaving a number of shallow and well-defined marshes and ponds. Wildfowl gathers in great numbers to spend the hottest months in these friendly havens. There were ducks of a number of species, including tree-ducks that make a shrill, whistling noise as they speed by and then drop on the ground near the marsh, to stand motionless and on the alert for possible danger before plunging into the water. Great gray herons croaked and waded sedately among the rushes, spearing frogs and fish as they went along. The horned screamer—a bird the size of a large turkey—is also an inhabitant of the marshes. It has rather long, but thick legs, that enable it to wade into fairly deep water, but also swims to floating islands of succulent water-plants which form a part of its food. The bird's color is slaty black, the back being glossy; the belly is white; a horn, or caruncle, several inches long grows from the forehead and curves forward. The feathers are soft, and the tissues for half an inch under the skin are filled with air spaces; the natives say that this protects the bird from the bites of poisonous snakes, and it is not impossible that this pneumatic cushion could serve such a purpose, although it is hardly probable.

Figure 2. Cattle grazing in the Cauca Valley.

Figure 3. Port of Guanchito.

The most remarkable thing about the bird, however, is its voice. Usually a pair sing together; they walk slowly back and forth, throw the head over the back, and emit powerful hoots, booms, and long-drawn, clear, ringing notes that, while harmonious and not unmusical, are nevertheless touched with pathos and conjure in one's imagination a picture of some trammelled spirit of the wild yearning for redemption. Numerous small birds, mainly tyrant-flycatchers inhabit the thorny thickets growing out of the water, and build their huge grass nests within the safe barrier of thorn-armed branches.

The surrounding country of the Cauca Valley is fertile and productive of most of the things essential to the support of a contented and thriving populace. A great deal of the land is used for grazing cattle and horses, but it will soon become too valuable to use for this purpose on account of the limited amount available. A far greater revenue can be derived through cultivation.

We paid a visit to a large sugar estate called La Manuelita, near the town of Palmira. La Manuelita is a little world of its own; it comprises fifteen hundred acres of the most fertile and attractive part of the valley. The ranch-house, occupying a site in the centre, is a rambling two-story building of generous proportions and attractive appearance. The gardens, surrounding it with a riot of color, give it a quaint, old-fashioned charm; there has been no studied effect, no precision in the arrangement of plants or flowers; oleanders, roses, hibiscus, geraniums, and hollyhocks grow in matted profusion. Clumps of magnolias, chinaberries and oranges conceal the high stone fence. Immediately without the wall surrounding the house is the *peon* village consisting of some fifty-odd houses of uniform size and appearance, and the sugar-factory. The *peons* are of Spanish, Indian, and negro blood, or of a mixture of any two or all three, and require constant supervision to secure the best results.

All the land is under cultivation, mostly in cane, for the production of which it is well suited. The soil is a rich alluvial loam. Some of the cane-fields at La Manuelita had not been replanted in ninety years; others on the estate of William Barney, former United States consul in Cali, had been producing one hundred and twenty years, and were still yielding eighty tons or more of cane to the acre. It was said, and all indications substantiate the report, that the entire region was at one time covered by a great lake! This accounts for the continued productiveness of the soil.

Cane grows to a height of fifteen feet, there being a dozen or more stalks to each hill. It requires eight to ten months to mature. The fields are divided into sections and cut at different intervals so as to provide a succession of ripe cane for the mill, and furnish steady employment for the several hundred *peons*.

The factory is modern in nearly every respect; its capacity is from five to eight tons of sugar daily, of good quality. It required a number of years to bring the heavy machinery over the mountains from Buenaventura. The more cumbersome pieces were slowly drawn up the steep slopes with the aid of block and tackle and oxen; the apparatus was so arranged that the animals could walk down-hill as they pulled, adding greatly to their efficiency. It is necessary to carry a complete stock of duplicate machinery to use in case of an accident; otherwise the factory might have to shut down a year or two while some badly needed article was being secured from abroad.

Nearly all machinery is ordered from London, as it can be had more quickly and better packed than from the United States. I heard this same statement in various parts of

South America. Although manufacturers were beginning to realize that in order to do business successfully in South America, they must first make a study of general conditions, they have not done so in the past, with the natural result that the bulk of Latin-American commerce has been done with the Old World. It is frequently necessary to ship merchandise on mule-back, or in small river-craft a distance of many days after its arrival at a port and before it reaches its destination; it is exposed to varying weather conditions—great heat and heavy rains; the treatment it receives is of necessity very rough. All this means that packing must have been done with great care and in a special manner. The fact that we have not adopted the metric system, and that there have been practically no American banks to discount bills, have been further drawbacks to the establishment of extensive trade relations between the two peoples.

Perhaps the most attractive thing of all about the Cauca Valley is its climate. A record of the temperature kept at La Manuelita during a period of ten years shows the greatest uniformity. The difference in the average weekly temperature is only 6° the year around.

A belt of tall bamboo entirely surrounds the *hacienda*; the giant stalks of steel-like toughness are armed with long, murderous thorns and form an interlocking mass that is absolutely impenetrable to man. Contrary to our expectations, birds were not plentiful in this land of tangled verdure. A few nighthawks dozed on the ground in the deep shade, and an occasional yellow-headed caracara (*Milvago chimachima*) that, perched on the tip of a swaying stalk, gave vent to its feelings in a succession of shrill, long-drawn screams.

Farther away, where clumps of woods grew, birds were more plentiful. There were many red-fronted parrakeets nesting in holes in dead stubs. Red-headed woodpeckers (*Chrysoptilus p. striatigularis*) in numbers hammered on hollow trunks; the strokes are so rapid that the sound resembles the roll of a snare-drum. Pigmy woodpeckers (*Picumnus*) no larger than a good-sized humming-bird, worked industriously on the smaller branches. They are obscurely marked mites of feathered energy, of a dark olive color with a few red dots on top of the blackish head. When the nesting season arrives a tiny cavity is excavated in some partially decayed limb in which two round, white eggs are deposited. These birds are nearly always found in pairs, and when the young leave the nest they accompany the parents, forming small family parties that forage for minute insects among the crevices of rough bark and in decayed wood.

Occasionally it seemed as if we were not so far from our northern home after all; for along the edges of the numerous marshes ran an old acquaintance—the spotted sandpiper. In the reeds yellow-headed blackbirds chirped and fluttered; but they are slightly smaller than the North American birds and have even been placed in a different genus (*Agelaius*). By walking quietly it was also possible to surprise a deer that had been tempted far from cover by the prospects of a luscious breakfast in some little plantation. These animals are so greatly persecuted that they make off at the first sign of danger.

Chapter 2

POPAYÁN AND THE CERRO MUNCHIQUE

After spending a few weeks in and about the Cauca Valley, Richardson and I started southward, while the two other members of the expedition began the homeward journey. I had looked forward very eagerly to my visit to southern Colombia because I knew that the country, towns, and even the people were different from those we had seen heretofore. But, above all, because ahead of us lay a vast region little known zoologically, and we hoped to penetrate into at least the mountain fastnesses west of Popayán in our insatiable search for the rare and interesting wild life that haunted that remote wilderness.

We left Cali at noon, May 13, well provided with riding and pack animals, and half-breed *arrieros*, and started on the well-beaten trail that leads toward the south.

At first there was no appreciable change in the valley, but by degrees the stretches of absolutely level-appearing land increased in size; instead of extensive cultivated areas there were pastures of large size, covered with a luxuriant growth of grass. Thousands of head of cattle were sprinkled over the velvety turf. We rode an hour through one of these ranches just before reaching the river Jamundi. This estate is the property of one Angel Mario Borreo, who is reported to be one of the most influential men in the Department of Cauca, and is only one of his sixteen similar holdings.

The Jamundi is not over one hundred and fifty feet wide at the point of crossing, and is spanned by a steel and brick bridge; dense jungles of bamboo line both banks. Just beyond lies the little town bearing the same name.

A tent-show had been billed to appear here at sometime within the near future, and the arrival of our pack-train was mistaken for that eagerly awaited event. The news spread rapidly and before long the populace had turned out en masse in the hope of getting a glimpse at the wonders our trunks and duffel-bags were supposed to contain. Not until we had taken refuge in the little *posada* or inn could they be convinced of their error and induced to return to their homes; but another surprise was in store for us.

The many and enervating tasks of the day called for our early retirement, and eight o'clock found us in our cots. Great was our surprise to be awakened an hour later by the sound of music at our very door. One of our men was sent to the door to learn the cause of the serenade and was told that the mayor of the town, with a delegation of the chief officials and the band, had come to pay us a visit. Of course, there was but one thing to do, and half an hour later found us out on the special seats that had been prepared, in full view of the visitors and perhaps half of the villagers who had accompanied them. Then followed speeches, singing, music, and a few native dances, interspersed with short intervals for smoking, drinking (a goodly supply of *aguardiente* had been brought along), and conversation. The visitors remained until one in the morning; a rather lengthy call, to be sure, but a pleasant one, and quite characteristic of the friendliness of the Colombians.

The next day's ride of ten hours' duration brought us to Buenos Aires, a very pretty little town nestling among and almost obscured by gardens of flowers and orchards of fruit.

A heavy rain during the night had filled all the sink-holes in the road with water, making progress slow on the following day. We rounded Mount Saint Ignacio early in the morning, and shortly after had our first view of the volcano Purace; we were to learn more of this mountain in the not distant future. Soon after, the *lomas* or great barren hills appeared; they form a kind of connecting-link between the Coast and Central Ranges. These gently rounded mounds are bare except for a kind of worthless, wiry grass that in some unaccountable way draws enough sustenance from the red-clay soil to maintain its meagre growth. These hills gradually increase in height, but the ascent is by such slow degrees that one is scarcely conscious of any rise at all. There are few houses, and the small number of inhabitants seem to be as sallow and lifeless as the hills themselves. A party of people had gathered at one of the Philippine-like structures near the roadside; they were chatting excitedly and drinking a good deal of *chicha*. When we dismounted we found that a child had died and was being prepared for burial. It sat propped up in a small, rudely made chair, covered with a piece of white cloth. No one seemed greatly concerned over the death, least of all the parents; on the contrary, they were proud of the *angelito*, and of the attention the event attracted from the people of the neighboring country.

In perfect accord with our expectations, there was little bird-life on the cheerless *lomas*. A few blue tanagers and *Veinte-vi* flycatchers (*Pitangus*) lived in the bushes that lined the infrequent rivulets trickling through narrow gullies between the hills. The *Veinte-vi* was an old acquaintance; its cheery call is one of the first bird-notes to greet the ear of the visitor to tropical South America. Its local name varies with the locality, and is an attempt by the natives to imitate the bird's cry. Thus it ranges from *Kiss-ka-dee* and *Veinte-vi* to *Dios te di* and *Christi fui*. This flycatcher is of a rather vivacious disposition, and pairs of them frequently may be seen singing together and beating their wings on the branches.

As a general rule these birds are of peaceful habits, except when nesting; but I have frequently seen them in pursuit of a carrion hawk at which they darted viciously and continued to follow until lost to view.

The diet of the *Veinte-vi* is varied, and the bird is most versatile in capturing its prey. Thus it will sit on a perch above a brook and plunge in after small fish or tadpoles, somewhat in the manner of a kingfisher; it may hover over a field and drop upon an unsuspicious mouse, lizard, or small snake; beetles, grasshoppers, and other insects are overtaken and captured on the wing. When a victim of some size has been captured, it is beaten rapidly upon a branch until its life is hammered out. It also hops about in fields looking for worms and grubs.

The nest is a huge domed structure, made of grasses and often wool, and placed in the branches of a tree six to fifteen feet up. Entrance is gained through an opening in one side, near the top. On account of the great size of the structure, being about twelve or fifteen inches high and eight to ten inches thick, it is very conspicuous; the exterior is carelessly made, with grasses and streamers of nesting material hanging down on all sides.

The eggs, two to five in number, although four seem to constitute the usual set, are long and pointed, cream-colored, and lightly spotted with chocolate-brown and purple blotches—mostly on the larger end.

Besides these species, there were ground-doves, lapwings, and an occasional sparrow-hawk. The latter is so similar to our common little terror of the air that it is hard to distinguish between the two.

Shortly after noon we encountered one of the most terrific tropical storms imaginable. Hour after hour a perfect deluge of rain poured down upon us from which rubber *ponchos* afforded little protection. Flashes of lightning pierced the semiblackness with blinding shafts of light, followed by deafening crashes of thunder—an indication that we were approaching the high zone of bleak mountain slopes and paramos.

That night we reached Morales, at an elevation of five thousand nine hundred feet. Fortunately there was no demonstration of any kind to interfere with our much-needed rest. Early the next morning, however, we experienced the thrill inseparably linked with the sudden display of one of those hidden forces of nature that forever and inalterably control our destiny.

From out of the gray and penetrating mist that seemed to envelop all the world there rose a low, ominous rumbling, distant, yet of thunderous volume; and the mud-walled, grass-thatched inn shuddered violently in unison with the trembling earth.

Through the open door of the adjoining room I heard the scratching of matches and saw the flicker of yellow light reflected on the whitewashed wall. A moment later the pious *señora*, surrounded by her little ones, was kneeling before the shrine of the Virgin, chanting a litany in low, monotonous tones. Two tapers flickered hazily. The gaudy tinsel flowers that decked the image gleamed in the uncertain light, but the pitiful squalor, ignorance, and general misery of the surroundings were mercifully left in darkness.

Without, all was silent, save for the barking of a pack of stray mongrels which had been asleep on the door-steps of Morales. The village again slumbered, and the chill, damp fog clung to the earth.

Alone I made my way up the only street, through the mud, to the eminence on which the adobe church stands, overlooking the valley and affording a view of the tremendous range on each side; for it was nearly the hour of daybreak and the sun rising above the lofty peaks of the Andes presents a scene of matchless beauty.

With the first faint glow of light in the east the banks of vapor became dissipated and gradually disappeared. Peak after peak reared its head above the ocean of snowy whiteness. First of all was Purace, the hoary monarch that dominates the southern part of the Cordillera Central and spreads terror through the land with threats and warnings similar to those we had just experienced. This great volcano has been active for untold ages. A huge column of smoke and vapor ascends continually straight into the clouds, and this, reflecting the light of the rising sun, makes a magnificent picture. Occasionally at night the eternal fires within the gaping crater may be seen tinting the low-hanging clouds and the snow that crowns the summit, fourteen thousand five hundred feet high, with rosy red. All about, the great barren lomas are strewn with black boulders, some of immense size, that serve to remind the wayfarer of the cataclysms of bygone ages. Everywhere they dot the hillsides and tower above the trail that winds among them.

Just below rises the silent mass of Sotará, crowned with the snow of centuries; the precipitous slopes are seamed and worn by the frequent slides of ice and stones from above, and deep, snow-filled gashes extend far down below the glittering dome in a ragged fringe. At night the moonlight steals softly up the frigid heights and reverently bathes the ancient head in a halo of dazzling splendor.

As the sun mounted higher and higher the peaks of the Western Range appeared one by one, like islands in mid-ocean, led by the awe-inspiring Munchique and followed by his lesser satellites. Between the two ranges, in the fruitful valley of the Cauca, Popayán still slumbered beneath a blanket of billowy softness.

By six o'clock the *arrieros* had corralled the mules and riding-horses, and half an hour later we were on the march.

Replacing the dry and barren *lomas*, we now found a bush-covered country with occasional long strips of low forest in the hollows; but the trail was an exceedingly difficult one, owing to the rocky nature of the country and the great boulders that obstruct the way. Frequently a small stream had to be crossed, such as the Rio Piendano, which is spanned by an arched bridge built of large, hand-made bricks, a curious relic of olden Spanish days. Down goes the trail five hundred feet or more at an angle of forty-five degrees, and then up again on the other side, the mules snorting and puffing as they creep along at a snail's pace. All the rivers seem to flow through deep gorges. Only sure-footed mules are of service on this trail, each carrying not more than two hundred pounds.

The distance from Morales to Popayán is not great; without cargo-mules it is an easy day's ride, but with a caravan of tired, heavily laden animals that have come all the way from Cali it is the part of wisdom to spend the night at the little *posada* La Venta and ride into the city early the next morning. Here a room and a good meal can usually be had on short notice, but one must carry his own cot and bedding, as luxuries of this kind are not furnished in Colombian inns except in the larger cities.

We were up and on our way early the next morning, for it was market-day—the day when the inhabitants from miles around flock to the city to buy and sell and to have a good time generally. It was our first visit and we could not afford to miss such an interesting and typical sight.

While still several miles distant from Popayán we began to meet small parties of Indians that dotted the trail, slowly wending their way toward the Mecca of the Upper Cauca. By the time we had reached Belen, a settlement of about twenty houses, the trail had widened into a beautiful thoroughfare and was crowded with oncoming hordes. These Indians are probably descendants of the ancient Guanacas, while some are doubtless the offspring of the tribe of Paeces which inhabits the Cordillera Central to the north. Many, no doubt, still preserve the original purity of the old stock, but the vast majority have mingled and intermarried with the native Colombians until one finds every possible stage of intergradation.

Before us passed the motliest crowd imaginable, each bearing the fruit of his toil, to be appraised and sold in the public plaza. There were small family parties, the man leading a decrepit mule that threatened to collapse at every step, laden with fruit and vegetables, fire-wood, hemp ropes and bags, calabashes, pottery, or any one of a hundred different things. The wife, acting as auxiliary beast of burden, carried the surplus. A band passed over the forehead supported the heavy pack; usually a small child was carried in a sling at her side, while several larger children clung to her skirt or trudged behind. As she walked she worked, spinning from a bunch of wool or cotton tucked under her arm, the spindle, a sharpened stick with a potato stuck on the end, dangling from her hands. The most characteristic occupation of the women is the making of small fibre bags, or *muchilas*, from hempen cord. They are meshed entirely by hand as the overburdened worker trots along, and when completed somewhat resemble a lady's shopping-bag. If the meshes are close it requires weeks to finish one which would fetch forty or fifty cents.

The men are dressed in loose white-cotton trousers that come below the knee; then there is the inevitable square of homespun woollen cloth, usually brownish, gray, or blue, called *ruana*; the head is thrust through a hole in the centre so that it drapes down to the waist, the corners often touching the ground and giving the same effect as the toga of a Roman senator. At night the *ruana* takes the place of a blanket under which the whole family sleeps. A broad-brimmed, high-crowned straw hat completes the outfit. The women are fond of dark-blue skirts (also the product of their industry), pink waists, and shawls of almost any color so long as they have fringes. Their hats are similar to those worn by the men. The feet of both sexes are, of course, bare.

Half an hour after leaving Belen we were cantering across the great brick bridge that spans the Cauca and forms the entrance to Popayán. This bridge is really a marvel of ancient Spanish architecture, five hundred feet long, forty feet wide, and supported by a series of arches.

Popayán is one of the oldest and most picturesque of Spanish-American cities, though by no means the largest. I doubt if its population exceeds ten thousand. The early history of the city is full of interest, and from it one gains an insight into the conditions attendant upon the conquest and colonization of a large part of South America. Spurred on by the love of adventure and the lust for treasure, the *Conquistadores* overran vast portions of the continent, establishing depots here and there from which they could start anew in search of El Dorado, which they were destined never to find. In this manner Popayán was founded in the year 1536 by Sebastian de Belalcazar, the son of a peasant from the border of Estremadura and Andalusia, in the south of Spain.

After founding Popayán, Belalcazar extended his raids down the river and formed the settlement which to-day is Cali, the largest and most important city in the Cauca. Being a fair example of the usual type of Conquistador, he showed no mercy toward the Indians, but nearly exterminated them; the country which had been a fruitful province was turned into a famine-stricken waste. In the meantime Pizarro had sent an officer, Lorenzo de Aldana, to arrest his erstwhile lieutenant; but Belalcazar, satisfied with his conquests, set sail for Spain in 1539 for the purpose of securing a charter before he could be apprehended.

The city lies high up on the level plain, more than six thousand feet above the sea, surrounded by rugged peaks, some snow-capped, others unbridled as yet by the hand of time, presaging catastrophe and disaster; and still others covered with impenetrable growths of virgin forest, untrodden by human foot, and known only to the wild creatures that lurk within the dark recesses. Above all hang the fleecy clouds that encircle the lofty pinnacles, dip low to meet the earth, and then vanish again into space. About the city prevails an air of calm repose; an air of sanctity and mysticism that radiates into every nook and corner, permeating every fibre. The city is famous as a centre of learning. Its colleges and university, conducted by the Order of Maristas, attract the youths from all parts of the country. There are numerous old churches, all very ancient, the gilded interiors rankling with the damp of untold years. Bells of antique workmanship, and covered with verdigris, dangle in open niches in the walls or in the low, square towers, and hourly call the faithful to prayer in monotonous cadence. The cathedral was completed in 1752 after many years' work. In one of the streets a delightful view may be had of three successive chapels, one above the other, and of the streams of pious penitents wending their way up the rocky path. There are also the overgrown ruins of a house of worship, but I could never quite decide whether the edifice had fallen into decay or whether the medley piles of bricks and rubbish between the four crumbling walls were still waiting to be placed in position. The streets, crooked and narrow, are paved with cobblestones. The buildings are of the old adobe type, one-story and whitewashed, with red-tile or sod roofs. Glass is not used except in the churches, but the windows are heavily barred. Recently a few modern brick structures have been erected. A look into the corridors and inner courts, of which there may be several in one house, conveys an insight into the domestic life of the people. The front courts are very attractive with their flowers, shrubbery, and trees, but the rear ones are anything but inviting, the dungeon-like enclosures reminding one of the stories of atrocities and persecutions carried on here in the turbulent times of the Spanish Inquisition.

On an average, the people are of a higher class, both intellectually and physically, than in most Colombian cities of equal size; comparatively few negroes are seen, and the good health and bright looks of the inhabitants are the natural result of a cool climate and pure mountain air.

One day, at noon, as I was photographing in the vicinity of Popayán, after having ridden perhaps five or six miles from the city, I was accosted by an elderly woman who invited me to stop at her humble cabin, where she had prepared a really palatable lunch. Her reason for doing this was that she had recognized me as a foreigner. During the course of the meal she tearfully related that she had had a son, of about my own age, who had gone to the States many years before. Had I met him, and could I give her any tidings? I could have, but I did not. By a strange and inexplicable coincidence I knew that her son had not left the country. Instead of going to the coast he had engaged in one of the revolutions common enough at that time and had been captured and shot; but what right had I to remove the only support that maintained the spark of life in her aged body? It was only the hope of seeing her boy again that gave her the strength to resist the onslaught of advancing years. Doubtless, she still waits, hoping against hope for the message that will never come. Hers is the mother-love that never despairs. How clearly it shows that human nature is very much the same the world over, even among the lowly!

On June 23 I was fortunate enough, while in Popayán, to behold one of the religious celebrations formerly all too numerous in Latin America. It was the *Fiesta del Sagrado Corazón de Jesús*. Troops of soldiers and bands were lined up in front of the cathedral; all were quiet and orderly while the sacred rites were being performed within. Suddenly the doors burst open, bells boomed and jingled, and the contents of the vast church poured through the portals in a steady stream. First came the altar-boys in white surplices and red cassocks, carrying gilded crosses on long poles and lighted tapers in silver holders, followed by the small children, the girls with tinsel wings, resembling tiny angels. Then came the governor of Cauca, the prefect of Popayán and their staffs, each bearing a standard. Next in line were the maidens, covered with large black shawls, or *mantas*, with folded hands and downcast eyes which, however, they were not averse to raising to meet the admiring glances cast by some of the onlookers. The students from the seminaries and a choir of singers preceded a life-size statue of the patron of the feast, borne aloft on the shoulders of stalwart youths; then came the archbishop and the higher ecclesiastics in tall mitres and gorgeously embroidered and glittering robes. Those of the general public who chose to march fell in line behind the bands that followed, chanting prayers. The remainder knelt in the streets with bowed, uncovered heads as the procession passed. All the buildings, even the trees, were gayly decorated with banners, a mixture of the papal and national insignia. Colombia is perhaps the only remaining country in the New World in which religion still dominates the government.

If we examine a map of Colombia we will find that the Cerro Munchique, the highest of the mountains in the Western Range, lies directly west of Popayán. There is an exceedingly difficult pass across the Cordillera at this point, leading to a place called the Cocal, still far distant from the coast. A trail was also being opened a short distance to the south leading to the Rio Micai. When this is completed it will require a four days' journey on mules to the river; then two days in canoes on the Micai, said to contain many rapids and to flow through country inhabited by savage tribes, before the coast is reached.

A day's ride from Popayán took us to El Tambo, and at noon the following day we were in the Indian village of Chapa at the very base of Munchique. A heavy electrical storm delayed our departure until noon the next day. There were but a dozen or fifteen adobe huts in the village, and during the height of the tempest one of these suddenly collapsed into a heap of mud and straw; the occupants barely escaped by fleeing into the deluge when the buckling walls apprised them of their danger.

After the agitation had subsided the people erected an altar in the plaza for the celebration of a mass of thanksgiving. Each one brought some trinket—a few paper flowers, a picture, a bit of tinsel, or a candle—with which to embellish the sacred structure. Then they all knelt, with bared heads, and in deepest devotion assisted at the religious service; that is, all but a plump Indian woman who boiled *chontaduros*, or palm-nuts, in a huge kettle, in back of one of the huts and sold them to the worshippers the moment devotions were over.

It required fully a half-day longer to reach the end of the mule trail, and by that time we had reached an elevation of eight thousand feet.

From this point up the mountains are covered with a dense growth of primeval forest. Below this elevation there are occasional strips of woods and patches of brush interspersed with clearings. Maize grows splendidly up to an altitude of seven thousand feet; this was proven by the few small fields cultivated by the Indians. The slope was also dotted with areas planted in rice.

The ascent of Munchique is very abrupt; there are no streams near the summit, as the top of the mountain is composed of solid rock that sheds rain as soon as it falls. The highest pinnacle is a flat, bare rock, about ten thousand feet above sea-level.

Figure 4. Cerro Munchique.

Figure 5. A deserted Indian hut on the Cerro Munchique.

Robert Blake White states that from this spot one may "obtain a view over more than fifteen thousand square miles of country. The whole of the Central Cordillera, from the frontier of Ecuador to the confines of the State of Antioquia, with the valleys of the Cauca and the Patia, were visible to the north, east, and south; whilst, on turning to the westward, the Pacific coast from the bay of Tumaco to the mouth of the San Juan River seemed spread out like a map before us.

"A more gorgeous panorama cannot well be imagined. The belts of bright-colored vegetation, marked by the valleys with their winding rivers and streams, were backed with masses of the Cordillera with their varied tints and snow-capped peaks. On the other hand, the dark-hued vegetation of the virgin forests of the Pacific slopes stretched down to the ocean's margin, which with its thousand bays and inlets and fringe of foam which was quite visible, looked like an edging of lace. The island of Gorgona could be distinctly seen.

"The Cerro Munchique should be visited in the dry season, for its peculiar prominence makes it a grand lightning conductor, as we clearly saw from the shattered rock on the summit."

We discovered a deserted Indian hut in the centre of a large, overgrown, abandoned plantation, and made it our headquarters for a week or more. The site was ideal. Tall forest hemmed in the clearing on all sides, and a rivulet of clear, icy water flowed near the shack. The elevation was eight thousand two hundred and twenty-five feet. Obviously, the place had been unoccupied for a number of years, doubtless owing to the fact that maize and rice would not thrive at this high altitude. However, these same conditions were most congenial to a host of other vegetation. Blackberries and rhododendrons, with lilac, red, white, pink, and yellow flowers formed a solid tangle, acres in extent, and creepers entirely covered the tall, dead stubs, and crowned them with a thick canopy of green leaves from which clusters of orange and scarlet trumpetflowers drooped.

At night the temperature went down to about 45°, but this did not deter giant hawk, owl, and sphinx-moths from appearing at dusk to feast on the nectar of the myriads of flowers. The little stream was the rendezvous for numberless frogs. One hardly suspected their presence during the daytime unless a careful search was made of the rotting wood that littered the ground, and of the tangled, snakelike stems of second-growth sprouts and leaves; but at night the concert was always sure to begin in startling volume. Some of the notes reminded me of our own spring peeper; others were sharp and metallic, like the twanging of a banjo-string; and others were low and mellow like the murmuring of a 'cello. They all blended into a deafening chorus of unflagging animation and unvaried monotony. At first the din was rather disconcerting, but gradually there came to us the realization that this was but the bubbling over of care-free little hearts rendering a song of happiness and thanksgiving to nature for the pure, unsullied joy of an unfettered existence.

Birds were not particularly plentiful in the forest. There were, however, a number of interesting forms, particularly among the tanagers. One species (*Psittospiza riefferi*) was about the size of a robin and of a deep grass-green color, with a chestnut-colored face and abdomen; these birds live singly and in pairs in the tall trees and are of a wary disposition. Another tanager (*Sporathraupis*) has a bright-blue head and olive-green back; the breast is deep, dull blue merging into golden yellow on the legs. The natives called this bird *jilguero*, a name applied to the solitaire in other localities. It lives in the lower branches of trees, travelling in pairs or small flocks and feeds on fruit; the song is not unpleasant, but cannot compare with any solitaire known to me.

While collecting one morning my attention was attracted by a chorus of chirps and screams, and following up the sounds I reached a tall tree where a peculiar bird drama was being staged. A number of California woodpeckers (*Melanerpes flavigularis*) had drilled numerous holes in the tree-trunk, from which sap trickled in small streams. A dozen or more buff-tailed hummers (*Boissonneaua flavescens*) had apparently come for their daily jag, for the sap very evidently had an intoxicating effect. Arriving in a bee-line, newcomers landed against the trunk, where they clung like so many moths, the buff-colored tail spread wide and against the bark for support. Their antics as the different stages of hilarity were reached were most amusing. They twittered, fought, turned, and tumbled in the air; others dozed on small twigs, and several fluttered toward the ground in an exhausted condition. This performance continued daily for a week, until the sap suddenly ceased to flow; then the tree was deserted and silent, the capricious band having no doubt sobered up from their debauch and gone back to their normal and more profitable pursuits in life.

In getting water from the brook, one of our men discovered a narrow trail under a giant log. We justly surmised that animals of some sort used the runway in journeying to and from the water. A trap was set in the path, and next morning a fine white opossum of large size had been safely ensnared. In the days that followed we secured an even dozen of the animals. They proved to be a form unknown to science that now bears the name *Didelphis paraguayensis andina*. The cook said that they were delicious eating, and prepared for us an unusually fat individual; but we found the meat of rather strong flavor, and not very palatable. A solitary weasel (*Mustela affinis costaricensis*) was also taken in the same spot. It would be interesting to know whether this animal came down to drink, or was in pursuit of some of the other creatures that frequented the runway. Weasels are courageous, active, and bloodthirsty little animals; their eyesight is poor, but their sense of smell is keen and they will tirelessly follow their intended victim until it falls into their clutches. I have frequently heard that they attack and kill small deer by clinging to the neck and doggedly chewing their way through the skin until the jugular vein is severed; this does not seem probable to me, however, and it is far more reasonable to believe that rats, mice, frogs, and other small creatures form the bulk of their prey. On account of their slender proportions, they can trail the quarry through small holes and crevices; in addition, they are also expert climbers. On one occasion, while "squeaking" to attract a bird, a weasel came instead, looking for the supposed helpless creature, and ran over my feet without suspecting the fraud.

They will fight savagely to protect their nest, usually made in a hole in the ground or hollow stump, and I know of one instance where one of the animals sprang into the face of a native who had trapped its mate at the mouth of a burrow.

Nearly a month had passed since we left Popayán, but the time had been spent so pleasantly and profitably that it seemed scarcely longer than a week. Our scheduled time for the region had been exhausted, however, so we reluctantly retraced our steps to Popayán.

Chapter 3

THE ANDES SOUTHWEST OF POPAYÁN; CRUISE OF THE *CALDAS*

After our return to Popayán we spent a few days wrapping and packing the large collection of birds and mammals that had been secured on the Cerro Munchique; and, during the odd moments when this work became arduous, we sought information about the mountains south of the place we had just visited. Eventually our quest took us to the governor's palace, where we had the good fortune to meet the executive of the province of Cauca, Doctor Alfredo Garcés.

The first thing that attracted our attention was a framed poem hanging above his desk; the title of it was "Death to Foreigners"; but the kindly and sympathetic governor soon dispelled any doubts we may have harbored as to his feelings. He was a man of great refinement and education, and had travelled extensively in the United States. Our mission interested him greatly. He brought out maps and charts, and then, with the aid of a pair of powerful field-glasses, showed us the places he had pointed out on the drawings.

Doctor Garcés proved to be one of our best and most highly valued friends, despite the rather alarming notice on his office wall. He called on us at the inn several times each day, and admired the birds and mammals we had collected. Our rooms were always in the state of greatest upheaval with packing-cases, equipment, provisions, and a hundred other things occupying each available place; but the governor took it all as a matter of course, enjoyed delving among our possessions in search of things unknown to him, and probably considered himself fortunate if he could find his silk hat and cane in the place they had been left when he was ready to leave.

June 24 found us again upon the trail, heading south-westward. Both the Central and Coast Ranges were visible for many miles, the snow-covered Purace and Sotará dominating the former, with Munchique standing unequalled in the latter. Barren, rolling hills stretched away in the distance like the waves of a storm-tossed sea; this undulating country is the first indication of a connecting-link between the two ranges.

For two days there was no perceptible change in the country; but on the morning of the third day, shortly after leaving the settlement of Monos, we entered virgin forest at an altitude of seven thousand five hundred feet. A shelter-house, known as San José, is just a thousand feet higher up, and at ten o'clock we were up ten thousand one hundred and forty feet. From here one has an unrivalled view of thousands of square miles of country. The magnificent valley, appearing greener and more level from our height than was really the case, lay below, and stretched far to the north. The paramos and volcanoes flanking the far side were abreast of our station. Frequently, while in similar positions, there recurred to me the sentiment so aptly expressed by Hudson: "Viewed from the top of a lofty mountain, the world assumes a vastness and varied beauty that revive the flagging spirit and refresh the soul." And quite as certainly there is forced upon our recognition the infinitesimal smallness of man when compared to the immensity of nature—a mere atom existing by virtue of a benevolent force that has so ordained, but that reserves the power to crush the whole fabric of life at a breath.

The top of the ridge is ten thousand three hundred and forty feet high, and the vegetation is typical of the temperate zone; low, dense bushes, mingled with the gnarled branches of stunted evergreen trees and shrubs, burdened with clumps of red and lavender flowers. Numbers of low grass shacks had been built along the trail; some of them were very long and housed the peons working on the road to Micai. Although these structures were of comparatively recent origin, many small rodents had been attracted to them by the corn that formed the principal article of food of the *peons*. A large brown rat (*Oryzomys pectoralis*) was very abundant; apparently this rodent had formerly existed in small numbers only, for it was rare out in the open; but the artificial conditions created by the settlement of the region had proved so congenial that it increased rapidly. The same is true of several other species of rats that almost overran the houses.

In riding or walking along the trail, I frequently encountered a species of snake resembling in coloration a coral snake; however, it was not unusual to find an individual five feet or even more in length, and two or three inches through in the thickest part. They appeared exceeding sluggish and even refused to move although almost trodden upon. We never molested them, as they appeared to be harmless, and were really of striking beauty. Unfortunately, we had no way of preserving any. A species recently discovered in Nicaragua by Mr. Clarence R. Halter, which is similar to the one we saw almost daily, belongs to the genus *Coronella*.

The birds belonged to a typical temperate-zone fauna. Among them was a new species of beautiful honey-creeper (*Diglossa gloriosissima*); it is black with blue shoulders and a deep-rufous abdomen. They clambered about over the clusters of gorgeous flowers, feasting on the nectar they contained. Another common bird was a tanager (*Iridosornis*) the size of our redbird, but of a bright-violet color that merges into greenish blue on the wings. The head is black excepting the crown, which is deep orange.

It is a vivacious creature, travels in small flocks that frequent the taller shrubbery, but possesses limited singing powers only.

During our stay we had occasion to witness a christening ceremony performed by a priest who was travelling through the region for the purpose of ministering to the people's spiritual needs. The fact had been widely advertised, so early on the appointed day many natives appeared, bringing small children to be baptized. About thirty had been brought by noon, when the priest commanded the god-parents to line up, each holding his gaudily dressed and probably fretful little charge. The priest began at one end of the row, dispensing one part of the sacrament to each child as he passed; then he went back and began all over again, giving the second part to each of them, and so on until the rite was completed.

It so happened that there was a small child in the hut we had chosen for our several days' sojourn. To honor the *gringos* who were stopping under her roof, the señora asked Richardson to be the little one's godfather, while I was permitted to suggest the name. We naturally felt as if we should give the baby a present, but a thorough ransacking through my effects revealed only a can of talcum powder, which I promptly presented to the mother. A few days later she came to me in distress: "The baby has a slight fever," she said. "I gave it some of the white powder you made me a present of, but it did no good. How much is it necessary to take at one time?"

The western slope of the range is very abrupt. Large forces of men were engaged in cutting a narrow ledge for a footpath into the face of the steep mountainside. The trail wound back and forth continuously; looking over the rim of the narrow shelf one could count six or eight loops underneath, one below the other.

The work of cutting such a way is hazardous for the men engaged in it. The soil is loose and saturated with water, so landslides were of frequent occurrence; and whenever the irresistible avalanche swept the precipitous terrain, it usually carried one or more of the laborers with it and buried them under tons of rock and débris. It was invariably hopeless to try to find the body, so the survivors simply erected a cross on the spot. Frequently there were several crosses together, and in one place I counted seventeen standing side by side.

Late in the afternoon we reached a lone hut in a small clearing—the hastily erected shelter of a group of *peons*. The men invited us to stop, and as the locality looked interesting we accepted the invitation; but we erected our tent and lived in it in preference to the hut.

All the surrounding country was covered with virgin forest. It had never been trodden by man, at least not within many years; there were no side trails of any kind, so that it was impossible to penetrate very far.

Among our first mammals were a doe and fawn of a little forest-deer (*Mazama setta*). They are commonly known as brockets or spike-bucks, as each horn consists of a single prong. I believe that these deer are not so rare as generally supposed, but they are seldom collected on account of the difficulty of hunting them in the thick jungle. The *peons* shot the ones we secured. They had discovered a path leading to a small stream, and concealing themselves on the opposite side, waited until the animals came down to drink; then they shot them. The men also brought in a huge bushmaster that they unearthed while clearing away underbrush. The deadly reptile is known as *equis* to the natives on account of the black X-shaped marks on its back.

The mountainsides were scarred with deep fissures and ravines filled with the darkest and most impenetrable of forests. It was possible to look across from one side to the other, but crossing them was impracticable. Each morning I could see a flock of some thirty-odd swallow-tailed kites soaring just above the trees on the far side of one of the ravines. The magnificent birds resembled huge white-bellied swallows, or swifts, as they circled majestically over the dark forest; they uttered shrill screams all the while. Apparently they fed on the wing, and specimens collected by us later had eaten quantities of large beetles and flying ants. At about nine o'clock the band always resorted to the top of a tall tree that towered far above its contemporaries for a short rest; this was rather unusual, as the birds are rarely seen perched, and the natives said they never alight during the daytime. Live birds, or specimens freshly killed, have the glistening black back covered with a beautiful grayish "bloom" or powder that gives it a soft, velvety appearance. However, this disappears soon after death.

A species of pigmy squirrel (*Microsciurus*) lived in the forest, but we seldom saw any of the little creatures scarcely bigger than one of the larger kinds of mice. I have always found these animals much rarer than the ordinary squirrels; usually they live in pairs. They seem to prefer forests abounding in palms and to feed on the various kinds of palm-fruits and nuts. They frequently evince a great deal of curiosity and can be approached to within a short distance before taking fright and scampering out of sight among the leaves.

The minute creatures move rapidly and gracefully and make long, daring leaps. In running over the leaves and branches they follow the lateral stems, and on reaching the ends ascend through the tree-top or thick foliage by leaping crossways from twig to twig, as if ascending the rungs of a ladder.

I kept one that was given to me by the natives as a pet for some time. It made its home in the pocket of my flannel hunting-shirt, where it was always sure to find a bit of cracker or sugar, and to which it could retreat when frightened. Whenever anything of an unusual nature occurred, the bright, inquisitive little eyes always appeared suddenly so as not to miss a single thing that might be of interest. It never attempted to bite or run away, and seemed perfectly contented with the company of the friends that fed and protected it.

Eventually we started back toward Popayán. We crossed the high brush-covered divide July 4; a violent electrical storm had been staged on the wild mountain-top as if in noisy celebration of the day. It began with a dark mist that covered everything. Then rain and hail fell uninterruptedly for more than an hour, while lightning flashed and blue-green flames seemed to leap toward the blackened rock. Sometimes the bolts came from above, and again they were below us so that we were enveloped in a weird, ghastly light. The thunder was terrific and kept up a continuous crash and rumble. It was impossible to see any other member of the expedition on account of the thick haze—there was no shelter of any kind—only the narrow shelf-like trail that snaked its way along the steep slope. After the storm we made camp; the men and mules kept straggling in until a late hour; they were very wet and much bedraggled, but apparently none the worse for the nerve-racking experience.

Popayán was reached without further incident. Richardson's contract having expired, he determined to leave the country, so we returned to Cali to rearrange the equipment and pack the collections; then he left for Buenaventura to take a north-bound steamer.

In the meantime Doctor Chapman, who had reached New York, arranged to send down a man to fill the vacancy left by Richardson. Doctor Arthur A. Allen, of Cornell University, was selected for the place. He reached Cali about the middle of August and accompanied me during the succeeding eight months.

In compliance with instructions received from Doctor Chapman, I immediately planned an expedition northward, then toward the east to make a zoological exploration of the forests bordering the Quindio Pass and of the high paramo of Santa Isabel. The first stage of the journey was down the Cauca River.

Regular steamboat service is maintained between Guanchito and Cartago during the rainy season. The *Sucre*, a boat of small size, makes frequent voyages, requiring about three days' time each way. On this vessel one may travel in comparative comfort—if one is not too squeamish. We had to be content with a smaller craft, however, as there was not sufficient water to float the *Sucre* over the numerous sand-bars.

The *Caldas* is a little steel launch of not over fifty feet from stem to stern, with a beam of fifteen feet and drawing eighteen inches of water. When the river is full the *Caldas* is used to carry freight only, for which purpose she doubtless serves admirably; but at other times she assumes the double responsibility of carrying both cargo and passengers. Of course there is the alternative of going overland; but the trip takes twice as long, and after having spent some time on the muddy trails, the novelty of a river trip is likely to make a strong appeal, whatever the odds.

On the announced date of sailing thirty-seven individuals of all sizes, ages, and shades of color gathered on the river's bank, each impatient to be the first to cross the narrow plank and board the small craft. There also waited a huge mound of boxes, bags, bales of hides, and other freight; this was loaded first and piled in the front and rear. The engine occupied the centre of the boat, as did the kitchen. When the people were finally permitted to go aboard, there was a wild scramble to the top of the heap of boxes and bundles. To sit up straight under the sheet-iron roof was impossible; fortunately the sun shone intermittently only or we should have been suffocated.

From the very beginning there was enough of interest to keep one's nerves tensed to a high pitch. The crumbling banks, great chunks of which settled gently into the water as the waves, caused by the launch's propeller, washed away the last bit of restraining sand; the numbers of bamboo rafts laden with bananas, plantains, and other tropical fruits on their way to the port of Guanchito; the dark-skinned fishermen who cast their nets into eddies and quiet pools, and the washwomen, each smoking an enormous black cigar and beating the clothes upon stones until one expected to see them fly into shreds, were very interesting. There were also hundreds of cormorants and anhingas that swam and dived or flew up into the trees; some of them sat on snags drying their wide-spread wings.

The banks of the river are very high and abrupt in most places, and the stream runs through a tortuous channel. At each bend the current dashes with great force against the bank, and then rebounds on down-stream. The little *Caldas* could not hope to battle against the rushing torrent, so she would head straight for the bank; frequently her nose struck the soft sand and held fast; then the current swung her around and back into midstream, where after spinning around a few times she regained her poise and was swept along. As wood was burned exclusively, stops had to be made every few hours for a new supply. The launch, in her crowded condition, had little room for fuel, but the brief pauses gave those on board an opportunity to go ashore—a welcome respite from the cramped position made necessary by the limited space available on the boat. At one of these spots an extensive cacao-plantation lined the bank, the tall *madres de cacao* reaching high up into the heavens above their lowly but precious protégés. The "mother of cacao," it might be said, is a species of *Erithmas* planted to protect the delicate cacao-trees from the sun. A colony of snake-birds or anhingas had selected this grove for a rookery. Thousands of the birds sat on the topmost branches while other countless numbers were flying back and forth in endless streams, each bird a component part of a whirling, living mass. The slender body, long thin neck, small head, and sharp bill give the bird a peculiar appearance; when swimming under water with only the neck protruding it greatly resembles a snake—hence the name snake-bird. Each tree within an area of several acres contained a number of nests; they were clumsy structures made of sticks. The eggs, three or four in number, are white and as long as a hen's egg but only half as wide. Later in the day a flock of scarlet ibises approached from down-stream, flew past, and then disappeared like twinkling bits of flame.

In the early afternoon the *Caldas* struck a sand-bar with full force. The greater number of the passengers had eaten their luncheon—brought by themselves in small parcels neatly done up in banana leaves—and were quietly dozing. There was a harsh, grating sound, a shock, and the water swirled around and past the boat, which moved not an inch. The engines were reversed and the crew sprang into the river and pushed, but it availed nothing, so after repeated efforts these attempts were abandoned. Luckily, the craft carried a small dugout canoe, into which the passengers were unloaded, three or four at a time, and taken ashore by two husky negroes who waded to the bank, one pulling and the other pushing the canoe. There was no break in the abrupt banks for perhaps a quarter of a mile, so it was some time before all on board had been landed. The crew then began to dig away the sand that held the launch fast.

The spot where the passengers had been landed was an open, treeless plain with not a shelter in sight. At first the heat of the sun was insufferable; then it began to rain as we had never seen it rain before. No one had a poncho, so there was nothing to do but stand quietly and endure the drenching downpour.

When the sand had been dug away the launch, suddenly freed, shot down-stream a half mile before a landing could be effected. This of course necessitated a long tramp through deep mud and tall, wet grass, which added to the cheerlessness of the luckless, half-drowned victims of backward methods of transportation. The banks were as steep as ever, but a capybara runway, resembling a giant muskrat slide, had been discovered, and down this we slid, one at a time, into the arms of two negroes who acted as a back-stop below.

The delay prevented the launch from reaching Buga, so as soon as darkness settled, she was tied up for the night. A great tree-trunk, embedded in the sand with huge branches swaying high above the water, lay nearby. We swung our hammocks between the sturdy limbs, covered them with mosquito-nets, and spent a miserable night; those who attempted to sleep aboard had a harder time of it by far.

We were off with the first streak of dawn, startling flocks of muscovy ducks and herons from near the banks. A faint blue mist was rising slowly from the water, and the air was chill and damp. The mantle of silence that falls over tropical South America at nightfall had not yet been lifted. For some little time we glided on, farther and farther, it seemed, into a great vacuity that led to some vaguely defined sanctuary of everlasting peace and oblivion. Then, without warning, a sound so terrible rent the vast solitude that it seemed as if some demon of the wilds were taking a belated revenge for the few hours of quiet in which the earth had rejoiced.

At first there was a series of low, gruff roars that would have done credit to the most savage of lions, and made the very air tremble; but this was not all. Added to the majestic frightfulness of the jungle king's voice was a quality of hate and treachery, of unfathomed rage and malicious bitterness. Then followed in quick succession a number of high-pitched, long-drawn wails or howls of tremulous quality that gradually died, ending with a few guttural barks. This uncanny performance lasted a number of minutes; but having perpetrated this outrage upon a heretofore peaceful world, the weird chorus suddenly stopped.

The mists of night had lifted, revealing clumps of tall bamboo and the beginning of heavy forest. In the top of the very first trees sat a group of large monkeys, red, with golden backs, properly called howling monkeys; they were the authors of the terrific chorus we had just heard.

Figure 6. The Caldas fast on a sand-bar in the Cauca River.

Figure 7. Bamboo rafts on the Cauca River.

How an animal that rarely attains a weight of thirty pounds can produce such loud sounds is most remarkable; the hyoid bone is developed into a huge cup which gives resonance to the voice. The howlers are rather sluggish and seldom descend from the trees. Their roaring, which can be heard several miles, resounds through the forest morning and night; whether it is merely a form of amusement with them, or is used to intimidate enemies, seems to be unknown. Very little is known about the habits of howling monkeys, despite their abundance and wide distribution. They are usually found in small family parties, including young of various sizes; but I have noticed, on various occasions, that the females desert from the troop when their babies are males and do not rejoin it until the young are half-grown, perhaps fearing that the old males will kill them; but I do not know if this is always the case.

C. William Beebe, in the course of a lecture at the American Museum, stated that he had on several occasions watched troops of these animals feed, in British Guiana. The older ones sent their small young to the tip of the slender branches that they, themselves, could not venture upon on account of their weight, to pick fruits; then they pulled the little ones back and robbed them of their food. This was repeated a number of times.

The second night we tied up near a heavy growth of forest, at a place called Riofrio. This is one of the few sections of the Cauca Valley still retaining its original stand of virgin jungle. We slung our hammocks between the trees. The nets furnished ample protection from the mosquitoes, but not from an army of foraging ants that chanced our way. From across the river came the whine of an ocelot, and the sharp snort of deer, while more than once we were awakened by the pattering and shuffle of cautious feet close at hand, some light, some heavy as if belonging to a large animal.

Contrary to her custom, the Caldas steamed on after dark on the third night of her voyage. A train of bright sparks trailed far behind, and when the wind blew it carried them into the boat where they set fire to clothing and baggage alike. Within a short time we had reached the port of Cartago, found the arriero who was awaiting us with the animals, and were off for Cartago a league away. The town was enveloped in inky blackness, and fast asleep, notwithstanding the early hour. A stray dog barked and a mule whinnied, but there were no other signs of life.

Chapter 4

CARTAGO TO THE PARAMOS OF RUIZ AND SANTA ISABEL

Dawn revealed the fact that Cartago was not materially different from Cali. It was not so large, however, and the temperature was much higher. Upon our arrival the preceding night we had finally succeeded in arousing a sleepy landlord, who admitted us to a dusty, bare room in the Hotel Colombia. We had learned long before this time that the word "hotel" usually meant a roof only over one's head and perhaps food, so we at no time travelled out of sight of our baggage, with which we could make ourselves fairly comfortable under almost any circumstances.

The country surrounding Cartago is level and of a dry nature; at any rate, it does not compare at all favorably with the Cauca Valley at Cali. We saw few evidences of cultivation and the number of cattle and mules grazing on the scanty vegetation was very small.

The outskirts of the city are picturesque. The huts are low and lightly built of slabs of flattened bamboo; fences made of split bamboo neatly woven in a basket pattern surround them, and cannon-ball trees rear their slender, awkward branches laden with great glistening spheres of green fruit, high above the narrow, muddy sidewalks. When the huge calabashes are ripe they are collected and used as containers for water, wash-basins, bowls, and a variety of utensils; narrow sections that have been split carefully and resemble miniature pointed barrel-staves even serve the purpose of spoons.

A small marsh lies just in back of Cartago. It was filled with several species of aquatic plants—mostly water-hyacinths and wild lettuce on which cattle fed, half submerged in the murky water. Swarms of mosquitoes issued from the stagnant borders and invaded the town at nightfall, but this was by no means the only breeding-place of the obnoxious insects. Drinking water, kept in uncovered kegs and pots, teemed with larvæ, and glasses full of it set before us in one of the shops were fairly alive with wrigglers.

One thing impressed me as being really appalling, and that was the number of infirm beggars in the streets. In most Colombian towns beggars are permitted to ply their profession only one day each week and are required to wear a cardboard license tag suspended from a string around the neck; but it seemed as if they were out in full force every day in Cartago. Some of them presented an offensive appearance; they were suffering with leprosy and other terrible diseases, and were in such a pitiful plight that one was literally touched at mere sight of them. They always asked alms in the name of the Virgin and all the saints, and if results were forthcoming heaped a copious blessing on the donor; but in the event that nothing was given the benediction was in some instances replaced by such a tirade of profanity that one quickly realized what a good opportunity to acquire merit had been neglected.

We left Cartago as soon as possible and after a half-day's ride over gently rolling, brush-covered country reached the Rio Viejo. A good-sized dwelling known as Piedra Moller stands near the river; there one may obtain men and dugout canoes with which to cross the stream.

Beyond the river the trail passes through a little valley or depression about four leagues across. Tall brush, some first-growth forest, and extensive jungles of bamboo flank the narrow passageway. I counted no less than forty species of birds during the afternoon and heard the notes of several others that I did not recognize. Small green parrakeets (*Psittacula conspicillata*) were exceedingly plentiful. They always reminded me of English sparrows—not in appearance but by their actions. Flocks of them sat on telegraph wires or house-tops, chirping and chattering incessantly, or fed on fruits or seeds in the bushes. They are also abundant in towns and villages and nest under tile roofs, in hollow posts, and in holes in walls. The people are very fond of the little "love-birds" as they are called and keep them in their patios as pets.

At Balsas, which served as the first night's stopping-place, we discovered a whippoorwill's (*Stenopsis ruficervix*) nest in a clump of bamboo. The single egg had been deposited on the leaves near a bamboo sprout that was rapidly pushing its way upward like a huge stalk of asparagus. The incubating bird fluttered away as we approached, but we returned the next morning and Allen secured a photograph of her on the nest.

Noon of the next day found us at Finlandia, an inviting village with a population of about four hundred, and situated at an elevation of six thousand four hundred feet. All this country is the foot-hills of the Central Andean Range. Rounded hills follow one another in a succession of gentle billows, the sides of which are so gradual that one hardly realizes there is a steady ascent. The forest that covers the ridge on the other side of Finlandia is of a heavy, subtropical character—the first of its kind we had encountered on this trip. Red howling monkeys were roaring in the ravine below, but the birds of the forest belonged to a fauna different from the one we had just left.

The palm-filled valley of the Boquilla had been reached by night. Salento, with its low, whitewashed houses, was clearly visible on top of the next ridge. It required just thirty minutes next morning to reach the town after a climb of nine hundred feet. We did not stop at the settlement, but continued up the time-honored trail leading to Quindio Pass; within a short time forest of the most promising kind had been reached and camp was being made in a sheltered spot about half a mile above a lone house called Laguneta. The pack-animals were sent back to Salento, where there was an abundance of pasturage, until they should be required again.

The woods at Laguneta were rather open and there was little underbrush. The trees, however, were burdened with moss, bromelias, orchids, and other epiphytes. Climbing bamboo and creepers filled the few clearings with impenetrable thickets. Most of the vegetation had small, harsh leaves, and the stems were gnarled and stunted. Clusters of fruit resembling pokeberries, on which numerous species of birds fed, grew on tall bushes near the forest's edge. Begonias covered with red and white flowers filled the hollows.

Figure 8. The town of Salento.

The Laguneta region was remarkable for the number of ant-birds found there (*Grallaria*, *Chamœza*, etc.) that are rare in collections on account of the difficulty of collecting them. We secured fifteen different species in the neighborhood. As they live in thickets and on the ground, the only knowledge one has of their presence is their strange whistling notes, distinct in each species, that come from some gloomy spot deep in the tangled vegetation. *Grallaria squamigera* was to me the most interesting species. It is a huge, heavy-bodied bird, olive above and tawny barred with black below. From a distance the coloration reminds one of a large immature robin, but the tail is very short and protrudes only about half an inch beyond the lower coverts, and the long legs measure fully five inches. The plumage is long and full. Occasionally we saw the shy creatures as we worked in front of our tent in the afternoons; we always made it a point to be very quiet and the reward came in the way of shadowy forms that unconcernedly pursued their lives among the logs and brush without suspecting our presence. This shows the advantage of camping in the midst of the wilderness, where one is sure to see and hear wild things at the most unexpected times—experiences that are lost if one does not spend his entire time in the very heart of their environs.Squirrel Woods is the name we applied to a spot below Laguneta and several miles nearer to Salento. On the upward journey the place had been singled out as being unusually attractive for a week's collecting, owing to the number of birds and particularly of squirrels seen from the trail. This, however, proved to be the one place in all Colombia where we were not welcome, and in this regard it is unique in my two years' experience in that country.

After leaving the Quindio trail we followed a narrow path through fields and forest for nearly a mile. It led to a neat, new cottage surrounded by pastures in which there were cattle and horses. The owner and his wife, middle-aged Colombians of the mestizo class, but of better appearance than the average, did not seem overjoyed to see us; they had no room, they said, for strangers. Explanations and the display of credentials bearing flaring, important-looking seals were of no avail; the people did not care to have the drowsy tenor of their ways disturbed by a couple of gringos. The region, however, was too alluring to forego, so we camped beside the house and took possession of the veranda for sleeping-quarters. There we remained a week, much to the displeasure of our unwilling hosts.

We had supposed that the presence of a wheat-field surrounded by primeval forest had led to an increase in the number of small mammals indigenous to the region, but this assumption proved right in so far as squirrels only were concerned. A granary had been built in the centre of the clearing, which was of considerable extent; bundles of grain were piled in it from floor to roof. Squirrels of three species came from the woods, and ensconcing themselves in the structure feasted on the wheat. They ran the entire distance between the forest and the house on the ground, taking advantage, however, of any logs or branches that littered the place. They were especially plentiful in the early morning and just before sundown. If one crept cautiously to the border of the field he was sure to see dark little forms scamper over the ground and disappear in the storehouse. The animals were very tame at first and did not leave their shelter until one was but a few yards away; then they appeared on all sides and ran quickly to the protecting woods. Later they posted a sentinel or remained on the alert, for no sooner did we reach one side of the clearing than all the squirrels hurried away on the opposite side, being careful to keep the granary between themselves and us. There were many stray dogs in the neighborhood; they pursued the squirrels while making their pilgrimage across the open space, and devoured any they succeeded in catching.

There were also other marauders that exacted a heavy toll in grain from the farmer. Yellow-throated woodpeckers (*Melanerpes flavigularis*) and green and yellow jays (*Xanthoura yucas*) were always about and frequently came to grief in our traps set for small rodents.

A species of pigmy opossum (*Thylamys caucæ*) lived in the woods. It is the size of a mouse, but has a longer tail. The slate-colored little animals prefer small cavities in tree-trunks for their homes, where they spend the days curled up in sleep; if disturbed they are very sluggish and may be taken in the hand, their only concern being to find a dark spot where they can snuggle up to one another and go to sleep again. At night they are more active and go on foraging expeditions for fruit, insects, and almost anything of an edible nature they can find.

The camp pet at this time was a young sloth (*Cholœpus andinus*). The slow-moving little beast reminded one of a "Teddy Bear," and when it clambered among the branches of a tree it always recalled to me Hudson's description to the effect that he "hugged the branches as if he loved them." Our pet had been brought in by a native hunter who had shot the mother and found the young one clinging to her long, gray hair. It was easy to handle owing to its inactivity, but occasionally it struck viciously with its front feet, each armed with two formidable claws, and also snapped suddenly in an attempt to bite, its strong teeth enabling it to inflict severe injury. It ate quantities of tender green leaves at regular intervals, but it was always necessary to first sprinkle them liberally with water and then feed them to the little creature one at a time and in quick succession. I have kept a number of sloths at various times and found that they thrived on young shoots and buds of many trees and plants, such as cacao, cabbage, lettuce, and almost any succulent vegetation.

I know of no animal that appears more stupid and lifeless than a sloth. They move with great difficulty and in a sprawling posture on the rare occasions when they descend to the ground, on account of the peculiar formation of the feet; nor do they attain any great speed while moving in the tree-tops, where they always maintain an inverted position except when climbing up or down a trunk. When resting they roll up into a ball, and as a species of green alga not infrequently grows on the fur, they are very inconspicuous among the leaves and moss-covered branches of their home—at least when viewed from below. But from above they do not always escape the sharp eye of the harpy-eagle, which is their chief enemy.

In spite of its lifeless appearance, it would be difficult to find a mammal more tenacious of life; in this respect it resembles the reptiles. Sloths will withstand the most frightful wounds and frequently make their escape after having been shot many times. The natives are very fond of the flesh and not infrequently capture the animals when cutting down trees in clearing land; a favorite way to kill them is by drowning, but this is a lengthy and barbarous process, as it requires a long submergence before the creatures cease struggling and life is extinct.

People of the lower class attribute peculiar powers to the sloth. They say that when one of the animals finds it necessary to descend to the ground it is unable to climb back to its lofty perch; but a friendly cloud is always hovering near by which envelops it and carries it back to any desired station in the tree-tops. In some localities they also attribute the wild call of the giant goatsucker to the sloth. The only time I heard the latter utter any sound was when a mother called to her young that was a few feet away; she gave a fairly loud ‘peep’ and her offspring at once went to her.

After a time our work at Squirrel Woods was completed, so, much to the relief of the inhospitable couple, we left the place and returned to Salento, where we had better fortune and were well cared for by one Colonel Martinez; his wife had come from Bogotá, was a well-educated woman, and, what interested us more just then, was a splendid cook. The family conducted a fairly good *posada* and shop and had various other business interests, including several worthless mining claims along streams flowing into the Quindio River just below. A few excavations had been made into the hillsides; the largest was known as *La Mina del Gallo* and had yielded hundreds of tons of rocks and earth; but as not a speck of the elusive yellow metal they so eagerly sought had been forthcoming, the mine had been abandoned, and owls and bats inhabited the dark tunnel. The greater part of the mining population had deserted Salento for a place about ten miles distant, where extensive cinnabar-fields had been discovered. They expected to acquire fabulous riches extracting the mercury from the deposits. Some Englishmen headed by a man named Lloyd-Owen were also interested in the enterprise, but I learned later that the prospect failed.

At dusk we occasionally had a brief view of the Nevado del Tolima far to the east. The snow-capped summit is over eighteen thousand feet high, but we could never see more than a small portion of it on account of the ridges that surrounded it. At night the snowy dome gleamed white and frosty beneath a brilliant moon, and chill winds blew from the frigid heights and roared through the town. The paramos of Ruiz and Isabel, composed of high, cold valleys, plateaus, and snow-covered peaks are south of the Tolima. We straightway resolved to visit that region, and as the rainy season with its severe electrical storms was fast approaching, no time was lost in starting on the expedition. My experience on the Cerro Munchique was still too fresh to make me want to duplicate it or expose any other members of the party to a similar ordeal.

September 12 found us wending our way along the Quindio River toward its headwaters. The valley floor is covered with grass that is kept close-cropped by cattle and horses. Low shrubbery grows along the river-bank; the stream—not over one hundred feet wide—is clear and swift and the icy water rushes over a boulder-strewn bed. A scattered growth of tall palms dots the entire valley and extends up the mountainsides to an elevation of about nine thousand five hundred feet.

The trail is so indistinct that Allen and I, who were riding in advance of the pack-mules, lost it and spent two hours in a vain endeavor to recover the way; then we saw the cargoes and peons far below, resembling moving black dots, and hurriedly rejoined them just as they were leaving the valley for the abrupt slope. The trail from here onward was steep and rough. Before us stretched a seemingly endless succession of ridges, farallones, tall rocks, and high precipices that reach a climax in the brown paramo of Santa Isabel, backed by walls of gleaming snow. In looking back over the way we had just come we could see the Quindio and the thousands of palms growing in its valley spread before us like a map.

The lower slopes were barren, having but recently been burned over; fire was still raging in a number of places and the hissing and popping of burning vegetation could be heard frequently with distinctness. Tall, smouldering stumps were clustered here and there like blackened chimneys from the tops of which wisps and columns of smoke ascended into a hazy sky. The pungent odor of burning green plants was at times almost suffocating.

Forest begins at nine thousand five hundred feet. It is at first somewhat open and reminded us of Laguneta. The rich mould of the forest floor was very deep and caused us much anxiety lest some of the pack-animals be lost, for they sank into it to a great depth, and there was constant danger of their floundering and pitching headlong down the mountainside. The arrieros took the utmost precautions, but even then one of the mules became overbalanced and fell off the trail. Fortunately the trees grew close together and one of the packs became wedged between two of them and halted the rolling creature a short distance below. It struggled there with feet in the air until the peons released it and led it back to the trail.

Toward evening we reached a native hut—the second since leaving the valley. The elevation of the place was ten thousand five hundred feet. A large clearing in which white clover grew abundantly surrounded the house. The inhabitants also had other clearings farther down, where they planted corn and wheat. They were all suffering with colds and the dreaded *dengue*, from which I was fortunately able to give them some relief with the aid of our medical kit. In return for this service they treated us most courteously and placed one of their two rooms at our disposal, although it happened that a score or more of chickens occupied the same quarters. The night was cold and damp. Next morning the wretched people gave us milk and cheese and we purchased several dozen eggs—certainly a great luxury in such an out-of-the-way place. They also showed us the skin and feet of a tapir one of the men had killed in the forest above. The hide had been used to make bottoms for chairs and was of a black color. They reported the presence of two species of bears, one entirely black and the other the tolerably well-known spectacled bear. Although the latter is the only species of bear supposed to exist in South America, I have been told repeatedly by the people that a large black bear is found in the high Andes and have seen skins that appeared to bear out their statements.

After leaving the house next morning we soon reached heavy mountain forest. A deserted hut stood near the border of it, so on our return from the paramo we spent several days there. The chief attraction about the place was the abundance of white-throated sparrows (*Brachyspiza capensis capensis*). Their cheerful little song cannot fail to endear them to any one with even a limited æsthetic nature. Whether one hears it in the hot, tropical lowlands or on a bleak mountain-top twelve thousand feet above sea-level, the happy little melody is always the same. Nor is the music confined to the hours of daylight only. I have frequently heard it in the darkest hours of night, ringing clear and sweet from somewhere out in the all-pervading blackness. These birds are fond of the proximity of man and are most abundant where he has chosen to break the soil and erect his abode. As a general rule they are not gregarious, but I have seen them congregate in flocks of many thousands to spend the night in some particularly attractive spot in places where sleeping sites were limited in number. Farther south these sparrows also gather in flocks of varying size during the winter season.

The nest is a neat, cup-shaped structure made of fine grasses; it is placed in a low bush or on the ground. Two or three pale-blue eggs thickly spotted with brown are laid and not infrequently two broods are reared in a season.

During our stay at the solitary house on the edge of the great forest a white-throat or *chingolo* came daily and perched on the bannister of our porch to pour out its overflow of happiness. We grew very much attached to the confiding feathered mite and eagerly awaited its frequent visits. After a short time I discovered the runway of some small rodent under the porch and set a trap to catch the animal. Not long after we heard the dull snap of the spring, and upon investigation found the limp body of the unfortunate songster. The place seemed deserted without the sprightly little bird and we never ceased to miss it.

The belt of forest through which we penetrated before reaching the paramo was magnificent. A species of orchid bearing long spikes of yellow flowers was in full bloom; there were many hundreds of the thick-leaved plants, some perched on lofty branches, others growing from crotches but a few feet above the ground, but all surmounted by a glorious halo of golden blossoms.

We left the forest with its giant moss-covered trees, ensnaring creepers, and breathless silence that suggests a thousand mysteries, at about noon. It ends abruptly and is replaced by a narrow strip of low, dwarfed trees and bushes with small leaves that are either very stiff or are covered with thick down. There were also clumps of blueberry-bushes, but the fruit was woody, bitter and inedible for human beings. Lupines and gentians grew in the hollows and numerous composites thrived on the slopes; among the latter was one with showy purple flowers that the peons called "arnica."

After a stiff climb of an hour we gained the summit of a rise; the whole panorama of the paramo was spread out before us—a marvellous series of brown plateaus, sunken valleys with tiny rivulets meandering through them, and stern ridges dotted with blackened, rocky peaks. The snow-fields of the higher altitudes were entirely obliterated by banks of cold, gray clouds.

The word *páramo* means an elevated plain, barren of trees, uncultivated, uninhabited, and exposed to the icy blasts of wind from the higher elevations. This description exactly fitted the country before us. We descended into one of the valleys, at the head of which lay a placid lake of small size, and made camp at the base of one of the protecting walls of rock that flanked it. The elevation of the valley is about twelve thousand seven hundred feet, and the main peaks of the range hemming in the paramo rise to a height of sixteen thousand feet or more.

Long, wiry grass covered the valley floor; the top was bent over, forming a billowy expanse of brown, variegated here and there with a diminutive patch of green. Lifting any one of the tufts disclosed a labyrinth of tunnels and runways apparently made by small mammals; but, strange to say, we saw a small number only of rabbits, and few rats came to our traps. If the network of tunnels harbored other creatures, they effectively succeeded in evading our every effort to discover them. Probably the denizens of this underworld had learned the value of extreme caution and wariness because numbers of eagles (*Lophotriorchis*) were always soaring overhead ready to pounce down on any of them that for an instant relaxed their vigil.

Figure 9. The lake on the paramo of Santa Isabel.

Figure 10. Snow on the paramo of Ruiz.

A large part of the soil was springy beneath our step; it was undermined by numberless rivulets which trickled from the slopes and made their way to the stream in the centre of the valley. These wet places were covered with extensive areas of daisy-like plants having clumps or rosettes of stiff leaves; the squat, green hummocks were strong enough to support one's weight, but walking over them was always accompanied by the feeling that they might give way suddenly and precipitate one into the deep mire. Sphagnum flourished along the edges of the marsh where it was not too wet.

The peculiar, gray, mullein-like plant called frailejón thrives in rocky places that were sheltered to some extent; but clumps of the plants also braved the open, wind-swept slopes and grew to the very edge of the snow-fields.

The heavy, orchid-laden forest through which we passed just before reaching the paramo encroached upon the valley's lower end, but for a short distance only. There were well-worn trails made by tapirs and deer that came nightly to feed on the abundant grass, for despite the dry and withered appearance of the upper layer there was a deep carpet of tender green shoots underneath.

There was an abundance of birds on the paramo, especially along the bush-grown banks of the streamlet; but all were of dull colors—slaty blue, gray, black, or deep brown, that harmonized well with the bleak surroundings. Their habits reminded us of open-country birds of the northern United States. Gray flycatchers ran over the ground; at frequent intervals they mounted high in the air, like horned larks, for which we at first from a distance mistook them. A small wren-like bird, black with brown flanks (*Scytalopus sylvestris*), lived in the taller herbage. It had a piping note that could be clearly heard fifty yards away, but the agile bird was hard to see on account of its obscure color and mouse-like habits that kept it constantly in the thickest cover. Numerous marsh-wrens (*Cistothorus æquatorialis*) inhabited the sedges, scolding and nervously flitting about.

More interesting than the foregoing, however, were large Andean snipe (*Gallinago nobilis*) bearing at least a superficial resemblance to the American woodcock. Single individuals or pairs of these birds were found running over the bogs and drilling in the soft earth. In many places the ground was perforated with dozens of the deep, symmetrical holes where the tireless workers had labored diligently for a meal. Shooting them was good sport. They sprang into the air with a piping bleat and then sped away in a zigzag course for fifty or a hundred yards, dropped back to earth and instantly merged into their surroundings so completely as to be invisible.

The finches were perhaps better represented than any other family of birds. A few goldfinches, in small bands, frequented the flowering shrubs. A kind of slaty finch (*Phrygilus unicolor grandis*) was far more abundant and fairly evenly distributed over the entire paramo. We discovered a nest of this species among the grass at the base of a frailejon; the structure was beautifully made of down taken from the leaves of the plants that sheltered it. It contained two pear-shaped eggs of a greenish color heavily speckled with fine dull-brown dots.

From a distance the small lake at the head of the valley appeared to be a promising field for investigation. It yielded, however, but a solitary Andean teal greatly resembling the gadwall (*Chaulelasmus*), that was swimming on the unruffled water, and when this had been taken our work in that particular spot was completed. The bottom of the pond was covered with a solid mass of long algæ far out as we could see; these concealed any aquatic life that may have existed in the chilly depths.

The weather was usually agreeable during the greater part of the day, the thermometer registering in the neighborhood of 76° at noon, and dropping to 30° at night. It rained little, but banks of clouds rolled in frequently and precipitated a superabundance of moisture.

One day Allen and I undertook an exploration trip to the snow-line. We started at daybreak, taking with us our guns, an abundant supply of ammunition, cameras, and a small parcel of lunch. We made straight for the head of the valley, passed the lake, and had soon reached the top of the weathered ridge that formed the first barrier to our progress. From the summit, fourteen thousand four hundred feet up, we could see numerous other isolated depressions like the one we had just left; in one of them was a newly made trench—probably the work of some venturesome miner who had drifted to this lonely place in search of gold. So far we had had not a glimpse of snow on account of the heavy mist. We followed along the top of a hogback running northward and gradually leading to higher country that flattened out into a marshy plateau on its farther end. Progress was difficult. At each step the bog quivered within a radius of several yards and the clumps of matted vegetation depressed by our weight were quickly covered with water that oozed from below. This was an ideal spot for snipe and several sprang up as we painfully picked our way over the treacherous ground; but the great exertion and high altitude had a demoralizing effect on our aim, with the result that we were relieved of a good deal of ammunition without securing a single bird in return.

A high wall of bare rock rose just beyond the confines of the bog, and gaining the top of it we were up fifteen thousand feet. It was covered with blackened rock fragments—mostly the result of weathering, but some of them probably detached from the many towering crags and columns by the shattering force of lightning. The highest point in the wall is fifteen thousand two hundred feet. As we rested a moment to recover our breath, a procedure necessary every twenty steps, the fog suddenly lifted and disclosed the snow-bound slopes of Ruiz a short distance away. Between us lay a valley flanked by perpendicular walls of rock and hundreds of feet deep. The snow apparently extended down two hundred feet lower than our station, making its lower limit fifteen thousand feet.

We stood lost in admiration at the marvelous spectacle that unfolded itself before us. The hurrying curtains of clouds revealed ever-changing scenes. One moment miles of slopes covered with a white mantle of snow stood out in bold relief; the next, they were whisked from view and bare pinnacles of dark rock, like the spires of a cathedral, appeared momentarily high above, their ragged outlines softened by a veil of thin blue haze. Again, the lower edges of the panorama came into view, revealing glaciers and avalanches of snow and rocks perched on the brink of the wall ready to plunge with a boom into the deep valley.

The floor of the valley was a series of ponds and morasses. Ducks disported in the cold water, all oblivious of our presence, and apparently safe in their, at least to man, inaccessible retreat. A raging torrent tore along the base of the wall, adding its roar as a fitting accompaniment to the general awe-inspiring character of its desolate and inhospitable surroundings.

A whisp of vapor borne on a chill wind hurried across the intervening chasm and blew into our faces. Time had passed faster than we realized and we discovered that half of the afternoon was gone. Hurriedly we began to retrace our steps along the wall of rock and through the treacherous bog. By the time the sharp ridge was reached, clouds in such volumes had rolled in over the paramo that everything was obscured outside of a radius of a few yards from us. There was no trail of any kind, and even the most familiar rocks assumed strange shapes swathed in the dank vapor. A compass is useless under such circumstances. Before long we reached the interlacing mass of ridges and, after holding a consultation, followed along the top of one that seemed to lead in the right direction. We stumbled along for two hours or more, and then realized that we were lost. Darkness was fast approaching and a raw wind swept down from the region of perpetual ice and snow. We began to look for a sheltered spot in which to spend the night, for it now seemed certain that each step was only taking us farther from camp. Just then a rift in the clouds appeared, and before it again closed we caught sight of a faint glimmer far below and to the right. That could mean but one thing: it was a reflection from the lake at the head of "our" valley. For more than an hour we had been travelling in exactly the opposite direction. We gave up the thought of a bed of frailejon leaves without regret and stumbled down the steep slope straight for the spot where the lake had flashed into view. After many collapses from thirst and fatigue we reached the brook with its crystal, ice-cold water; then progress was easier, and within another hour the glow of the camp-fire appeared through the haze, and soon we were snugly ensconced in the depths of our blankets.

A few days after our journey to Ruiz the weather changed greatly. Low-hanging fogs covered the paramo day and night; lightning flashed among the clouds, and frigid gales roared over the plateaus. These were signs of the coming winter and warned us to leave the paramo before it was too late. Soon there would be only snow and ice, penetrating mists, the reverberating roll of thunder, and blinding displays of electricity. The elements would be unleashed and in all their grandeur, and awe-inspiring frightfulness take possession of the upper world. Life would then be unendurable, so we accepted the warning and returned to Salento.

Chapter 5

THE CHOCÓ COUNTRY ON THE WESTERN COAST OF COLOMBIA

Upon returning to Cartago from our expedition to the bleak paramo of Santa Isabel, we began preparations for a visit to the notorious Chocó, which lies along the western coast and within the San Juan River watershed. This section of the country presents the other extreme in climatic conditions. It has been rarely visited by naturalists on account of its inaccessibility; and the few who have succeeded in forcing their way within its inhospitable borders have found it impossible to remain any length of time. Malarial and yellow fevers are endemic among the natives, but quickly sap the vitality and life of newcomers into the region; rain falls daily—four hundred inches being the average precipitation for one year—and the heat is so intense that when the sun appears during the intervals between showers the whole jungle is converted into a steaming inferno. Small wonder, then, that the fabulous wealth in gold and platinum of the Chocó has been little more than touched.

Our plans called for an overland trip to Nóvita on the Tamaná River; after reaching that point local conditions would have to guide our subsequent movements. Trail there is none, but a footpath, often so faint that it loses itself among the vegetation or in the beds of streams, serves the purpose of partially guiding the stalwart negro who carries the mail to Nóvita at infrequent intervals, as well as others who undertake to cross the Western Range into the tropical lowland.

The townspeople of Cartago had heard a good deal, in a general way, about conditions existing in the Chocó, but they could give no information of practical value. We haunted the market and other places where *peons* congregate in numbers in our endeavor to secure porters for the trip. The few who reluctantly expressed a readiness to go did not seem physically fit for such strenuous work, so I rejected them. One day a caravan of oxen arrived from the settlement of Salencio, and I hastily engaged them for the return trip, as these animals, while slow, are sure-footed, and can pick their way through mud and jungle that horses could not penetrate.

Leaving Cartago, we crossed the arid Cauca Valley; the land west of the river is more rolling than on the opposite bank, but the character of the plant life is much the same. Within an hour Ansermanueva, a cluster of twenty or thirty adobe hovels, was sighted in the distance, but the trail divided just before reaching the village and we followed the southern branch. Beyond this fork the climb into the mountains began; there are two ridges, six thousand eight hundred feet and seven thousand five hundred feet high respectively, with a ravine of five thousand eight hundred feet between. The "cloud" forest does not begin below the top of the first ridge; then there is an abundance of mosses, ferns, bromelias and other epiphytes forming a growth that is both rank and beautiful, and equalling in density that found in any other region. The greater luxuriance of the flora on the western slope indicated a heavier rainfall on that side; this is accounted for by the fact that the summits of the various ridges stop the moisture-laden winds from the Pacific, to a large extent, and cause them to precipitate the water on the ocean side of the divide.

Within two days we arrived at Salencio, small, dilapidated, situated on a little plateau between the peaks, and inhabited mainly by half-breeds. We were advised to wait until the weekly market-day, when many people from the surroundings would come to town, and it would be possible to secure porters for the continuation of the journey. In the meantime we made short excursions into the neighboring forest; they yielded several novelties, among them a splendid example of the military macaw (*Ara m. militaris*). This gorgeously colored bird is rare, indeed, and we have never seen more than two at the same time. Spectacled bears were said to be common and to come to the clearings when corn is ripe; the number of pelts exhibited by the inhabitants amply verified their assertion.

When Sunday came, and with it the gathering of people always present when market is held, we had no trouble in engaging the required *peons*, each of whom agreed to carry a pack of seventy-five pounds. Early the next morning they appeared, eager and ready for their undertaking. They shouldered their loads and started away at a fast gait, while we brought up the rear of the column to prevent straggling.

The way lay across a low, forested ridge, and then adhered closely to the bamboo-covered banks of a small stream called locally Rio Cabeceros or Rio Vueltas, but which is really the headwaters of the Sipi River. At one time we waded in the knee-deep water a distance of over three miles, as it was easier than to force a way through the matted plant life on either side. I soon discovered that the porters did not possess the endurance of those we had previously employed on other expeditions, and I believe this was due to the fact that the use of coca leaves is unknown in this part of Colombia. Whenever our *peons* had an abundance of coca to chew they seemed tireless in the performance of their work; those not given to the habit required large and frequent meals, ate *panela* all day long as they marched, and were capable of covering a short distance only in the course of a day's walk. We were compelled to halt early and chose the top of a knoll for a camping site.

A steady downpour of rain had fallen the entire afternoon, which continued throughout the night, and this, coupled with the severe cold (the elevation being seven thousand two hundred feet) and the desirability of preparing hot food, caused us to long for the comforts of a huge camp-fire. Dry wood was out of the question, but the men cut down a tree, the green wood of which burned readily, and had soon started a fire adequate for working purposes. Their ponchos, which had become saturated with water, were of no service in keeping them warm, so they sat up the entire night, singing, telling stories, and drinking hot coffee in their endeavors to remain cheerful and keep warm.

On the following day the vegetation was far more dense, and advantage was taken of numerous narrow fissures in the mountainside roofed over with logs and moss; through these tunnels we crawled on hands and knees, but that was easier than forcing a way through the tangled mass of plants growing above. When camp was made that night the base of a tree was selected for a fireplace. At first glance it seemed that the diameter of the vine-covered trunk must be at least ten feet, but this was a delusion. After the men had vigorously plied their machetes on the creepers, moss, and ferns, a stem not over two feet across was revealed; they cleared away the lower tangle, leaving a protecting umbrella-like canopy overhead that shielded the entire party from the rain while they cooked their food.

We crossed three ridges in all, the elevation of each being slightly in excess of seven thousand feet, with depressions of from two thousand feet to three thousand feet between them. All are heavily forested, the growth above four thousand feet being subtropical in character, while that lower down is typical of the tropics and comparatively open.

At the end of the third day we heard the welcome roar of water, and not long after halted on the bank of the Hávita River. A naked negro came from the far side in answer to our calls, and ferried us across the stream in a huge dugout canoe. There we found a settlement of half a dozen bamboo huts filled with lazy negroes clothed in scanty attire. The place is called El Puente. About one hundred yards below the group of hovels, the Hávita is joined by the Rio Ingara. The water of both streams is swift, cool, and of a bluish-gray color. Each of the streams is about seventy-five yards wide just above the junction.

After crossing another ridge which required two days' time, we reached Juntas de Tamaná, on the south bank of the Hávita, a stone's throw above the point where this stream empties into the Tamaná, and but four hundred feet above sea-level. Excepting only the little clearing in which the fifteen dilapidated negro abodes stand, the entire country is covered with a forest of tall trees; there is little undergrowth, but many of the lower branches are covered with epiphytes, and long vines or "forest ropes" dangle down from the interlocking tree-tops to the very ground.

The negroes of Juntas are a miserable, sickly lot. They suffer from lack of food, for the simple reason that they are too indolent to grow in sufficient quantities the plantains, yuccas, and other plants that thrive with a minimum of attention in such a favorable location. Instead of making clearings and cultivating the fertile ground, they prefer to lounge in their hammocks and take a chance at starving to death. At irregular intervals, when the pinch of want is too great to endure longer, the men paddle in canoes to their *fincas* to cut sugar-cane, gather plantains, and to pick palm-nuts in the forest. Upon their return the family gathers about the food and eats until not a vestige remains. So effectively do they attack the mound of provisions that one might easily imagine a swarm of locusts had paid the region a visit.

Figure 11. Native of Juntas de Tamaná with trail-haunting blacksnake.

Figure 12. The author with natives of Juntas de Tamaná.

A day or two after our arrival at Juntas a two-year-old child belonging to one of the families died. The news spread rapidly and by night the entire neighborhood had turned out for a wake. We followed the crowd. The baby, in a white dress, with bright red and green ribbon trimming, lay in a wooden box on the table. A canopy of muslin had been erected above the bier which was strewn with wild flowers. The room was packed to suffocation with the black forms of the populace, which glistened in the dim, flickering candle-light. At first bottles of *aguardiente* were distributed, and every one had a number of liberal-sized drinks. Then the older folks withdrew against the four walls and, squatting on the floor, sang or lamented as fancy dictated. The younger people divided into two parties and played games around the coffin. One of them was a kind of charade and, when the guessing side solved the riddle, they pursued and caught the others, amid loud shouts and laughter. I feared constantly that they might upset the coffin. Occasionally someone would stop long enough to pet or caress the dead little form, and address a few terms of endearment to it, such as *pobrecito*, *angelito*, or *tan lindito*. The revelry lasted until daylight; then a procession slowly wound its way to a newly dug grave and deposited its burden, leaving the only little mound visible that side of the Tamaná.

Christmas was drawing near. We were surprised to see the women apparently making preparations for a celebration, which is most unusual in South America. They worked several days cutting the weeds around the village and cleaning up the place. When we asked about it, they said it was not on account of the approaching *fiesta*, but a form of penance they performed annually in atonement of their sins. Apparently the men were without blemish, for they gazed upon the workers and addressed jocular remarks to them from the comfortable retreat of their hammocks, even enumerating particular misdeeds and suggesting special forms of penance that might be effective.

The next stage of our journey had to be performed on the river. We secured a huge *bongo* and stalwart negro paddlers, and December 21 found us speeding down-stream toward Nóvita. The Tamaná is a rapid stream, varying between one hundred and three hundred yards in width. Its bed is strewn with boulders, causing rapids easily navigable on the downward voyage, but difficult and dangerous to negotiate when bound up-stream. Then there are deep passages between high, crumbling banks, where the water glides silently onward like an olive-drab stream of molten glass. The densest of tropical jungles lines both banks; its matted walls facing the river are interrupted by small clearings at infrequent intervals, where low hovels stand surrounded by the rich foliage of banana and yucca plants. Chonta-palms, with bristling, spiny stems, rear their plumed heads above the other forest-trees, or droop over the water in a graceful manner, forming a dainty filigree against the brazen sky. The brassy, merciless sun blazed down with unrelenting vigor, and we were glad when dark storm-clouds obscured the sky and provided a greatly needed respite.

It was possible to proceed only to a point called Cabeceros, below which rapids of a formidable character obstruct further navigation. The few negroes living on the river-bank can usually be induced to assist in making the portage, men and women alike undertaking to carry packs to Tambito at the foot of the rapids. Here it was necessary to secure another *bongo* and the trip was resumed.

The Tamaná grows wider constantly. Cataracts are of more frequent occurrence and present greater hazards in their navigation. The *bongo*, made of a huge tree-trunk and measuring thirty feet in length, and a yard in width, was most seaworthy; but frequently it shipped water in alarming quantities, and scraped and bumped over the hidden rocks until we expected the craft to be rent asunder and flounder.

During the greater part of the afternoon we were in sight of a high, isolated mountain, appearing on the map under the name Cerro Torra. So far as I can learn no explorer has ever succeeded in gaining its summit, and when I beheld the vast stretch of impenetrable jungle extending from the river to apparently the very top of the mountain, I could readily understand why the few men who had attempted this piece of exploration had failed in their undertaking.

Late in the afternoon we landed at Nóvita. I was somewhat surprised at the size of the town, which consists of about fifty hovels. The white population, which was very small, consists mainly of traders, and is more or less transient. I was told that they remain in the region a year or two to buy gold and to sell their stock of provisions and merchandise at exorbitant prices, and then return to a more healthful climate—to suffer many years afterward from the effects of their sojourn in the Chocó.

Nóvita is essentially a mining town. A good deal of gold and platinum are washed out of the small streams that form a network in the surrounding country. The negroes and Indians bring in the precious metals in small quantities—wrapped in leaves—and trade them for tinned food and cloth. However, the town seemed to be on the decline in favor of Condoto, Pueblo Rico, and Quibdó, where richer mineral deposits had been located.

The forest contained comparatively little wild life, and that was typical of the Pacific tropical faunal zone. We daily took long tramps and discovered numerous things of more than passing interest. Among them was a colony of nesting black-and-yellow orioles (*Icterus*). The birds had selected a solitary ceiba-tree standing in the centre of a banana-field. It was seventy feet to the lowest limbs and the trunk was so thick and smooth that no predatory animal could climb it, which insured the safety of the colony from such a source of danger. The nests, like huge pears, dangled from the tips of the branches; I counted one hundred and four, and there must have been many others concealed by the foliage. The adult birds were busy and excited, and were coming and going in steady streams, keeping up their noisy chattering all the while. We found numerous bits of egg-shells, white with black dots, on the ground, indicating that the young were just hatching.

One evening as we were returning from a long hunt, we noticed lines of bats emerging from the little church standing on the edge of the village. Next day (Christmas) visited this rendezvous accompanied by several negro assistants. The bats were all concealed within the board walls, so that it was impossible to get at them, but the negroes unhesitatingly tore away the slabs of flattened bamboo and soon had the room filled with a squeaking, fluttering swarm which they attacked with sticks. This method of attack proving too slow, they grabbed guns and fired into the masses amid wild shouts of merriment. When the pandemonium was over and the heap of slain had been collected, they respectfully removed their hats and in passing out of the church reverently bowed the knee before the altar.

We had been cautioned to be on the alert for snakes. The deadly bushmaster or *verrugosa* was said to be particularly abundant. While hunting one day, Allen shot a hawk and placed it in the back pocket of his hunting-coat. To all appearances the bird was dead; while crawling through a thicket a short time later he felt a sudden sharp sting in his back and, throwing up his hands in terror, yelled, "Oh, Lord! one got me at last," thinking, of course, that he had been struck by a snake. Hurriedly removing his coat, the discovery was made that the supposedly dead hawk had been stunned only and, reviving, had promptly dug its talons in the first thing that offered a firm hold. One may well imagine the unpleasantness of such an experience.

Occasionally we saw a species of blacksnake that grows to a length of more than twelve feet. It is perfectly harmless, but has the disagreeable habit of haunting trails and footpaths near the villages. When a pedestrian approaches it rears its head several feet above the ground and calmly gazes into his face. The first few times this happens, the sudden, upward lunge of the big head, the rapidly playing tongue and the beady eyes give one a decided shock and provide ample cause for flight. Later, one becomes more or less accustomed to it. This snake was also plentiful in tropical Venezuela and Bolivia.

Figure 13. Nóvita, the largest town in the Chocó.

It was impossible to secure fresh meat at Nóvita; salt beef was imported in barrels, but it was of such poor quality that we could not eat it. We therefore depended on toucans and parrots for our meat-supply, and found both species very palatable.

The paper money used throughout the greater part of Colombia is not recognized by inhabitants of the Chocó. It rots in the wet, hot atmosphere and for that reason is valueless. Neither are gold coins wanted, but some of the shopkeepers accepted them at a twelve per cent discount. The money that finds favor is composed of silver coins from Mexico and practically all the other South and Central American republics; it is valued according to size, the "dollars" passing for forty cents, the halves for twenty, and so on. I found a number of United States half-dimes circulating at two cent, and dimes at four cent values, and "collected" all that came within reach.

After a few days' hunting around Nóvita we secured another *bongo* and resumed our journey down-stream. The Tamaná empties into the San Juan, about ten miles below Nóvita. The latter river is wider and deeper, but there is no change in the country bordering it. All day long we glided steadily onward, stopping at noon only for a brief respite from the burning sun. At dusk we landed to spend the night near a negro hut. The floor was raised five feet from the ground and the ragged, thatched roof nearly touched it; there were no walls. Altogether it was a most primitive dwelling, in which the dusky forms of the occupants moved like shadows against the dim light of their cooking fire. Noanamá was reached the next day. It is not quite so large as Juntas de Tamaná, and stands on a bluff overlooking the river. The inhabitants are all negroes; the males wore breech-cloths only, while the costume of the women consisted of a narrow cloth fastened around the waist with a string. Both men and women spend a few hours each day washing gold on the river-bank, securing enough from this work to pay for provisions brought from Buenaventura. When they have accumulated a small quantity of the fine, sparkling flakes they embark in their canoes and make their way to the seaport in three days, there to do their trading. It was impossible to hire them for any kind of work; one woman had flour, but could not bake bread for lack of fire-wood, because no one would carry it from the forest one hundred yards away. Indians came to the village daily. They wore many ornaments of beaten silver about their necks and wrists; some of them also had earrings made of the same metal, the size of doorknobs; they were so heavy that a framework of sticks placed at the back of the head had to be used to support their weight. I was greatly amused by the actions of one stalwart young brave who, with his wife and baby, came to the settlement each day. While in town, where he might be observed, he paid no attention whatever to his family; he walked several paces in front of the woman, who, of course, carried the baby, and not once even condescended to glance in their direction. However, when they reached the river-bank or some other secluded spot where he was safe from prying eyes, he snatched the infant from the mother's arms, kissed it, tossed it into the air and acted exactly like any other fond parent. If any one approached, he hastily returned it to his wife and resumed his taciturn expression.

At times a small steamer, the *Fluvial*, from Buenaventura, visits the settlements on the lower San Juan. We waited in vain ten days for her appearance. However, a launch belonging to a miner, a Mr. Stapleton, chanced to pass, and the owner kindly offered to take us to the coast.

The San Juan grows constantly wider. Its banks are dotted with the conical huts of Indians; the floors are always raised on poles, high above the ground, to escape the floods and insects.

As we sped down the river many of the naked, painted savages rushed out in their canoes, paddling and yelling like demons in attempts to overtake the launch. I do not know what object they had in mind as we always outdistanced them. We also saw others catching crabs in places where the high, sheer banks were honeycombed with holes made by these crustaceans. They had slender, sharpened sticks with a barb on the end, which they inserted in the burrows and then withdrew with the struggling victims impaled on them.

We reached the mouth of the San Juan in two days' time. The river is very wide at this point and dotted with low mangrove islands. A sand-bar almost completely blocks the estuary, and when we left the next morning we had great difficulty in finding a passage. Then followed a wild, careening dash of forty miles in the open ocean. The launch was but twenty-one feet long, and we were compelled to go out of sight of land to avoid rocks and reefs; but dusk found us well within the confines of Buenaventura Bay, ploughing through the placid water at great speed and frightening up innumerable flocks of brown pelicans that much preferred to float comfortably on the unruffled surface, and took wing only as a last resort to escape being run down.

Buenaventura had never seemed attractive or inviting to us before, but after a month in the steaming coastal land, with its almost constant downpour, insect pests, and terrific heat, it appeared to be altogether delightful. We returned to Cali and spent weeks on our backs suffering from the fevers with which we had become inoculated. Allen's attack was so severe that he was compelled to return to the United States two days after reaching San Agustin on our next expedition, and just before the discovery of some of our most valuable material.

Chapter 6

In Quest of the Cock-of-the-Rock

On my fourth visit to Popayán we had to remain in the city the greater part of a week, arranging for the continuation of our journey across the Central Andes to the headwaters of the Magdalena. Hereafter we were to travel on foot, partly due to the fact that some of the trails were impassable, both to riding and pack animals, and partly to enable us to be in a position better to study the wild life of the region we traversed. I was accompanied on this particular expedition by Doctor Allen and Mr. J. T. Lloyd, of Cornell University.

On February 27 we left Popayán on foot, the mule-train following some little distance behind. The route lay through undulating country, rather well cultivated, where there were numerous huts at which we found shelter for the nights. At one of these stopping-places the natives were engaged in thrashing beans. The pods had been heaped upon a straw mat and the family were beating them with heavy flails. Wheat was thrashed in the same manner, but after the grains had been beaten loose from the chaff large pans full were held high above the head and poured out in a thin, steady stream; the wind blew the chaff from the falling column and the wheat dropped upon the mat. At another hut men were manufacturing "cabulla" by stripping off, between two sticks, the fleshy part of the leaves of the yucca-plant. The tough fibres remaining were mixed with horsehair and braided into strong ropes. Food was scarce, the natives subsisting upon the inevitable "sancocho" of boiled green plantains, and cornmeal "jarepas." However, we managed occasionally to pick up a fowl, some green corn, and once, we succeeded in purchasing a live sheep; this, in addition to the provisions we carried, enabled us to fare passably well.

On March 7 we had reached the top of a ridge ten thousand three hundred and fifty feet high, having passed the little villages Timbio, San Miguel, Santa Barbara, and La Vega. La Vega means "fertile plain," and the surrounding country fully justifies the name. Far as the eye could see the gently sloping mountainsides had been divided into a network of small, irregular plots by rows of high, thick hedges. Wheat, corn, cabbage, and rice flourished under the cultivating hand of the Indian; there were also small flocks of sheep, and occasionally a few head of cattle. Small mud-walled huts, singly and in clusters, dotted the maze of green landscape, and over all breathed an air of quiet and contentment.

The trail had gradually led upward, though often descending into gorges and ravines a thousand feet deep. We had passed through patches of barren country, and then entered a wilderness of lovely flowering rhododendrons. The masses of red wild oleanders were beautiful, but the lanes of a species of shrub covered with small waxen blossoms of purest white, mingled with deep-green foliage and the fronds of monstrous subtropical ferns, surpassed any picture that pen can describe or the imagination conjure. From afar we could hear the steady buzz of bees and other insects that swarmed about the flowers, and frequently a humming-bird whirred into the arena, hovered a few moments, and then sped away; myriads of nocturnal insects appeared at night, and great sphinx-moths took the place of the hummers.

The top of the ridge is covered with tall, magnificent forest. We saw numerous signs of bird and animal life. Toucans of several species yelped and clattered their bills in the tall trees above. There were also yellow-shouldered troupials, blue and yellow cotingas, brown creepers, bright-colored hummers, and many dragon-flies. The latter possessed a special interest for Lloyd, who immediately erected breeding-cages and began to study their life history. The larva of the dragon-fly resembles a good-size black beetle and lives in water. It is the possessor of a voracious appetite, feeding upon aquatic insects, the larvæ of mosquitoes, and even upon members of its own kind. Finally it rises to the top, hatches, and continues the cycle of its existence as an aerialist, the terror of the winged insects upon which it preys. Penelopes, small turkey-like birds, were abundant, and proved to be excellent eating. One day we succeeded in taking two specimens of a rare, beautiful tanager (*Serricossypha albocristata*) that lived in small flocks in the tall tree-tops. It was as large as a robin, of a velvety blue-black color, with a white crown and breast of deep scarlet. With such a display of lovely colors one might expect harmony in song; but apparently the vocal ability of the gorgeous creature was limited to a few shrill "peeps" like those of a strayed pullet. Deer also were abundant, and one day we caught a fine cat of the ocelot family.

We pitched camp in the heart of the forest. The vegetation was really wonderful. In spots the lower growth consisted entirely of climbing bamboo, so dense as to be impenetrable; the moss carpeting the ground was often knee-deep, and the trees seemed to be breaking under the weight of the creepers, orchids, mosses, and lilies that burdened every trunk and branch. It rained a good deal, and when the downpour stopped there was always the drip, drip of water that had been absorbed by the spongy masses overhead.

The forest zone extends along the top of the ridge for three or four miles and down about one thousand five hundred feet on the other side, but the slope immediately below this line is either bush-covered or cultivated, and bears every evidence of having been cleared.

Figure 14. Threshing wheat.

Figure 15. Indian hut in the Valle de las Papas.

Fifteen hundred feet lower down we came upon the little settlement Almaguer, which boasts about one hundred adobe houses and two severely plain little churches, but all are whitewashed and present a clean appearance. The main industry is the making of Panama hats of a rather coarse kind. Many Indians visit the town on market-days, bringing coca leaves, lime, and *sera*, a kind of vegetable wax, obtained from a berry that grows in the mountains and used for making candles. Pigeons are very fond of the berry, and as they ripen the great band-tailed species congregate in flocks to feed upon them, becoming so fat that they finally pay with their lives for the short season of feasting. The candles made of *sera* are green, but burn well and are generally better than the ordinary tallow dip. The lime, or "mambe," is used for chewing with the coca leaves, which is a confirmed habit in this part of the country.

As elsewhere, the weekly market at Almaguer is a day of great activity and is looked upon almost in the light of a fiesta. Early in the morning, usually at four o'clock, a cow is killed in the plaza and all the inhabitants gather around to watch the skinning of the carcass.

At eight o'clock the plaza is filled with tradespeople, usually women, squatting on the ground with their wares spread before them in wooden trays, bags, or baskets. All that these simple people deem necessary to existence, and even some luxuries, may be had. There are rows of venders of bread, cakes, and dulces; others with vegetables, rice, coffee, corn, and cheese; occasionally peaches, apples of an inferior quality, oranges, and a few plantains are brought up from some sheltered valley; but the greatest space is always taken up by the coca merchants, who unquestionably do the most thriving business, as everyone takes advantage of market-day to have their "mambero" replenished. Sometimes a buyer of hats visits the market. On such occasions the day is ushered in with an unearthly hammering noise that proceeds from all the houses, and investigation will disclose the women industriously pounding the Panamas into shape on a wooden block. Later they carry them to market on their heads, where the buyer, after a casual examination, makes an offer which varies from forty cents to a few dollars, according to the texture of the hat.

At night the temperature falls rapidly as the cold winds sweep down from the mountains and howl through the streets. We have every reason to remember our night's experience in Almaguer. The pack-animals had failed to catch up and we carried nothing with us, so we spent the long, cheerless hours until sunrise shivering in our bare, dusty room in the *posada*.

The first night from Almaguer was passed at an old mill on the banks of the Caquiona, built by monks many years ago. They had thoughtfully provided a large room to house the Indians who formerly came to have their wheat and corn ground, even to the extent of providing rough bunks; and just outside stood a massive stocks, doubtless also provided for the use of the Indians, but it must have detracted somewhat from the effect of the hospitality extended by the good monks. There was plenty of tender, luscious grass for the mules. Near the river large numbers of butterflies settled on the moist sand to drink; the boulders on the bottom of the clear, cold stream had many houses of the caddis-fly cemented to them—little pebbly mummy-cases in which the owner lay snugly ensconced in the silky lining and quickly repaired the break if we opened them. The next day we passed San Sebastian, the last settlement, and climbed steadily higher toward the cold, bleak paramo that marks the dividing-line between the Cauca and the Magdalena.

After four days we reached the marvellous Valle de las Papas, just below the mist-enshrouded paramo, and took refuge in the pretentious house of old Pedro, a full-blooded Andaquia, while preparing for our final dash across the great barrier.

The Valle de las Papas is a great level stretch of marshy land covered with a growth of tall grass and small clumps of forest, between ten thousand and eleven thousand feet up. The tops of the ridges hem it in on all sides and somewhat protect it from the icy winds. It is said that the ancient Indians cultivated the potato in this valley; hence its name—"The Valley of Potatoes." An elaborate network of canals or drains runs through the valley, but the climate and soil are such that I doubt if cultivation could be carried on to any great extent. Often, for many days at a time, rain and hail fall steadily and the mist is so thick that one cannot venture far on the treacherous boggy soil. Yet, strange to say, cattle thrive wonderfully on the high plateau, and their rearing is the occupation followed by the few Indian families who live on these heights. Beautiful orchids abound in the trees, especially in the forest that reaches up to the valley; we saw many of yellow, purple, and snowy-white. Some of the trees are of the evergreen family, including a kind of holly. There were many indications of deer and tapirs, although we shot none. Large snipe and ant-thrushes were plentiful, and on the streams we saw a number of peculiar little torrent-ducks, or merganettas; large white gulls, which the Indians say are old birds that come up from the sea to die, soared high overhead.

At one end of the valley lies a small lake, of which we had an occasional short view when the clouds drifted up the slopes. All about grew clumps of frailejones. Two streams leave the grassy borders of the lake, mere rivulets ten or twelve feet wide, through which we waded daily; one flows down the extreme eastern slope and develops into the mighty Caquetá that helps to swell the yellow flood of the Amazon; the other breaks through the ridges to the northeast, and dashing down the mountains in a series of rapids and cascades forms the Magdalena, which empties into the Caribbean many hundreds of miles away.

Allen was suffering considerably from the fever contracted in the Chocó four months before. Instead of being benefited by the high, cold climate as we had hoped, his condition grew steadily worse, so we found it necessary to continue our journey sooner than we had anticipated. I hastened back to San Sebastian to engage Indian porters, as mules are unable to carry packs beyond this point, and was assisted in my mission by the schoolmaster, who took a sympathetic interest in our undertaking. He was a pathetic example of a man who might have accomplished great deeds had the opportunity presented itself. One of his most highly cherished possessions was an old magazine containing illustrations of an aeroplane and an article on wireless telegraphy.

With a great deal of difficulty I succeeded in arranging with a dozen Indians to carry our luggage across the cordillera the following week. They were of splendid physique and as fine a looking lot as I had ever seen. The price agreed upon was about seventy-five cents per *arroba* of twenty-five pounds, each man carrying from two to four *arrobas*. The journey would require five days, and each man was to carry his own food for the trip in addition to the pack. The charge was high, judged by local standards, but on account of the rainy season the trail was all but impassable; also, it was the *Semana Santa*, one of the greatest fiestas of the year, when all good Indians should roam the streets, dulling their senses with an excessive use of coca leaves and *guarapo*, and fighting, while the women spent the greater part of the days in church acquiring grace for themselves and their delinquent husbands. A small advance was made to each man to enable him to purchase a supply of ground corn, cane-sugar, and coca. Acceptance of this advance is considered equal to signing a contract, and they rarely, if ever, go back on the deal.

On Wednesday, April 3, the day set for our departure, the men appeared, each provided with a board and strong cords. The packs, consisting of boxes, steamer trunks, and bags, were tied to the boards which fitted the men's backs; a broad band was passed over the forehead and two bands across the chest. Each man carried in his hand a forked stick, or "mula," as a means of aiding him in going up and down the slippery inclines and in walking the logs that crossed the streams.

After a short, steep climb we were out on the bleak paramo, in the midst of the rain, hail, and mist. The wind blew a gale and the cold was intense. Through an occasional break in the banks of fog we had glimpses of the valley on each side filled with dense clumps of frailejones. We continued on in the face of the blinding storm for several hours, but with the coming of darkness the trail left the wind-swept zone and started downward, winding along the canyon of the Magdalena; in the failing light the scenery was bewitchingly beautiful. High, rugged peaks, sheer cliffs, and black masses of forest towered above the sparkling stream that bounded from rock to rock in a succession of falls. Allen and Lloyd had gone on ahead, and after dark I came upon them camped in a unique spot. They had thrown their blankets on a ledge in the face of a cliff that towered several hundred feet above them. A tiny waterfall dashed over the edge of the precipice, cleared the ledge, and joined the greater torrent below. The regular night's stopping-place is known as Santa Marta, which the Indians reached at nine that night.

Immediately after arriving at the camping site the porters boiled corn-meal, which they ate with brown sugar. Each man had brought a sheepskin to use as a bed, and these were dried beside the fire while their food was cooking. Before starting in the morning they had another meal of mush and sugar. During the gruelling day their mouths were kept well filled with coca and lime, and the apparent amount of sustenance and endurance derived from the herb is extraordinary; nor does it seem to have any bad aftereffect, though in Almaguer I saw a number of shaky old women with bloodshot eyes and blackened lips and teeth, said to be due to the result of excessive indulgence in coca.

The second night we failed to catch up with the men who had gone on ahead. We had waded streams and knee-deep mud the greater part of the day as the result of the steady downpour which rendered the trail indescribably bad; everything was drenched and it required more than an hour of hard work to start a small fire. However, the day dawned bright and sunny, and we lingered to watch the tribes of feathered folk that began feeding and chattering in the tree-tops. The ripening fruits had attracted great black guans, trogons with rose-colored breasts and metallic green backs, and wonderful curve-billed hummers with long white tails. Along a stretch of bamboo we saw scores of large, pearly butterflies flapping about lazily, the iridescence of their wings flashing like bits of rainbow in the sunlight; but not a glimpse did we have of the main object of our long wanderings—the rare and elusive cock-of-the-rock.

In the afternoon the rain again fell in unrelenting torrents, and we camped beneath a wall of rock hundreds of feet high, which the Indians called the Peña Seca, or dry stone. Great vines with bunches of scarlet flowers drooped a hundred feet below the top, like gigantic serpents, but not a drop of all the downpour reached us. The base of the cliff was blackened from the numerous camp-fires kindled by Indians on their way to Tolima in quest of salt. By way of divertisement our Indians gathered incense, which is a kind of gum that collects on certain trees, and which they intended to take home with them for use in the *santa iglesia*. I watched the social bees that live in company with termites building tubular entrances that may extend out eighteen inches or more like a coiled pipe-stem to their apartment in the nest; apparently the two different inmates of the common domicile never clash.

The third night we reached the hut of an old Indian who called himself Domingo, and who was as surly a creature as ever walked the earth. As he refused us the hospitality of his hut, we camped outside his gate.

We now occasionally passed through a cleared spot where grain and vegetables grew; cattle grazed on the long, tender grass, and dark-brown, wild-eyed children peered at us from under the fringed, low grass roofs of shambling Indian huts. On the top of every knoll was a row of tall wooden crosses, some newly erected, others decaying and ready to topple over; it is the custom of the natives to erect a new one each year on Good Friday, permitting the old ones to remain standing. We had reached the frontier of Huila.

On Easter Sunday we had our first glimpse of San Agustin, which was decidedly disappointing. All that we could see as we descended the last steep slope was a cluster of some fifty-odd mud huts protruding from the centre of a wide, barren plain; there is no forest within a mile in any direction, and very little cultivation is carried on in the immediate vicinity. The town is very old; the inhabitants are mainly of Spanish descent, but scattered throughout the surrounding country can be found small clearings, or *fincas*, cultivated by full-blooded Indians. These latter are of a reticent though friendly disposition, emerging from the seclusion of their forest-bound homes only on market-days to dispose of the products of the soil and of their flocks.

In recent years the name San Agustin has come into prominence on account of the prehistoric ruins and monoliths that are found in its vicinity, and which are supposed to be of very great antiquity, dating back to a culture that has entirely disappeared and of which nothing definite is known. Even the Indians who to-day inhabit the region have no traditions or folk-lore of the vanished race, and scientists who have examined the ruins have, up to the present time, been unable to account for their origin. It has been suggested that they may represent the work of the tribe of Andaquias, but this statement is disputed by Carlos Cuervo Marquez, who points out that the mute reminders of an ancient civilization already existed in the same unknown condition at the time the Conquistadores overran the empire of the Chibchas.

Figure 16. The village of Santa Barbara.

The thing that first attracted our attention was the row of twelve stone images that stand in the centre of the plaza facing the village chapel, which vary in height from two to eight feet and are carved from sandstone and granite. Gigantic heads, with round faces and staring, expressionless eyes, are set upon short, square bodies. Some are crowned with hats or head-coverings that range in pattern from the Turkish fez and sugar-loaf to curious curved caps that may have been intended to simulate the rainbow. Many of the figures are quite naked, while others are clothed in a narrow band, or loin-cloth. The teeth of many of the human beings represented are prominent, and each has two pair of great pointed canines like those of a beast. This row of images was placed in its present location by order of the priest who had charge of the parish; we may imagine at what cost of labor when we realize that many of the stones weigh several tons. Of course, there are no trails, and the only way was to drag them out of the forest with ropes.

Figure 17. A corner of San Augustin.

One of the monoliths represents a woman with a small child in one arm and a club in the other hand raised in an attitude of defense; on one is carved a woman meshing a *muchila*, and on another a man is holding a fish. There is the hewn figure of a large monkey crouching over a smaller one, and some distance away stands an owl holding a snake in its beak. A flat slab in a recumbent position bears the engraved figure of a woman and possibly served as the covering of a coffin or a grave. Then there is the statue of a woman with a mallet in one hand and a chisel in the other, thought to represent the goddess of sculpture. It seems not improbable that the greater number of the images represent idols which were worshipped by the ancient people.

The most interesting examples are to be found in the forest above San Agustin. Under the giant cedars and tall cecropias that cover the slopes one finds works of a more pretentious nature, scattered among the dense low palm growths and covered with creepers and epiphytes. There a huge stone tablet may be seen, supported on four richly carved stone columns six feet high, which probably served as an altar for the offer of sacrifice; or it may have been the entrance to a temple. Near-by is an underground gallery leading to two large caves in which are carvings of the sun and moon with rays darting in all directions. There are many other statues within a radius of several miles, and doubtless a systematic search of the region would reveal rich archæological treasure-troves. Numerous mounds and caverns furnish abundant evidence of the existence of ruined temples and the remnants of works of art that have yielded to decadence with the passing of the centuries. Most of the known statues have been undermined by fortune-hunters and have toppled over; others have been broken by the excavators in their mad search for the small gold replicas or ornaments that are found in the graves, while several have been demolished by order of the clergy. The only thing that prevents the removal of the stones themselves is their great weight and lack of transportation facilities.

The ruins about San Agustin possess none of the ornate massiveness of those found in Guatemala and Yucatan, but rather has the work been executed along severe lines and in bas-relief; nor are they nearly so well preserved, which might tend to show that they date back to an earlier period. Hieroglyphics are almost wholly wanting. Doctor Karl Theodor Stoepel, who spent some time in San Agustin previous to our visit, has traced a similarity between one of the monoliths and an example found in Pachacama, Bolivia. In one or two instances the work resembles that of the Aztecs.

Just how to account for the advance of civilization to a point where art and architecture were encouraged, and which supported a well-organized form of government, and then to explain its complete extinction, is a question on which students of the subject are at variance. Religion in some form or other has always wielded a powerful influence upon the life and customs of primitive nations; one evidence—almost invariably the deities and the temples erected for their veneration represent the supreme efforts of the ancient artists and alone have withstood the weathering of ages. This points strongly to the supremacy of a sacerdotal order; but whether the reigning classes who withheld their knowledge from the common people for selfish purposes were annihilated by an uprising of the servile hordes or by an outside invasion, or whether some great cataclysm of nature extinguished the progress of ages at a stroke, may forever remain a secret.

The bird life around San Agustin was varied and abundant. Trees were in blossom, especially one with a feathery, pinkish flower (*Mimosa*), and to this scores of hummers came. One species had a slightly curved bill and was green in color, with a patch of deepest purple on the throat; another of a blue color had tail-feathers six inches long. In the ravines there were many chachalacas that kept up a demoniacal cackling. The bushes were full of finches and lovely velvety red tanagers, while honey-creepers came to our table daily and gorged themselves on sugar. In the forest we saw many large, woolly monkeys, some bluish, others silvery gray. There were kinkajous, agoutis, and peccaries. The two-toed sloth was abundant; the flesh of all these animals was greedily eaten by the natives. Numbers of large lizards or iguanas prowled about the town and feasted on the tiny chickens and ducklings. A flight of locusts covered the entire upper Magdalena, and for days the air was black with the pest; millions would rise from the ground in a steady cloud in front of us as we walked along through the fields. In a few days not a speck of green remained. The hungry, insatiable hordes moved on, but behind them remained a wide, brown desert, filled with sorrow and desolation, for the crops of corn, yuccas, and bananas had been destroyed and there would be famine for many months to come.

Figure 18. A mountain stream, such as the Rio Naranjos, where the cock-of-the-rock spends its existence.

We scouted the forests daily, confining our search to the untrodden ravines of the Rio Naranjos, a turbulent, wicked stream that joins the Magdalena a short distance below. Great precipices flank its sides and the water rushes through dark, narrow gorges. Everywhere the river-bed is dotted with great boulders against which the water dashes with a force that sends clouds of spray into the air. The slopes of the mountains and ravines are covered with a dense palm jungle, the trees laden with bunches of purple berries. It is in places such as these that the cock-of-the-rock spends its existence. After several weeks of the most strenuous work our efforts were rewarded: we came suddenly upon a flock of male birds in the top of a palm, the bright scarlet color of the wonderful creatures flaming among the deep-green fronds in a dazzling manner as they flitted about, and with outstretched necks and raucous "*eur-rr-ks*" surveyed the disturbers of their time-honored solitude. We were the first human beings to penetrate their jungle fastness and excited curiosity rather than fear. The mere sight of these beautiful birds in their wild surroundings was worth all the discomforts of the long journey. In size they are no larger than domestic pigeons, but the color is of a most intense and brilliant scarlet, with wings and tail of black; the upper wing-coverts are of a light shade of gray, and the eyes and feet are golden yellow; a flat crest an inch and a half high completely covers the head and hides the yellow bill. The female is of a dull shade of brown.

We wanted to find their nests and to study their home life, of which little was known; also to secure material for the museum group. With the aid of Indians, and ropes made of creepers, we began to explore the face of the cliffs, some of which were a hundred feet high. On many of the steep slopes the palms grew so close together that we utilized them as ladders. As it rained nearly every day the footholds were very slippery, and many times one or another of the party fell, being saved from being dashed on the rocks far below only by the rope that bound us together.

One day, as we crept along slowly and painfully, we flushed a bird of sombre brown from a great boulder that rose from the centre of the stream. We waited breathlessly while she fluttered about in the palms and then returned to the rock. She flew many times back and forth, carrying food in her bill, and at last I discerned a dark object against the face of the rock upon which the bird centred her attention. There was no longer cause for concealment, so we moved to the edge of the torrent and saw the grass and mud nest plastered against the face of the rock; below raged a whirlpool, and on each side there was a waterfall. A more inaccessible spot could not have been chosen by the bird, whose haunts had never been violated.

After a consultation the Indians decided to build a raft, and accordingly cut down trees and lashed the trunks together, but no sooner had the craft been launched than it was caught by the raging swirl and spun about until the creepers parted and we found ourselves struggling in the whirlpool. A great liana which had been securely tied to the raft and fastened on the bank swept past, and this proved to be our salvation.

A tall tree was now felled, and its course so directed that the top should fall across the inaccessible rock island, but it fell several yards short and again we were outwitted.

The sun was now directly overhead, and the fierce rays entered the narrow confines of the canyon so that it was stiflingly hot. Angry peals of thunder warned us of the approaching storm, and red howling monkeys, disturbed from their midday rest, roared dismally. Above, the river flowed like a greenish stream of molten glass; below, it dashed through the gorge with a dull roar, and to the towering boulder in the centre clung a treasure, to possess which men had risked their lives; but on the very verge of success we seemed likely to fail. Even the Indians, pioneers of the jungle, shook their heads doubtfully and wanted to return.

We tried the only remaining resource. With poles and lines two of the Indians and myself picked our way to a number of small rocks that jutted out of the angry flood at the very mouth of the gorge. The other Indian spliced together joints of slender bamboo and climbed out into the branch of the fallen tree which had lodged against some rocks. From this precarious position he made repeated thrusts at the nest; finally it fell and began its maddening career in the whirlpool. Around it went, many times, and then shot straight for the gorge, swerving toward the rock on which Juan stood. As we shouted encouragement Juan dived. In spite of the fact that he was a powerful swimmer we doubted if we should ever see him again, but after what seemed minutes he reappeared, battling furiously with the flood that sought to sweep him into the maelstrom. We threw him a line and dragged him ashore. In his mouth he held the precious nest, a young bird, drowned, still clinging to the grass lining.

Later, and under circumstances hardly less thrilling, we found other birds and nests with both eggs and young, but we took only those that were absolutely necessary. The others, and there were many, we left to the eternal mystery of the wilderness, to dance in the shadows and to woo their mates beside the rushing waters; to rear their young and to lead the life that was intended for them from the beginning.

Chapter 7

CROSSING THE EASTERN ANDES INTO THE CAQUETÁ

Of the many little-known places in South America, the least known lie eastward of the eastern base of the Andes. One such region is the Caquetá of Colombia. We had been considering the feasibility of undertaking a trip into this country, but the departure for home of my companion, Doctor Allen, and Mr. Lloyd, from San Agustin, left me alone in the field, and I doubted the advisability of taking the journey without their assistance. From all the information I could gather, the crossing of the Eastern Range presented great difficulties and would have to be accomplished on foot. The rainy season had set in, adding to the difficulties of travel. Also, the rivers were swollen to such an extent that there was danger of our being stopped at any one of them; or, far worse, of being unable to recross them upon our return. However, a nearer view invariably changes the perspective, so I determined to approach the region as near as possible, gather all the data available, and then follow the course that seemed best.

Accordingly, we bade a reluctant farewell to San Agustin one Sunday morning. The entire village turned out to see us depart and gave us numerous tokens of their good-will and friendship in the form of embroidered handkerchiefs, panama hats, food, and pets. An old Indian solemnly presented me with a small monkey, which he said could cry if spanked thoroughly; he offered to give a demonstration of the creature's accomplishment, but I assured him that his word was sufficient. A parrot was contributed by another person who said it would be good company, as it "conversed" well. The *Vaya con Dios!* of these simple, honest folk was touching, and we took away with us only the most pleasant memories and friendliest feeling.

After a three days' ride through level plains and gently rolling grasslands we forded the Rio Suaza and drew rein in the town of Guadaloupe. It stands at the foot of the Cordillera Oriental. A trail was being constructed from this point across the mountains and into Amazonian drainage; however, work had little more than begun, and the reports of the route we had from the villagers were not very encouraging.

There was nothing of particular interest about the village. We moved to a site known as La Danta three thousand five hundred feet up the slope. There was abundant woods all around in which we hunted with good results nearly three weeks.

One day a party of Indians made camp on the bank of a creek not far from La Danta, and immediately built a rock and mud dam across the little waterway. Then they crushed a great many leaves of the yucca-plant and threw them into the stream. The milky juice quickly mingled with the water, and soon scores of catfish came to the top, stupefied by the poison, and floating on their backs. They were gathered by the basketful and taken away by the Indians. These catfish, living in rapid mountain streams, are provided with a sucking disk which enables them to attach themselves to a rock to rest; otherwise they would be washed down stream, as they are not very powerful swimmers.

The cost of being married is so high in some South American countries that in many cases the ceremony is dispensed with. Occasionally, however, bands of missionaries visit a region and attempt to undo the wrong inflicted by the local *padres* by uniting in marriage free of charge all those who appear before them for that purpose. The *padres* are not always to blame; frequently the inhabitants are simply too indifferent or lazy to go through the formalities, or there may be no one in their midst to look after their spiritual wants.

While we were at La Danta a half-dozen priests came to Guadaloupe and urged the *paisanos* to take advantage of this opportunity to become united in wedlock according to the ritual of the church. The people listened to the exhortations, promised to heed the admonitions, and—failed to show up at the proper time. Then the *padres* lost patience and talked the matter over with the *jefe*. The latter sent out soldiers to scour the country and bring in all the offenders living together within a radius of many miles; the pairs were frequently brought in handcuffed together, all objections and excuses being promptly overruled or ignored by the officiating clergy. Then they were lined up and married.

Several weeks later I was the guest of a very high government official in another state. In the course of dinner conversation the *señora* asked me in the most casual way: "Tell me! In your country, do people get married, or *así, no más* like here?" The last phrase was accompanied by a dainty snap of the fingers. I am afraid I said: "*Así, no más!*"

From *peons* working on the new road we learned that their operations had extended to a point near the top of the range, and that a *tambo*, or rest-shack, had been built there to shelter the laborers. We immediately started for the place and by dint of hard travel reached it in one day's time. The shack bore the name Andalucia and was seven thousand nine hundred feet up. The *peons* gladly shared their quarters with us, and we divided our rations with them, which must have been a welcome change from their everlasting boiled corn and *panela*.

The weather at Andalucia was most severe; fog, strong wind, almost continuous rain, and a freezing temperature reminded us of conditions on a paramo at the worst season of the year. Also, the forest was dense, and the vast number of fallen trunks and branches rendered the greater part of it impenetrable. Birds were scarce and hard to find, but small mammals were plentiful.

The foreman of the work gang had cleared a few acres of land and sowed wheat, but the chances of harvesting a crop were very small, because it seemed as if all the rats and mice for miles around had located the spot and promptly migrated there to unearth the seed and cut down the tender shoots.

Water for drinking and cooking was secured from a deep pit dug in the slope. One of our first cares always is to investigate the water-supply of the region in which we are working; an inspection of the excavation near the *tambo* revealed a most astounding state of affairs; three earthworms, as large as good-sized snakes, make the reservoir their home. They resembled the well-known "shiners" that appear on our lawns after a shower; but the size! The largest, by actual measurement, was thirty-seven inches long and four inches in circumference. When I asked the cook for an explanation as to why he did not remove them and keep the water clean, he promptly informed me that they were *cojures* (cohoories) that he had dug up in the woods and placed there for safe-keeping until he had time to use them on a fishing trip in the low country. Needless to say, perhaps, his pets promptly disappeared; he always insinuated that they had met with foul play at my hands!

One day a person of distinguished appearance rode up the road and introduced himself as General Rafael Santos, of Bogotá. He had heard that we were in the locality and wanted to get into the Caquetá. Could he be of any service to us? As he was in control of the work being done on the new trail, he certainly was in a position to be of great help. He told us of conditions on the eastern slope and also of the country we were so eager to see; before leaving, one of his *peons* was despatched down the trail to inform his scouting-parties that we would follow within a short time, and for them to have camping places prepared for us.

We lost no time in starting on the trip. I had with me several natives who had been with the expedition some months, and their number was augmented by men from Guadaloupe who were eager to have a hand in the undertaking. Every one walked, the *peons* carrying the packs; but mules were driven ahead to test the trail, and also for use after we reached the level low country.

The heavy subtropical forest that begins at La Danta continues on to the top of the range, and down the other side in an unbroken mass of solid, living green. There were practically no signs of life, but the wind blew less violently and the cold was less intense and not so penetrating as at Andalucia.

The slope is less abrupt than on the western side. On the second night a palm-leaf lean-to called *El Paraiso* was reached. The elevation was two thousand four hundred and twenty-five feet. A number of bedraggled and discontented laborers had erected this shelter and said they would stay there without doing another stroke of work until their pay, several months overdue, should arrive. Perhaps they are still camping there, unless the prospect of starving to death forced them to move, as we had heard several times that the foremen were in the habit of drawing the money for all the men under them, and then decamping for parts unknown.

Beyond "the paradise" the way lay through a region that might well be called *El Infierno*. There was an unbroken succession of pools and sinks so that we struggled onward hour after hour through water and thin mud several feet deep. Contrary to our expectations, we had been able to use the mules for very light packs on parts of the previous day's journey; but now they floundered and caused so much trouble, that we heartily regretted not having left them behind.

On the following days the country was dotted with steep, rocky foot-hills, alternated with deep, muddy depressions. Rain fell almost continuously, but it served to keep away troublesome insects. The *peons* were cheerful withal and seemed to enjoy the experience in spite of the hard work. However, it was with a feeling of relief that we emerged from the mountainous country and entered a stretch of level forest, the elevation of which was one thousand feet. From the edge of this "plateau" we had our first view of the Caquetá—a perfect ocean of forest stretching out ahead as far as the eye can see, which on clear days is a distance of many miles. The sight is most impressive. Not a single rise is visible above the uniform expanse of green, as the trees appear to be all of the same height.

We stopped at the first native hut encountered, which was but a ten-minute walk from the settlement of Florencia. There was a clearing of considerable size; the greater part of it was overgrown with grass and weeds, but there were also fields of cane and plantains. The latter were the finest I have ever found in all South America—eighteen inches long and sweeter and better flavored than the best bananas. It was almost impossible to grow sugar-cane in any quantity; capibaras were abundant along the streams and made nightly inroads on the plantation, devastating large areas on each visit.

The great Amazonian forest extending on all sides was full of surprising sounds emanating from a fauna entirely new to us. For the first time we heard the clear, ringing whistle of the "false bell-bird" (*Lathria cinerea*). The penetrating *whoo-ee-whee-oo* filled the woods with music as the birds called to one another, but the obscurely colored singers were hard to see among the dark branches. The song contains several low, *churring* notes that are lost from a distance.

The abundance and variety of wild life was so great as to almost bewilder us and we worked day and night preparing the wealth of material that came into our hands. Working conditions were most unfavorable; it rained daily; sand-flies took away a great deal of the pleasure that each day brought in the form of new and interesting creatures, while mosquitoes and fleas insisted on gaining an entrance under the nets and making the nights disagreeable. Every member of the expedition suffered from malaria during our entire stay in the Caquetá region. Notwithstanding these handicaps, we lost not a single day, and the collections rapidly grew to record-breaking size.

It was, of course, necessary to depend to a certain extent upon native hunters. They were always carefully instructed as to the area they should visit and how to work it; from the results they obtained I could usually tell whether directions had been followed. One of these *cazadores* was a lazy, thoroughly good-natured half-breed named *Abrán*. He came in daily with a tale of woe, recounting in detail the great distance he had covered, the hardships of such a long tramp through the jungle, and—bringing few specimens. I pretended to believe his stories, knowing full well all the while that he had really selected a comfortable spot a mile or so away and then settled down on a log for a quiet day of smoking and day-dreaming.

Figure 19. Tree-fern, typical of the Andean forests.

When any animal came within sight he shot it. In this manner he secured many of the shy, ground-haunting species, such as rails, tinamou, and ant-birds that one seldom sees while moving about through the forest. This was exactly what I wanted. It is all but impossible to find a native hunter with patience enough to sit and wait for these things, so while *Abrán* thought he was playing an easy game, he was in reality the most valuable *peon* in the outfit. His brother *Moisés* was of the opposite temperament; he walked many miles each day and considered it beneath his dignity to shoot anything but large, brilliantly colored birds, such as parrots, macaws, cotingas, and tanagers, or monkeys—in short, game worthy of a man's efforts. The two brothers made an ideal combination. *Moisés* had spoken frequently about a marvellous bird called *tente* which he said was found in the region, and of which he was determined to secure one as a pet for the *patrón*. One day he brought in a queer, frightened little creature—all legs and neck—that he proudly introduced as the *tente*. It was a young trumpeter (*Psophia*). After being tied up a few days it grew very tame and was given full liberty about the place. It walked slowly and in dignified fashion, catching flies and pecking at insects on the ground or walls; but if a dog should chance to pass near by it darted at it with outspread wings, making a loud, rumbling sound deep down in its breast; the dog always fled in terror. The bird increased rapidly in size and before long the beautiful metallic-blue throat-feathers appeared. When we emerged from the hammocks in the early mornings it was always there to greet us with low bows, spread wings, and deep murmurings. In travelling, a large-meshed fibre bag served as its container; upon being turned loose when camp was made, it first carefully dried its plumage before the fire, then strutted around a while, and finally flew into the branches of the nearest tree to spend the night. We kept this interesting little pet until our departure from Colombia, and then gave it to an acquaintance in Neiva, where it was well cared for.

A colony of cultivator-ants had taken possession of a patch of young *cecropia*-trees near the house. They carried particles of earth to the branches and formed them into large balls in which the seeds of a succulent plant were sowed and cultivated. The earth was kept loose and moistened and the bunch of tender shoots resembled a clump of mistletoe. In this manner an abundant food-supply was assured.

Florencia was a small village of adobe and bamboo huts, built in anticipation of the opening of Colombian Amazonia, when the new trail across the Andes should be completed. The region is undoubtedly rich in natural resources, and there seemed to be a possibility that the dreams of these pioneer settlers might some day be fulfilled. However, five years later, while aboard the S. S. *Vauban*, bound for New York, I chanced to meet among the passengers a Colombian with whom I had become acquainted in Florencia. He stated that the climate there had proved so unhealthful that most of the people had died or gone away and the settlement was all but deserted. The elevation of the site, though thousands of miles from the Atlantic Ocean, into which its rivers drain, is only six hundred and seventy-five feet.

During our stay in the vicinity we had occasion to witness a celebration of the feast of San Juan. On the eve of the festival a pig was slaughtered in each hut; those who had none went into the jungle and shot a wild one. The dressed carcass was placed in an oblong wooden bowl, surrounded with plantains, yuccas, and yams, and then baked four hours in a mud oven. The roasts were delicious and every one ate until not a morsel remained, which was far into the night. Next day the *fiesta* proper began with a bull-fight, local talent, shirtless and in tattered drawers, supplying the places of the gorgeous *toreadores*, *banderilleros*, and *matadores*. This was a fine chance for the youths to display their courage to the weaker sex, which had gathered *en masse* to witness the performance, and, if one enjoys such spectacles, he would doubtless say that the showing made was quite creditable. The men charged the bull, flourishing their bright-colored ponchos, and when the animal turned the tables and chased them they fled to shelter, as is the custom of the profession. We did not remain to see the finish, but later in the day the women were roasting chunks of beef over open fires. The merrymaking continued for several days, and the latter part of the period consisted in drinking *aguardiente*, with the resultant fighting that always marks the wind-up of such affairs. The *alcalde* was a leading spirit in the activities of the festive occasion; he had been a priest at one time, but was excommunicated for preaching sermons of too liberal a nature. Then he married and was rearing a family. He told us that he owned a ranch called La Morelia, two days' distant from Florencia, and offered to send us there; so we accepted his courtesy with pleasure, as we were eager to see the country farther in the interior.

A faintly defined footpath led to La Morelia. The forest is comparatively open, that is, free from dense undergrowth. The trees are tall and there are a few tree-ferns and palms; many climbing lilies and other epiphytes grow on the trunks and branches. Moss is lacking; near the streams bamboo, wild cane, high grass, and briars, united by creepers, form dense jungles that are hard to penetrate. Streams and rivers are numerous and we were at once impressed with their size and depth. Crossings were effected in dugout canoes. While the current is swift, the waterways are so silent that one is not aware of their existence until reaching their very borders.

We saw little of the Huitoto Indians inhabiting this district. They seem to remain in seclusion in their forest homes and rarely venture into the path of the settlers. Those we encountered were of low stature, yellow in color, and had features so nearly resembling the Japanese that they might be easily mistaken for that race. They are of a shy and retiring disposition. Their ornaments were very elaborate, consisting of anklets, amulets, and necklaces of colored seeds and jaguar and monkey teeth, skilfully wrought into pleasing combinations.

The hut at La Morelia was of large dimensions, built entirely of bamboo, with palm-leaf roof. An unusual feature was that it contained two stories, the lower used to store grain and plantains, the upper serving as living quarters. A clearing about one hundred acres in extent surrounded it; most of it was grass-covered, providing pasturage for a few head of cattle, the remainder was under cultivation. The several acres that had been given to growing plantains produced so abundantly that hundreds of bunches were going to waste. If left to mature on the plant the fruit bursts and is destroyed by insects. The choicest clusters were cut green and then placed in a down-stairs room of the house to ripen. At night hundreds of small bats visited the enclosure to feed on the mountain of rapidly yellowing fruit. We desired some of the creatures for our collections, but found it difficult to catch or shoot them in sufficient numbers. Finally we evolved the plan of suspending a fish-net from the ceiling and tacking out the edges so that it formed a cone with a wide base. A choice bunch of the ripest plantains was placed in the centre for bait. Bats soon gathered about the trap in swarms. At first they were suspicious and circled around the net without attempting to alight; but as their hunger increased so their caution decreased in like proportion, and before long they were striking the conical arrangement from all sides and madly endeavoring to scramble through the small meshes. Some succeeded in forcing their way through the openings and immediately fell upon the bait with ravenous appetites; the vast majority, however, became helplessly entangled in the meshes. Newcomers arrived in a steady stream; they paid no attention to our presence nor to the lights we carried, but frantically hurled themselves into the midst of their struggling brethren, until the net was covered with screeching, scrambling masses.

The house was within a stone's throw of the Rio Bodoquera—a stream two hundred yards wide. One night a jaguar attacked the cattle and chased them on to a sand-spit that projected out into the stream. We heard the mad bellowing of the frightened animals as they stampeded past the shack, hotly pursued by the snarling jaguar. A few shots sufficed to frighten the big spotted cat back into the jungle, but the cattle refused to leave the strategic position to which they had retreated. The river was rising rapidly, endangering the panic-stricken creatures. Every hand turned out; we took lanterns with us and, manning the canoes, paddled to the far side of the peninsula and attempted to drive them back to the mainland. All our efforts were in vain. The work was very exciting, as enraged members of the herd charged the lights repeatedly when we approached close to them. Finally the water became so deep that the animals had to swim, and then they made for the far side of the river and disappeared from view. It took several days to round them up, but a number were never seen again.

One day a Franciscan priest stopped at the *rancho* for a short rest. He was engaged in opening a trail to Mocoa. About twenty *peons* accompanied him, carrying his outfit. His robe was in tatters and his feet were bare; he had spent months in the jungles and showed the effects of hard usage. Each of his men carried an animal of some kind on top of his pack. There were monkeys, parrots, macaws, and a curious little creature belonging to the agouti family (*Myoprocta*) that they called *tin-tin*. We had seen numbers of the latter along the river-bank, where they lived in burrows. The flesh is white and of fine flavor. In spite of the hardships the priest and his party had endured they were in the best of humor, and after an hour's halt shouldered their packs and resumed the march. No one will dispute the fact that men of this type have done a great deal toward exploring unknown parts of South America; usually they are the real trail-breakers and lead the way for the pioneer settlers who are to follow.

Figure 20. The high, flat-topped panorama of the Andes.

The bird-life of the Caquetá is typical of the Amazonian forest, and many of the species are found on the lower river two thousand miles away. This is caused by the uniformity of topographical conditions, and the lack of a barrier that would interfere with the range of a species. On all of our visits to the headwaters of the Amazon's tributaries, in Colombia, Bolivia, and Brazil, a large proportion of the mammals collected were new to science and differed greatly from those found lower down the river's course. Such large animals as spider-monkeys (*Ateleus*), "flying" monkeys (*Pithecia*), and cats represented forms heretofore unknown to science; the smaller mammals also were new in many instances. Of course, we must not lose sight of the fact that the power of flight gives greater mobility to the birds and accounts for the wider range of some of them, but not for the equally vast distribution of the ground-inhabiting and almost flightless species.

After a strenuous three weeks at La Morelia we returned to our first stopping-place near Florencia. The rainy season was at its worst, and low clouds covered the forest day after day, while torrents of water fell almost continuously. The journey back to Guadaloupe was far more difficult than had been our entrance into the region, for the greater part of it lay up-hill and mud and water had accumulated in spots until it was waist-deep. The cold grew more intense as we neared the top of the range. We were never warm or dry until we reached our destination.

The maximum time allowed for work in Colombia had expired. Although I had spent over eighteen months in the republic, they had flown all too rapidly, and I heartily regretted that it was not possible to visit the numerous other places that invited exploration. The next best thing was to hope for a return trip in the future—a hope that was realized several years later in our expedition to the Antioquian Highlands.

The homeward trip was accomplished without noteworthy incident. At first there was a ride of five days' duration down the desert-like valley of the Magdalena to Neiva. The river is not navigable in this part of its course on account of rapids and shallow water. At Neiva a *champán*, or flat-bottomed freight-boat, was secured. The crew of twenty men rowed it down to Giradot in three days; it takes them thirty days to pull the craft back up-stream to the starting-point.

The remainder of the journey to Puerto Colombia was merely a matter of travel on river-steamers and train, and required two weeks' time.

In summarizing the work of the expedition to the Caquetá, Doctor Chapman, in "The Distribution of Bird Life in Colombia," writes as follows:

This "work during the rainy season in the humid Amazonian forests of the Caquetá, where with only unskilled native assistance he secured eight hundred and thirty birds and mammals in thirty days, is a feat in tropical collecting." And "this locality ... was one of the most productive of any visited by American Museum expeditions, and many species were secured which have not heretofore been recorded from Colombia."

Chapter 8

ACROSS THE ANTIOQUIAN GOLD-FIELDS TO PUERTO VALDIVIA ON THE LOWER CAUCA

Puerto Berrio is not the most attractive spot in Colombia, but it is nevertheless of a great deal of importance. All steamers plying on the Lower Magdalena stop at that port, the up-going ones after a six days' voyage from Barranquilla to discharge freight for Medellin, and those bound down-stream to take aboard gold and other products of the Antioquian highlands.

The arrival of the steamer always causes a great deal of confusion. Debarking passengers are required to look after their own luggage, which is not a simple matter, as it is invariably covered with mountains of boxes and bags on the lower deck; and after it has been located it is necessary to secure *peons* to convey it ashore, the ship's crew invariably refusing to render this service.

There is always a rush for the little hotel "Magdalena," built on a slight bluff overlooking the river. Accommodations are limited, and those who arrive first naturally have the advantage of selecting the cooler rooms in the upper story. However, the advantages gained are partly imaginary, at best. The climate is insufferably hot in the daytime, and mosquitoes filtering through rents in the nets protecting the beds are most annoying at night. Nor is it possible to seek the cooling comfort of a bath; a small, corrugated iron building in the garden is supposed to provide for this need, but a tank containing water for the shower is placed on the roof in the full glare of the tropical sun, and the water becomes heated to such a degree that it is almost scalding.

The town of Puerto Berrio is situated a few hundred yards below the landing. It contains about a hundred low buildings, many of which are utilized for shops where merchandise and, more important at least to transients, a great variety of fruit may be had. All the buildings are low—some constructed of adobe with red tile roofs, others of nothing more substantial than bamboo, and grass or palm-leaves.

Beyond the town is a low, rambling shed used as a slaughter-house. When one tires of watching the blue tanagers, orioles, and yellow warblers quarrel in the cocoanut-palms near the hotel, he may tempt his æsthetic taste by walking to the pavilion of bovine death, and look upon the hundreds of black vultures sitting on the roof, strutting and hopping over the ground, or tearing at the hides that have been stretched out to dry. These birds are so typical a part of most towns and villages of tropical Colombia that one soon learns to accept them as a matter of course. They act as scavengers. Without them the settlements would reek with foulness.

Puerto Berrio marks the beginning of a narrow-gauge railway, and each morning at six a passenger-train leaves the station for Cisneros, covering the first stage of the journey to Medellin. Almost immediately after leaving the port, the road plunges into the finest type of Magdalena Valley forest. We therefore debarked at the first settlement, called Malena, only fifteen minutes after leaving the starting-point. My assistant on this expedition was Mr. Howarth S. Boyle, of Elmhurst, Long Island.

At Malena the tropical forest reaches the height of its development. There is a clearing large enough only to provide room for the village of some twenty houses, and the stately living wall of trees hems it in on all sides. The people are most obliging, and while there is no *posada*, or inn, of any kind, a Mestizo family volunteered to permit us the use of part of their dwelling.

A short tour of inspection confirmed our first impression of the region; it was a naturalist's paradise. One had only to go to the outskirts of the town to find birds in greatest abundance. A number of tall dead trees had been left standing in the clearing, probably because it was easier to merely girdle them and let them die than to cut them down, and many blue and yellow macaws and Amazon parrots were nesting in cavities high up in the trunks. They had young at the time of our visit (March), and screamed and fluttered about the nests all day long. No one thought of disturbing them. Rough-winged swallows and martens nested in the same stubs, and apparently lived in perfect harmony with their noisy neighbors.

A shallow, narrow stream of clear water flows through the clearing, and a belt of woods and low sprouts mantles each bank with dusky green. This was the favorite resort of many small birds; oven-birds and ant-wrens ran about in the deep shade, while night-hawks, aroused from their slumbers, flapped noiselessly into the air and dropped again a few feet away. Scores of parrakeets chattered in the branches overhead, while flocks of large, spotted wrens (*Heleodytes*) added to the chorus with their incessant scolding.

If we remained close to the stream we were sure to surprise herons of several species, and black ibises wading in the shallow water. A species of ani (*Crotophaga*) fluttered in the overhanging bushes; they were awkward though beautiful creatures, the size of a blue jay, with brilliant, black iridescent plumage; the mouth was pure white, while the eyes were of a pea-green color.

If our tramp led to the heavy forest, the character of the birds changed. Giant orioles (*Ostinops*), grackles, and chachalacas always remained near the border of the taller growth, and toucans in flocks seemed to prefer the protection of the more inaccessible cover.

Figure 21. The town of Valdivia.

Figure 22. The Cauca River at Puerto Valdivia.

The forest is magnificent, and is composed largely of ceibas with thick, white trunks and wide-spreading tops. Many *tagua*, or ivory-nut palms, grow beneath the tall trees; their fruit is one of the important articles of export from the Magdalena Valley and, during August and September, many thousands of bags are shipped down the river to Barranquilla. Wild life, however, was comparatively scarce in the forest proper, with the single exception of mosquitoes, which were present in unlimited swarms, even in the daytime; and small troops of brown marmosets that showed themselves at rare intervals.

While crossing the clearing one day a flock of blue and yellow macaws passed overhead; we needed a pair for the collection, so I took a quick shot at the birds as they flew by; however, I succeeded only in wounding one of their number, which flew to the ground in a long slant and alighted so far away that it was useless to try to follow. On reaching home at noon, I was greatly surprised to find the bird perched on a ladder in the very house we were occupying. It had dropped in the yard, and having been seen by some children, they tried to catch it, whereupon it took refuge indoors and kept them at bay with its angry screams and attempts to bite.

The evenings at Malena were fully as profitable as the mornings. We always spent a pleasant hour or two at dusk, walking along the railroad. Pools of water had collected in the hollows where earth for the road-bed had been excavated, and many water-birds came there nightly to fish or catch frogs. Great blue herons, bitterns, and occasionally a cormorant or anhinga were surprised at their nocturnal feasts. When we returned after dark we started numerous goatsuckers, which had settled in the open lane to catch insects and to sing; this habit of resorting to open places, especially trails and roadways, has earned for them the name *guardacamino* (road-guard) among the natives.

Malena was such an unusually interesting place that we expected to remain there several weeks; but, unfortunately, an epidemic of dysentery had invaded the Magdalena Valley, and the village was soon writhing in the throes of this fatal disease. Sickness and death in the family of our hosts made it necessary for us to continue on our way.

It requires exactly six hours to reach Cisneros, the end of the railroad, from Puerto Berrio. The altitude of the terminus is three thousand seven hundred feet above sea-level, and as one approaches it the heavy forest gradually disappears, to be replaced with a lower growth of brush and bushes; finally the hilltops are barren.

At Cisneros one may secure riding-animals, a carriage, or a motor-car, according to the mode of travel preferred, for the short ride across the ridge to Botero, from whence the journey may again be resumed by train. The road is splendid, and as the highest point, called La Quiebra, is only five thousand four hundred and twenty-five feet up, a canter on a spirited horse across the divide is most enjoyable.

Botero is very similar to Cisneros. There are two small hotels where the traveller may rest in comfort until the train leaves for Medellin, which is at 4.30 P.M.

Numerous villages are scattered along the railroad, which follows closely the course of the Medellin River. The country is green and apparently fertile. Thickets of wild cane grow near the stream, and the valley is dotted with clumps of tall, slender willows; so dense is the latter growth in some parts of the region that it forms groves and woods.

Two and a half hours after leaving Botero the train arrived at Medellin. Medellin is the third largest city of Colombia, and boasts of a population of seventy thousand. The city is not modern but very picturesque, and lies in a depression almost completely surrounded by mountains. We were fortunate enough to make the acquaintance of the American consul, Mr. H. B. Meyerheim, who rendered the expedition invaluable service during our entire stay in Antioquia.

The people of the state differ from the Colombians living in other parts of the country in that they possess more initiative and business ability; and for this reason they are frequently referred to as the Jews of Colombia. Some authorities go so far as to assert that they are really descendants of a colony of Jews that settled there many years ago. For this belief there seems to be very little foundation. The fact that the climate is bracing and that it requires a greater amount of work to gain a living in the semiarid country probably accounts for the increased degree of energy displayed by the inhabitants.

Our first expedition was to a point in the mountains southeast of the city, known as Santa Elena, and only a few hours' ride on mules from Medellin. After crossing the ridge we found ourselves on a high, wild plateau, which had at one time been covered with forest; but the trees had been felled on the greater part of the area, and only small, scattered patches of woods were left untouched. There are numerous little huts in this upland country, and at one of these we decided to remain for as long a time as the country proved a profitable collecting ground.

On account of the great change in the flora, occasioned by deforestation, a corresponding change had taken place in the bird life. But little remained of the subtropical fauna we had expected to find; however, there were black thrushes, several species of tanagers, toucans, trogons, and motmots, besides many commoner species. Weasels were abundant and occasionally blundered into our traps; these animals are very easy to call up, and if one sits quietly and imitates the screams and squeaks of a wounded bird, it is often possible to attract a weasel to within a few feet, and at times it will run across one's lap in search for the supposed victim. There were also squirrels of several species, and tiger-cats.

Many flowering shrubs dotted the roadside, imparting a blaze of color to the muddy highway; some of them were covered with brilliant scarlet blossoms, and others with snowy trumpet flowers of great size. In addition to this wealth of native flowers, the people cultivated plots of gladioli and roses, both of which attained great size and beauty in spite of the cold, wet climate.

We continued on across the highland from Santa Elena to a place called Barro Blanco, passing through the villages of Rio Negro and Carmen on the way; but the character of the country did not change appreciably! One of the products of the high, bleak region that immediately attracted our attention was a variety of maize; this thrived even on rocky ground. The ears were large and well-formed, and the huge, even grains were of a milky-white color and of splendid flavor. Large flocks of pigeons came to the corn-fields to feed and furnished splendid shooting; they fly down the mountainsides at terrific speed, and the rushing noises made by the wings can be heard at a great distance. On the edges of the fields grew small trees (*Ficus*) bearing quantities of white berries; birds of many species, including flycatchers, came to feed on them.

After completing our work in the Santa Elena region we returned to Medellin. Then we took the train as far as a station called Barbosa, and started overland for the Lower Cauca. We brought both pack and riding mules with us on the train, as it was difficult to obtain them at Barbosa, and when everything had been unloaded at the station, packs were adjusted and the mules started up the exceedingly steep ridge to the north. The altitude of Barbosa is four thousand six hundred and twenty-five feet above sea-level, but there was no break in the narrow, rocky trail until we had reached the summit, eight thousand one hundred feet up. A few miles beyond the top lies the village of Don Matias, almost concealed in a deep depression and surrounded by fruit-trees. The trail continues to wind across a rolling, arid country. Boulders of great size are strewn on the ground; they are of a most peculiar formation, consisting of concentric layers of stone, one or two inches thick.

Water is scarce, and we passed only one stream, and that of small size, called the Rio Porce.

Seven leagues is considered a good day's travel in Colombia, on account of the mountainous nature of the country and poor trails. However, on our first day out from Barbosa we covered only five leagues, and spent the night at a hut called Sabanete, nine thousand feet up. Early the following morning we reached Santa Rosa, the centre of the Antioquian gold-fields. The town is of considerable size, but stands in the middle of a bleak, arid plain, and is about as cheerless a place as one could find. The surrounding country is exceedingly rich in gold, and numberless mines pierce the flat, stony surface, and penetrate into the hillsides. The only drawback to mining operations on a gigantic scale is the lack of water. During the rainy season the inhabitants of Santa Rosa gather water in barrels and every available sort of container, and then wash gold out in the street in front of their homes, or in the back yards. Despite its many natural disadvantages, Antioquia is one of the richest states in Colombia, and produces a great proportion of that country's yearly output of gold, which in 1916 amounted to $5,400,000.

The country beyond Santa Rosa is practically uninhabited for a distance of ten or twelve miles; after that a growth of low woods gradually appears, and with it an abundance of bird life, such as California woodpeckers, green and yellow jays, black thrushes, warblers, and parrots. This was in great contrast to the arid country we had just left behind, where practically the only sign of life was an occasional hawk hovering in the air for many minutes at a time, in the hope of surprising an unsuspecting lizard or some small rodent among the rocks below.

It was in this forest that we again encountered a number of one of the most beautiful birds found in the entire region—the white-crowned tanager (*Serricossypha albacristata*). A flock of sixteen sat in the top of a bush and kept up a continuous shrill peeping.

The third night after leaving Medellin we reached Yarumal, a large town built on a steep, rocky slope. From a distance it seems as if the houses were standing one on top of another, and it is difficult to understand what prevents the whole town from sliding down the steep mountainside.

The "Hotel de la Madre" is one of the institutions of Yarumal. It is conducted by an old negress who looked us over suspiciously and found it hard to decide whether or not to admit us. While deliberating and fumbling about her shawl she scratched her finger severely on a pin; to this I immediately applied a few grains of permanganate taken from my snake-bite lancet. This won her favor, and we were given a room. Later she confided to us that two Englishmen had stopped there the week before. "We were frightened to death when we found out that they were Englishmen," she said, "because England is at war, you know. But what do you think? They paid their bill next morning and left without hurting anybody. However, we made up our minds to be careful about admitting strangers in the future."

One may ride from Yarumal to Valdivia in one day; but we broke the trip by stopping at a large wayside inn called La Frijolera. It was in the midst of a splendid forest growth, the elevation being five thousand feet. From a distance the forest looked most promising, but on account of the density of mosses, ferns, and creepers forming the undergrowth it was all but impenetrable.

We located a grove of guavas a short distance from the house, and this proved the most prolific hunting-ground. It was always possible to shoot squirrels there, as they came out at all hours of the day to feed on the ripening fruit. Many birds also flocked to the low trees for their daily sustenance, and even opossums lurked about the roots and brush to pick up the sweet morsels dropped by the furred and feathered flocks feasting in the branches.

At La Frijolera we engaged a native hunter who owned a famous hunting-dog named *Golondrina* (meaning swallow). Words can hardly be found to convey an accurate picture of the hunter, but the dog's name at once suggests its chief accomplishment. Day after day our man took his dog afield in search of agoutis, but he always returned empty-handed, explaining that while he had started a number of the animals we wanted, *Golondrina* could never *see* them, and so she failed to catch them. However, one day he saved his reputation as a hunter by making a difficult trip of ten miles to a steep, heavily wooded ravine, and shooting a number of red howler monkeys. A few days later the dog accidentally came across a peccary, which some native hunters were pursuing, brought it to bay on a rock, and kept it there until it could be shot.

This place presented rare opportunities for hunting by night. A road had been cut through the forest, dividing it in two clean-cut sections. However, the tips of wide-spreading branches from each side of the clear swath met in several places, forming an aerial connection above the road. These are known as "monkey bridges" because night monkeys and other animals utilize them in crossing from one section of the forest to another. As there was a full moon it was only necessary to sit quietly on a stump near one of the bridges and wait. Before long a rustling sound would come from the tree-top, so slight as to be scarcely audible, and occasionally a deep, low grunt; then silent, shadowy forms emerged from the blackness of concealing foliage and slowly made their way across the springy passage. Kinkajous also used these bridges, and as the natives prized the skins of these animals highly for making chaparejos, they conducted a regular business of hunting them on moonlight nights. After shooting in one spot for several nights in succession, it was necessary to leave it undisturbed for some time, as the animals became wary and sought other bridges.

The town of Valdivia is located on a little ridge four thousand two hundred feet up, about ten miles from Puerto Valdivia, which is on the Cauca River. All the intervening country is wooded.

We reached the port on a Sunday afternoon. The people from a distance of many miles around flocked to the spot on this day for the purpose of having a "good time," so that there were upward of a hundred natives in and about the one corrugated iron and bamboo building comprising the *puerto*, dancing, drinking, fighting, and trading at the little shop. The owner of the house received us courteously (and where in all Colombia was courtesy wanting?) and we soon made ourselves comfortable in the large wareroom which formed one end of the structure. There was no thought of work that day, for everybody crowded about to have a good look at and welcome the *gringos*, but we made the best of the occasion and secured a good deal of information concerning the surrounding country.

The Cauca, a swift, muddy stream four or five hundred feet wide at this point, is hemmed in on both sides by the steep slopes of the Western and Central Andean Ranges, the forest extending down to the water. It is navigable from here on down to a small settlement called Cáceres, but rafts and canoes only are employed in making this journey, which requires half a day going down and two days coming up. The natives are a careless lot while on the water, and numbers of lives are lost annually. About the first thing we saw was the body of a man floating down the river, with a vulture perched on it. We asked Don José, owner of the place, why he did not send some of his peons in a canoe to recover it. He replied that if he did he would be required to care for the body until a government official from Yarumal came to view it, and then he and every one present would have to go back with the coroner to give their testimony as to the finding of the cadaver. This entailed so much trouble that it was customary not to pay any attention to such occurrences.

In few places have I seen such an abundance of interesting fauna as at Puerto Valdivia. The forest was teeming with birds; mammals were plentiful; shoals of fish and even caimans swarmed in the river; there were also insects enough to cheer the heart of an entomologist.

Figure 23. A naturalists' camp in the forest.

In such a region the naturalist has no idle moments. When we tired of working with birds and mammals, which were of chief interest to us, we had only to step to the river-bank, where vast swarms of brilliantly colored butterflies settled in thick masses in the mud or rocks to drink. A single sweep of the net often ensnared several score of the insects. A species of *Urania* of a black and green color predominated, but a *Diana*, deep red above and spotted with silver dots on the under-side was not uncommon.

Fish could always be secured in abundance. If we attempted to catch them with hooks we usually landed catfish or small eels. It is unlawful to use dynamite in Colombia, but Don José had a goodly supply stored away and did not hesitate to use it when occasion required. The *peons* detailed for that purpose selected a spot in the river where logs and brush had grounded to form a drift, or where the water eddied against a sharp bend; they tied a rock to the explosive, lit the fuse and threw it into the water. After a few moments, during which the water hissed and bubbled as the gases from the burning fuse rose and escaped, a dull thud followed and, almost immediately, the surface was littered with numbers of dead and stunned fish. They were invariably a species of "Pacu" (*Prochilodus nigricans*), weighing from one to four or five pounds, and proved to be excellent eating.

Figure 24. A native hunter with a red howling monkey.

Not far from the port is an old cacao-plantation which has apparently been deserted for a number of years. The trees are tall and covered with moss, while the sheltering *cochimbas* or *madre de cacaos* form a high canopy of interlocking branches. To this cool retreat almost every species of bird common in the region came to feed or to pass the noonday hours. There were buccos and wood-hewers in abundance—the former dull, stupid birds, which sat quietly on the lower twigs in the hope that some insect would wing its way not too far from their ever-hungry mouths; the latter, agile and alert as they scampered up the moss-covered trunks, eagerly examining each crevice for a hidden grub or an ant. Gorgeous trogons with resplendent green backs and blood-red breasts flitted among the lower branches, and little parrots of bright green with gold-colored heads screamed and fluttered in the leafy branches high overhead. Where ferns and brush grew thickest, near the numerous ravines, flocks of yellow manakins (*Manacus*) sputtered and whirred in the semidarkness; they proved to be an undescribed form.

Mammals, too, were not lacking. Of chief interest were giant black weasels with white throat patches (*Tayra*). These are truly dreadful creatures—at least to the animals on which they feed. They are of powerful build, the neck muscles being particularly well-developed, and I can picture them as a dangerous antagonist even to a deer or a peccary.

The smallest of ant-eaters (*Cycloterus didactylus*) was also found in this region. This little animal, while not rare, perhaps, is seldom seen on account of its diminutive size and arboreal habits. It is of a beautiful golden color, and the fur is so fine and silky that could it be obtained in quantities sufficient for commercial purposes it would perhaps rival in value the highest priced fur in use to-day. The creature lives in the tree-tops and is diurnal in habits. It moves along the branches with great rapidity, either in an upright position or inverted like a sloth, the prehensile tail being used constantly. Ants form the food, and as these ascend even the highest trees, the little ant-eater has a never-failing and abundant supply; they are gathered up hurriedly as the little creature moves quickly along.

One day an army of carnivorous ants invaded our quarters while we were busily occupied preparing the specimens collected during the morning. The first intimation we had of the arrival of the ravaging host was when scores of cockroaches suddenly appeared and frantically ran up the walls of the room. Not long after, several centipedes eight inches long joined the fleeing cockroaches, and before long a number of scorpions followed in their wake, hotly pursued by the multitude of ants. There was nothing for us to do but follow the lead of the panic-stricken insects, so we hurriedly transferred our collections to a zone of safety outdoors, and waited a few hours until the ant army had completed its work and gone on its way. The natives welcome these visits as the ants act as scavengers and rid the house of vermin.

While at Puerto Valdivia we were presented with a young night monkey not larger than a good-sized mouse. It was a most interesting pet, and readily took to a diet of condensed milk, which it drank from a spoon. My companion, to whom the little animal belonged, kept it on the window-sill, from which point of vantage it took a lively interest in all that occurred within its range of vision. It so happened that there was a very small crack in the sill, and this proved to be a matter of the utmost concern to the tiny monkey. Hundreds of times each day it crept timidly to the crack and peered down into it anxiously, although there was only darkness below. When we held the pelt of an animal near it paid no attention whatever to it, with the single exception of the skin of one of its species, which it recognized immediately, and to which it clung tenaciously. When we left the hot climate of the Lower Cauca and started on the return journey to Medellin the little creature was unable to withstand the cold of the higher altitude and died.

The purpose of our zoological exploration of this section of Antioquia was to secure material that would throw light on the geography of the country farther north; for, beyond the general knowledge that the junction of the Cauca and the Magdalena mark the breaking down of the Cordillera Central, we knew comparatively little of a definite character about this part of Colombia. It was not until several months later that our work farther west—on the Paramillo and the Rio Sucio—provided the material which, viewed from a distributional standpoint, furnished the clews that aided very materially in solving our problem.

Chapter 9

ASCENT OF THE PARAMILLO—COLLECTING ON THE RIO SUCIO

The return to Medellin from Puerto Valdivia occupied five days. We again went to our former headquarters, the "Gran Hotel," and spent a few busy days packing the large collections brought from the Lower Cauca. Then we began to gather provisions and cargo mules for a second expedition.

Upon leaving Medellin we started northwestward, having in view an ascent of the Paramillo, a lofty spur of the Andes, jutting out of the Western Range slightly below latitude 7° south. This region, so far as I am able to discover, had never been explored.

At first the trail is wide and very good, so that within four hours after starting we reached the summit of the first ridge, eight thousand seven hundred and fifty feet up. A great cleft in the bare, rocky peaks forms a natural pass and saves a climb of at least an additional thousand feet. The slope on the other (western) side is more gentle.

We were immediately impressed with the barren nature of the country, for, with the exception of a few patches of low brush, and clumps of withered grass, there was no vegetation. An occasional glimpse of the Cauca River, far below, presented the picture of a broad yellow ribbon lying upon a brown, rocky plain.

That night we reached San Geronimo, a small town well down in the valley. Limited plots of ground are irrigated in the vicinity of the settlement, where rice, corn, and pasturage are cultivated by the inhabitants. Yellow-rumped tanagers, anis, and finches (*Sycalis*) make this little oasis their home, and add greatly to its attractiveness.

Next morning we were in the saddle before six o'clock. A few hours later, after crossing a low ridge, we came suddenly upon Sopetran, a beautiful little town completely hidden in groves of palms, mangoes, and other lovely trees. The cluster of some hundreds of snow-white houses with red roofs, wide, well-kept streets, and the abundance of multicolored birds fluttering and singing among the deep green foliage, render Sopetran one of the most attractive towns of its size I have seen in tropical America.

At noon we reached the Cauca and crossed that sluggish, muddy stream on a suspension bridge about eight hundred feet long. The cables are anchored in picturesque brick piers built into the face of the steep banks, and hundreds of swallows utilize as nesting-sites the small openings where the wires enter the masonry. Gravel flats flank the sides of the river, and bare, sandy islands divide the water into several channels. The elevation is approximately two thousand feet.

One league beyond the Cauca lies the town of Antioquia. If Sopetran is the last word in attractiveness, Antioquia must be placed at the extreme other end of the scale. The wide, arid valley supports no vegetation except occasional clumps of cacti and dwarfed mimosas, which rather add to its desert-like appearance. The heat is almost unbearable, as the Western and Central Andes, hemming in the valley between huge walls of pink clay and sandstone, shut off all ventilating winds.

Although it was still early in the afternoon, we decided to spend the rest of the day in Antioquia, as the pack-mules seemed nearly exhausted; but it was not long before we heartily regretted not having avoided the town and made camp out in the open plains. Our *arriero* had guided us to the little hotel, where a matronly *señora* received us with evident joy and a great deal of ceremony, probably because we were the first guests in some time; we soon discovered, however, that she was not the only one to whom our visit gave pleasure. Fleas in droves appeared from the cracks in the brick flooring and made their way through leggings, trousers, and all other wearing apparel as quickly and easily as the proverbial rat running through a cheese; and when we entered our room, vermin of a still more objectionable character rushed joyfully from the beds, walls, and chairs to gloat in hungry anticipation at their prospective victims. We erected our cots in the *patio* and spent a long, long night out in the open.

Buriticá was reached on the following day. Immediately after leaving Antioquia, a mere ledge of a trail begins the ascent of the Coast Range, and while a good deal of anxiety was felt for the safety of the pack-animals, it was nevertheless a relief to escape from the cheerless desert wastes and the intolerable heat of the low country. The altitude of Buriticá is six thousand two hundred feet. On account of the jaded condition of the mules, we spent a half-day in the town, and also lightened the cargoes by leaving at the inn all equipment intended for a subsequent journey in another direction. We had, of course, never visited Buriticá before, but I had not the slightest hesitation in leaving with perfect strangers a good deal of valuable material. The honesty of the Colombians is well known, and we did not lose a single thing by theft during the entire two years I spent in that country.

At Tabacal, a half day's ride from Buriticá, we lost sight of the Cauca River. Our view was shut off by an independent ridge of mountains several thousand feet high, which rises out of the valley between the range we were on and the stream. A slight change was also perceptible in the character of the country; extensive areas covered with brush now dotted the slopes, although at infrequent intervals; and on the extreme tops of both ranges a thin fringe of green was plainly discernible. The country is also very rough and broken, and there are a number of ridges to be crossed, many of which are two thousand feet high. Several separate mountains, not connected with the main ranges, stand here and there like giant, man-hewn monoliths, rising from a basal elevation of three thousand to eight or nine thousand feet, which magnifies their tremendous proportions.

On the fifth day we reached an altitude of eight thousand feet, and entered a fine strip of forest, the first we had seen on this journey. This is the beginning of the forested zone, and close scrutiny revealed the fact that it begins at precisely the same height on both the Central and Coast Ranges, and continues to the very top of the mountains, several thousands of feet higher up. We travelled along the top of the ridge for some miles, and then again descended abruptly to the barren valley where the little village of Peque is situated, and where our journey by mules ended.

Peque contains about fifty dilapidated mud huts, and its population is mostly of Indian descent, but includes some pure-blooded Indians. We had a letter of introduction to one of the latter, Julian David, who is the chief man in the town, and he rendered us every assistance. He called together a number of sturdy young half-breeds and requested them to join the expedition; in other words, told them to carry our packs to the top of the Paramillo. The men eagerly agreed to do this, for they had never before been in the service of strangers, and the trip to the high country and also the society of *gringos* promised interesting possibilities. We spent a few days investigating the neighboring country, while the men had their wives prepare the provisions for their use during the trip.

Some of the country surrounding Peque once doubtless bore a light forest growth, with heavier forest in the ravines; but by far the greater part is naturally barren or covered with brush thickets. I was told that at the time of the Spanish invasion forty thousand Indians inhabited the region, and as the several mountain streams supply an abundance of water, and the soil responds fairly well to cultivation, there seems to be no reason why it should not have supported an extensive population; at the present time only a few hundred people are left, the others having gone to swell the ranks of victims exacted by the lust of the conquerors.

The forested zone, beginning at eight thousand feet on the ridge we had just traversed, gradually extends its limits downward as one goes farther north, until at Peque it reached as low as five thousand feet in the deeper and well-watered ravines; and, as previously stated, at Puerto Valdivia it reaches the very edge of the Cauca.

One day an inhabitant of Tabacal rushed into our room and begged me to show him the wonderful diamond ring he said I wore while in his village; he had been away at the time, so had not seen it, but tales had reached his ears upon his return of the marvellous brilliancy of the stone which lighted up the whole street as we walked along. At first I wondered from what sort of an hallucination the man was suffering, for neither my companion nor myself carried any diamonds with us; finally I remembered that, in trying to find our way through the street at Tabacal, we had used a small electric flash-light to avoid falling over the pigs or into the mud-wallows; whereupon I demonstrated its mysterious powers to him, and he started back on his two days' walk a better-informed but nevertheless a most-disappointed man.

Figure 25. The porters en route to the Paramillo.

Figure 26. Cuña Indians at Dabeiba.

A stream of clear, cold water flows around one side of the hill upon which Peque stands, and to this we went nightly for a swim. Don Julian could not quite believe us when we told him of the purpose of our nocturnal prowls; so one night he accompanied us to the stream and, wonder of wonders, we actually did go into the water. I invited him to join us, but he said: "No, such a thing is unheard of; and, besides, an Indian is just like a cat; when either one gets wet it dies!"

When the half-breed porters who were to carry the equipment finally had their *charque* and *jarepas* all ready, they shouldered their packs and started for the mountains. As there was no trail, an additional man was engaged to go in advance and clear an opening with his machete.

A three hours' walk brought us to a point called El Madero, because a few trees had once been cut down there for their lumber, but the clearing was overgrown with blackberry-briars, brush, and guavas. Then we plunged into the unexplored forest.

It was our plan to follow along the top of an undulating ridge which one of the men said was the shortest and easiest route to the Paramillo. He knew from experience, having once visited the region some sixteen years before. It was during the course of a revolution; his father was pursued by the opposing forces and fled into the forest, taking his son, who was then a small boy, with him, and eventually reaching the Paramillo, they spent some time there in concealment.

At first the forest was fairly penetrable, but soon the moss-draped, liana-garlanded walls closed about us in a compact mass; ferns, palms, and arums sprang up from the ground in a matted jungle to join the heavily laden branches above. Then our trail-cutter was pressed into service, and plied his machete with deadly effect on the vegetation, with the result that a narrow tunnel was opened, through which we walked or crawled as occasion might demand.

On account of the long climb, having ascended five thousand feet during an eight hours' march, we made camp at three o'clock at an elevation of ten thousand feet. This gave us an opportunity of observing a few of the birds living in this untouched wilderness. There were wood-hewers and yellow-headed tanagers; parrots and blue-throated jays. A large harpy-eagle sat majestically on a low branch, surrounded by a flock of California woodpeckers, which screamed and scolded and darted at his head; but he sat perfectly motionless, utterly disdainful of such ignominious prey.

There was no water on the ridge, but a supply was secured from a ravine a thousand feet lower down; it was the last we had until we reached the Paramillo two days later.

The second day's march we hoped would be over a gentler slope, but it was soon discovered that our ridge consisted of a number of knolls rising from five hundred to a thousand feet above the mean level, and the forest grew denser constantly. Every foot of the way had to be cleared. In places we actually walked over the top of the vegetation; the branches were covered with a solid tangle of creepers, climbing bamboo, bromelias, and moss, and formed springy aerial bridges. More frequently it was easier to burrow underneath, so tunnels many yards long were cut, through which the porters crawled on hands and knees. The tops of some of the eminences were void of trees, their place being taken by jungles of bamboo, wild oleanders, shrubs, and clumps of tall, coarse grass with blades eight feet high and six inches wide, the edges of which cut like knives. That night we camped at eleven thousand three hundred and fifty feet up. The men eagerly cut down clumps of bromelias, hoping to obtain water from the bases of the leaves, but all they found were a few drops of vile, black liquid filled with drowned insects. Although we had travelled steadily for ten hours, I doubt if we had covered more than three miles.

A few hours after starting, on the morning of the third day, we emerged suddenly from the gloom of the forest. Instead of the tall, overburdened trees, there were extensive areas covered with brush, evergreens, stunted pines, and ferns. Beyond stretched the bleak, wind-swept slope of the Paramillo. At sight of this, the porters struggled on frantically, for the attaining of the goal meant a release from their heavy burdens—and water. That afternoon the last knoll had been crossed and the packs deposited on a rocky flat which was to serve as a camping-site. Each man started in a difference direction in search of a brook, and by dusk a pot-hole at the bottom of a ravine, and only a few hundred yards from camp, had been found containing several hundred gallons of pure, icy water. Never was a discovery more earnestly welcomed, and the men sprawled around the edges of the pool and drank their fill; then it was arranged that they should stay with us for the night, start back to Peque the next morning, and return for us after ten days. Our cook was of course to remain with us.

Figure 27. Our camp on the Paramillo.

The Paramillo region is composed of a series of sharply inclined peaks, the highest of which has an elevation of thirteen thousand feet, and is interspersed with ravines and deep fissures. The surface consists mainly of dark sandstone, so shattered over vast areas that a thin litter of particles covers the fundamental rock. Occasionally a thin vein of white quartz crops out to the surface, especially where, as often occurs, the strata stand in a perpendicular position.

At night the temperature dropped to 28° F., and ice half an inch thick formed on the reservoir; in the morning the ground was white with frost. The sparse vegetation on the slope consists of *frailejones*, blueberry-bushes and tall, tough grass; stunted trees and bushes, all covered with moss, grow in the deeper ravines. Hunting in these latter places was a never-ending source of delight; there was no water so it was possible to walk unrestrictedly underneath the green vault of brush which fringed the sides and met overhead. Many little mammals' trails zigzagged over the moss-covered rocks, and burrows opened into the steep banks; if we stole noiselessly along, or better still, sat quietly for a few minutes, the inquiring eyes of a paca, a large, spotted, tailless rodent, were sure to peer timidly out of some dark opening, to be followed later by the animal's entire body as it moved out stealthily to nibble on the tender sprouts. Numbers of woolly, yellow rats (*Melanomys*) also appeared to stare with beady, black eyes, and to nervously twitch their noses; sometimes they came out boldly to chase one another over well-defined runways and through mossy tunnels; but more often, they were content merely to gaze from the entrance of some safe retreat into which they vanished at the first suspicious move on our part. Deer, too, were seen occasionally, but they were not numerous; they grazed on the slopes in broad daylight, and had snug lairs in dense clumps of bushes which always commanded a view of the surrounding country. We saw no cougars or bears although we found the remains of several deer which had apparently been killed by these animals.

Birds were extremely scarce and, strange to relate, exceedingly wary. Collecting them was heart-breaking work; the slopes are so steep, that it was impossible to walk many yards without becoming utterly exhausted, and tramping through the high, wet grass chilled the lower extremities to numbness. The slaty finch (*Phrygilus*) so common at Santa Isabel, and two species of honey-creepers (*Diglossa*) were by far the most common; followed by a queer, wren-like little bird (*Scytalopus*) called tapacola, which lives among the densest ferns and mosses; it was seldom seen, but a cheery whistle apprised us constantly of its presence. There was also a gorgeous humming-bird, the whole body being of the most resplendent, iridescent deep rose and green colors; we located a nest of this species, a tiny moss cup scarcely an inch across, suspended from a creeper dangling beneath a bower of protecting leaves; it held two minute eggs, so fragile that the mere touch of a finger would crush them.

One day we ascended the highest peak in order to obtain a good view of the surrounding country. The Paramillo rises like a rocky island, out of an ocean of forest.

Clouds fill the depressions between the neighboring peaks, and surging, tumbling banks of white roll up the slopes or ascend in columns to spread out in funnel-shaped masses in the higher altitudes and become dissipated by the sun. To the southward rises the lofty Paramo of Frontino, many miles distant, the flat top dimly outlined in a grayish haze.

Toward the close of the tenth day, we heard loud calls and, soon after, our faithful porters dashed into camp. We were astonished at their number for, according to our agreement, only the original number was to return, there being no need for the trail-cutters; however, several additional men had arrived. Upon reaching the Paramillo, we had jestingly remarked that we should ascend the highest peak because we could perhaps see New York from the top; the extra men heard of this, and seriously explained that they had come to make the ascent in order to get a view of "Rome where the Holy Father lives!"

Early the next morning, we broke camp and started back. The homeward trip was much easier, for the packs were lighter, and the greater part of the distance was down-hill. After two days we emerged from the lower edge of the forest, and there was Don Julian and a delegation of natives waiting to convoy us back to Peque and welcome us home.

Don Julian provided horses for our return to Buriticá. They were unquestionably the poorest animals I had ever seen, and I disliked greatly to use them; but as no others were to be had it was a case of either taking the ones available or remaining in Peque for an indefinite period. However, they arrived safely in Buriticá after two days' time, and having secured a new pack-train we started northwestward toward Atrato drainage.

Leaving the little town and the semiarid country surrounding it, we proceeded straight to the top of a ridge eight thousand feet high, where a narrow strip of forest grew; and then descended on the other side into the valley of the Rio Canasgordas. At this point the stream is a mere rivulet, but it widens rapidly and the fertile banks are planted in sugar-cane, maize, and bananas. Huts built of mud and grass, half concealed by orange-trees dot the narrow valley; near them half-naked, dark-skinned children, pigs, and chickens ran about in a care-free manner or stared at us as we passed.

Lower down the river is flanked by wide belts of tall bamboo. Birds were not particularly abundant, but occasionally we caught sight of a yellow-rumped tanager as the bird darted through the foliage; or heard the familiar *kis-ka-dee* of a tyrant-bird perched on some high branch to sing, and to wait for insect victims to come within range of its snapping, insatiable beak. We spent the first night in the town of Canasgordas, and the second in a dilapidated house known as Orobajo. The family living here was in great distress owing to an epidemic of some kind of virulent fever which had appeared in the district. There was no food in the house, with the exception of a few beans, but after scouring the neighborhood our cook succeeded in purchasing a hen and a dozen *jarepas* which we divided with the infirm family.

While waiting for supper we went on a tour of inspection over the premises and located a house-wren's nest in the roof. It contained one young bird, and the people told us that the other had been killed by falling to the ground. Later we found several other nests of this species, but in no instance were there more than two eggs or birds in one nest. This fact is most interesting; in a temperate climate the house-wrens rear a large brood—eight being not an uncommon number of young; but near the equator two seemed to be the usual amount.

Below Orobajo the river is known as the Heradura. It flows past the village of Uramita, which was all but deserted. The fever that had invaded Orobajo had also visited this place and more than half the inhabitants had died. A few men were engaged in pumping salt water from shallow wells which was led in bamboo pipes to a battery of low pans where boiling evaporated the water and left the salt. So far as we could see there was no other industry in the town.

Figure 28. Dabeiba on the Rio Sucio.

Dabeiba, our first objective, was reached the third day after leaving Buriticá. As we gained the summit of the last ridge, a wonderful view lay before our eyes. The little town, composed of whitewashed houses with red-tile roofs, glistened in a flat valley carpeted with the softest green. On one side a river, called the Rio Sucio, raged and fumed over a rock-encumbered bed; fields of cotton dotted its banks, the snowy bolls and yellow blossoms almost obliterating the large green leaves. Forested hills enclosed the peaceful view as with a protecting hand which would shield it from the terrors of the frigid Andes on one side, and the steaming Atrato lowlands on the other. In this garden spot we decided to remain, but our arrival was nearly marked by a tragedy. On account of the noonday heat I had tucked a towel under my hat which, hanging down in the back provided, in a measure, protection from the hot sun. One of our *peons*, in a spirit of fun, told several small boys we chanced to meet that I was the bishop come to pay the town a visit; the urchins rushed into the road and prostrated themselves at my horse's feet, imploring a benediction. Fortunately the animal took fright at this unusual occurrence and bolted to one side before it could be restrained, narrowly avoiding trampling the kneeling forms in its path.

At Dabeiba we made the acquaintance of a tribe of interesting Indians—the Cuñas. They lived in banana-leaf huts, scattered over a wide area, but spent most of the time in town, looking into open doorways, begging for rum, or standing in silent groups on the street corners. They are a short, well-knit people of a dark-brown color. When in the forest they wear a breech-cloth only; but the priest has provided them with large muslin sheets that they promptly dyed a dirty-brown hue with *achiote* seeds, which they wear while in town. They also wore heavy necklaces of silver coins, and bunches of weeds tied about the neck for charms. At first sight it appeared as if they had no teeth, but further scrutiny revealed the fact that their dental equipment was perfect, though colored black from the juice of a fruit which they chew continuously. The body is liberally besmeared with grease—especially before they enter the river to bathe, so that the water rolls off as from a duck's back. One of the men was entirely covered with star-shaped marks of a deep-blue color which had been stamped on with a die made of wood. They spoke practically no Spanish, but were a friendly lot and enjoyed being photographed.

In order to reach the best hunting-ground, it was necessary to go to the other side of the river, but this was not difficult owing to the fact that a raft ferry was available. Birds were plentiful about the outskirts of the town, though of species common to open country and easy to observe in more accessible regions; we therefore spent the greater part of our time in the forest.

One of our first and most interesting discoveries was a species of pigmy motmot (*Hylomanes*). It is no larger than a sparrow and has a very short tail in contrast to the long "pendulum" tails of the better-known varieties. This little blue-and-green bird lived in the dense vegetation on the steep slopes, and when several flocked together they joined in a loud, cackling chorus at frequent intervals.

The cotton-fields sheltered a varied fauna. Hummingbirds came to the blossoms, and numbers of fat, red insects resembling potato-bugs lived among the drooping white fibre of the opened pods. Doves ran over the ground, and small rodents had their burrows at the base of the thick stems.

While at Dabeiba we met one of the most delightful Colombians—a type which I am afraid is vanishing, even as the forests and virgin wilds disappear before the onslaughts of civilization. He had but recently penetrated farther into the wilderness, cleared a few acres of ground and erected a humble cabin of bamboo and wild banana leaves; to this he urged us to come for as long a time as we should care to remain; so one morning we gathered together the most essential articles of our equipment and tramped through the intervening eight miles of jungle to his home. The beauty of the forest is indescribable; and wild life was so abundant that by the time our journey's end was reached we had attained such a stage of thrilling expectancy it was difficult to restrain our enthusiasm for the few hours needed to seek shelter indoors from an approaching storm. The shrill cries of parrots cleft the air; trogons *cooed* plaintively; toucans yelped and rattled; and from all sides came the *whush-whush-whush* of giant orioles' wings as the black-and-yellow forms hurried by to seek their pendent nests swaying dizzily from the branches of some giant ceiba towering regally above the unbroken forest.

While we waited for the storm to subside, the cook shelled corn and then, placing it in a wooden mortar together with a handful of ashes, began to pound it to remove the skins. This operation required about half an hour, so frequently she paused to rest; but no sooner had she deserted her post than a swarm of *cargador* ants invaded the receptacle, and the first intimation we had of their presence was when a file of white kernels began to descend the side of the mortar and cross the floor at our feet. How the small insects are able to carry the large, heavy grains is a mystery. The burden weighs many times as much as the ant which bears it, and almost hides it from view. Later, we saw swarms of the same species at work in the clearing; they cut sections from the edges of corn leaves by digging one mandible into the leaf for a secure hold, and then rip toward it with the other; the cut is always circular. Most of the insects worked from right to left, but one out of every five seemed to be "left-handed" and worked in the opposite direction. When the section of leaf is detached it is dexterously swung over the cutter's back, and away it marches with the green banner waving aloft. In addition to carrying this load, several small ants often mount on the leaf for a free ride to the nest.

That night another denizen of the wilds invaded the house; as we sat quietly in front of the hut listening to a shrill, uncanny *oh-ho-ho-ho-ho* coming from the forest, and which the natives said was the mating call of the three-toed sloth, but which we recognized as the song of a giant frogmouth or goatsucker, a cat owned by the family began to cut queer capers about the fireplace. A light revealed a good-sized bushmaster making its way across the kitchen floor. Whether the reptile had been attracted by the warm glow of the embers—for the rain had been followed by a decided drop in temperature—or had entered the structure to forage for mice, I do not know; but fortunately the cat had discovered its presence in time to prevent some one from stepping on it, and was striking at it playfully with its paws. After that the cook slept on a bench instead of on the earth floor, as had been her custom.

Our daily excursions took us far into the forest which invested the low, rounded hills in all directions. There were few trails, but a lack of undergrowth made walking easy. On one of our first hunting expeditions we found the rare ground-cuckoo (*Neomorphus*), a beautiful iridescent greenish-black bird which, on account of its terrestrial habits, has nearly lost the power of flight. Once before, I had seen this bird, and that was on the upper Orinoco, near the foot of Mount Duida. There the single individual was engaged in a curious game of tag with a tinamou; the birds chased one another about on the leaf-strewn ground, over logs, and through the underbrush, and jumped over one another's back as if playing leap-frog. We also found the flat-billed motmot in considerable numbers. These birds usually clung to the lianas drooping in festoons and loops above the small mountain brooks, and were exceedingly stupid and unsuspicious. They uttered no note, and sat motionless many minutes at a time, silhouetted like dark, ragged spectres on their perches. Among the moss or green leaves their color blended well with the surroundings, and we doubtless passed numbers without being aware of their presence.

Not all the birds inhabiting the forests at Alto Bonito are inconspicuously colored, however. There are gorgeous little tanagers, humming-birds, toucans, and trogons. The latter, especially, are creatures of such exquisite beauty that they seem to belong to a world more ethereal than our own; their brilliant scarlet or yellow breasts resemble a flower of dazzling color, for which the shimmering, metallic wing-coverts and back provide a resplendent setting. The bird is as fragile as it is beautiful, and was evidently not intended to be defiled by the touch of mortal hands. If a specimen is shot, many of the feathers are lost before the bird reaches the ground, and at the impact of the ground many more are shed. The skin is so delicate that it takes an expert to remove it, and even then the bird is the despair of field-naturalist and taxidermist alike.

There was also a splendid representation of the parrot family, ranging from noisy little parrakeets to huge, green amazons. This reminded me of an interesting provision of nature whereby three families of birds frequently found in the same locality are able to obtain their sustenance. They are the parrots, trogons, and toucans, all of which feed upon fruit, each seeming to secure its food in a different manner. The zygodactyl feet of parrots enable them to *climb* out to the tip of fruit-laden branches and to cling to them in any position while feeding; toucans, endowed with an enormously elongated bill are able to *reach* a long distance for a coveted morsel, which is grasped between the tip of the mandibles and tossed back with an upward jerk of the head, to be swallowed; a trogon has a very short beak and neck, and the delicate feet are not adapted to climbing, but the wings of the bird are so constructed as to enable it to hover, from which position the fruit it desires may be snapped off the stem, when the bird returns to its perch to devour it.

One day our host's son, aged thirteen, undertook to guide me to a distant part of the forest, where he said a large herd of peccaries had their feeding-ground. At first we passed through a part of the country well known to me, as I had taken a number of hunting excursions over the same ground; then we ascended a steep slope and, reaching the top, began to explore a vast stretch of heavy woods but rarely visited by any one. Although we had come for the express purpose of hunting peccaries, there were so many rare prizes on all sides that it was impossible to adhere strictly to our first intention; the temptation to add new treasures to our collection proved too great. Dainty little pigmy squirrels played in the top of the palms, or clung like lichens to the tree-trunks.

Some of the trees bore ripe fruit, and to them many animals came which are hard or even impossible to find under other conditions, thus making an ideal spot for the naturalist. A few seeds of the alligator-pear cast away by a hunter years before had taken root and grown into good-sized trees; the fruit dropped to the ground as it matured, attracting agoutis, which collected, apparently from some distance, to feed on the rich morsels. Other trees were laden with small berries. Although there was no sound to indicate the presence of a living thing, we usually discovered that first impressions were deceptive. If we waited a short time, a gentle patter on the leaves at our feet rewarded our patience; and then a close scrutiny of the leafy vault revealed silent, dark forms carefully moving among the tops of the branches and reaching out to pick the fruit upon which they were feeding. Gradually the shadowy forms assumed the shape of toucans, parrots, or macaws; the latter two birds are very wasteful and drop far more food than they eat.

The presence of an ant army is invariably advertised by the sharp chirp of the ant-wrens attending it. We encountered one, and spent an exciting half-hour securing two species of ant-birds, one black with white shoulders (*Myrmelastes*), and the other of a brown color with a white line running through the centre of the underparts (*Anoplops*); they had been feeding on beetles and spiders, and examination of the stomach contents revealed also a few ants. After shooting a bird it was necessary to enter into the thick of the voracious insects to hunt for it; but before the trophy could be recovered swarms of ants had climbed up our legs and clung with a bulldog grip.

Occasionally we saw a flock of manakins—brilliant little sprites of the forest, always found in the densest thickets. Some are black with golden heads; others, also black, have yellow breasts and long tufts of feathers on the throat, giving the bird a comical, bearded appearance; a third species had a vivid scarlet crest. The males only are brightly colored; the females are green.

There were signs of peccaries in abundance, but the constant shooting had frightened them away; so after inspecting an ancient Indian tomb consisting of a pile of carefully placed stones, overgrown with creepers, we started for home. Instead of retracing our steps over the many miles we had come, we followed a narrow gorge which we knew must lead to the Rio Sucio. Progress was slow and difficult, for the brook descended in a series of falls, and the rocks were covered with moss and were slippery; however, having started *via* this route, it was impossible to retrace our steps.

There was little of interest along the course of the treacherous little stream; but we discovered nests of a barred black-and-white wren (*Thryophilus*) swinging gayly above the water. The basket-shaped structures had been placed in the wildest, darkest spots, and each contained a single young bird, dozing peacefully in the entrance opening, lulled to sleep, no doubt, by the semigloom and the sound of rushing water.

As we picked our way along slowly and painfully, frequently wading through water three feet deep, a dark, shadowy form lunged from the blackness of a cavern among the boulders and clung for an instant to the cuff of my hunting-coat; then it dropped to the ground, and slowly disappeared among the rocks. My companion, who was a few feet in advance, had just turned to make some comment, and it was not until his frantic shriek brought me back to earth that I fully realized what had occurred. A bushmaster, apparently four or five feet long had become exasperated at our close proximity, and aimed a deadly thrust at the disturber of its diurnal slumber. This habit of the snake is well known; by nature it is sluggish; one person may pass close by without arousing its anger, while to a second individual, immediately following, it will show resentment, although it may not strike; but a third may consider himself fortunate, indeed, if he does not draw the full measure of the reptile's fury.

The exploration at Alto Bonito yielded such rich returns that we regretted the necessity of leaving; but a field-naturalist's time is not unlimited, and presently we found ourselves riding across the parched Antioquian desert, *en route* to Medellin.

The work at Alto Bonito provided the last link in the chain of facts regarding the forestation of northwestern Antioquia, and also throws some light on the extension of the mountain ranges. For information on the latter subject we were compelled to rely largely on data furnished by Señor Cspinas, director of the School of Mines, Medellin; Señor Ernesto White, an engineer who has made surveys in the region, and the reports of Indians.

The Western Cordillera terminates in the Cerro Aguila, just below 9°, near the Golfo de Urubá, and is less than one thousand feet high. The range breaks down, gradually, north of the Paramillo. In latitude 7½° the highest peak is known as Alto Esmeralda, four thousand feet high; and the Abibi, a few miles farther north, reaches an altitude of only three thousand six hundred feet.

A trail recently built (by Señor White) from Turbo on the Gulf of Urubá to Montana on the Rio Sinú crosses the very country about which we knew least; the elevation of its highest point is eight hundred feet, and every mile of the way was cut through heavy virgin forest.

PART II. VENEZUELA

Chapter 10

FIFTEEN HUNDRED MILES ON THE ORINOCO

It seemed as if the declining sun had set the quivering world aflame; all day long the *Delta*, well remembered but unbeloved by voyagers on the Master River, had struggled on against the yellow flood toward her goal two hundred and forty miles above the Parian Gulf. Not a ripple stirred the placid water which glided ever onward, and no breeze stirred the heavy, dark vegetation that lined the river's bank. It had been one of those days which only the traveller to tropical lands can adequately picture; when all the earth silently droops beneath the unrelenting glare of the lurid orb overhead, and eagerly awaits the coming of night which alone can bring relief.

As the last vestige of the sullen disk dipped into the forest, and only a faint pink and violet glow lit up the banks of vapors hanging low in the west, the nightly gales from the ocean sprang up with unrestrained vigor; soon a choppy sea was raging, and as each white-capped wave struck her wooden sides with a muffled boom, the fragile, top-heavy steamer shuddered and threatened to capsize. Morning, however, found her still battling bravely with the somewhat subsided elements, and, not long after, the *Delta* was slowly dragging herself alongside the high, sandy beach on which stands Ciudad Bolivar.

The first white man to ascend the Orinoco was Ordaz, who in 1531–2 went as far as the mouth of the Meta; and after him came the usual bands of treasure-seekers in quest of El Dorado; but instead of wonderful golden cities they found yawning graves in a hostile wilderness.

In the middle of the eighteenth century missions, founded by the Jesuit fathers, dotted the river-bank as far up as Esmeraldas; these have long since vanished. Humboldt made his memorable voyage to the Cassiquiare in 1800, and a number of other scientific expeditions followed in his wake at irregular intervals; to enumerate them all would be a tedious and unwarranted use of time. However, one remarkable fact must not be overlooked, namely, that even to this day the actual sources of the Orinoco have not been discovered.

To trace this huge artery to its very beginning, supposedly somewhere in the Serrania de Parima on the Brazilian frontier was not the object of our expedition; but rather to explore the regions north of the inaccurately mapped Rio Cunucunuma, more particularly Mount Duida, thought by many to be the locality described in a widely read book entitled "The Lost World." Of this country, and of the people and animal life inhabiting its virgin wilds, very little was known.

With the tying up of the *Delta* the first stage of our journey had been completed.

Ciudad Bolivar, formerly called Angostura, meaning narrows, on account of the narrowing of the Orinoco at this point to the width of a mile, stands on an eminence on the left bank, and is the capital of the Department of Guiana; it is the largest and only city of importance on the river. The red-tiled roofs and whitewashed walls of the houses can be seen from afar. On landing, one is confronted by a strange medley of low, thick-walled edifices; narrow, crooked streets, and swarthy, unkempt people. Practically all of the windows are heavily barred, a custom common in many parts of South America, and retained from the Moors.

Whatever beauty attached to the place is indoors. There are no green lawns or flowering gardens to cheer the eye of the passer-by; but a glimpse behind the sombre walls will invariably reveal an open court or *patio* filled with flowers and tropical shrubbery, and occasionally a fountain; but this is not all. In the *patio* of the hotel, which served as our headquarters, there lived in perfect harmony several large tortoises, a deer, two sheep, about a dozen tree-ducks, turkeys, chickens, guinea-fowl, and several pigs; fifteen species of birds, including parrots, orioles, and finches occupied cages hanging on the walls. The desire to keep caged animals is an inherent trait of the South American. Back of the city lies an extensive swamp from which, at least during the month of December, came great numbers of mosquitoes. As may have been inferred, the heat was very great; but regularly at nightfall the strong wind came up the river, causing a drop of several degrees in the temperature; then the town cast off its torpor, lights twinkled, the band played on the water-front, gayly dressed and painted women peered from behind the heavily barred windows, the streets were filled with a roving crowd of men and boys, and Ciudad Bolivar presented a wide-awake appearance.

On the opposite side of the Orinoco is the small town of Soledad; this village supplies a large portion of the sailors who man the boats plying on the river.

Our first care was to try to find a way of proceeding on our voyage. On account of the low stage of the water from the months of January to March, steamers do not ascend beyond Ciudad Bolivar regularly, and at best they go only as far as the Apure. It was, therefore, decided to charter a sailboat of shallow draft which would take us to the first great barrier to navigation, the cataracts of Atures. To secure such a craft was not an easy matter. We visited several of the large export houses, mostly German, but none of them had vessels at their disposal. Finally, we heard of a man named Guillermo Montez; he was a type frequently met with in South America; owning a small store which contained chiefly long ropes of garlic festooned on the walls, living in a mud hovel, and apparently poverty-stricken, he nevertheless possessed great wealth and knew how to handle his fellow countrymen. This "handling" consisted of keeping them constantly in debt to himself, so that he owned them virtually body and soul. Montez immediately sent to Soledad for one of his debtors, and within a short time we had secured the contract for the transportation needed.

On December 16 word reached us that the boat was ready. We had spent the intervening days adding to the stock of provisions brought from New York, and it might be added that the shops of Ciudad Bolivar were well filled with a splendid assortment of foodstuffs at reasonable prices.

The *Hilo de Oro* (*Thread of Gold*), for that was the name of the sloop impatiently bobbing near the bank, was a boat capable of carrying one hundred and fifty *quintales*, under the command of one Pedro Solano; her crew consisted of four men, and the captain's wife, whose position was that of cook. To properly load the equipment and provisions required half a day, and with the springing up of the evening wind we hoisted sail and, skirting the towering rocks protruding from the centre of the river, glided easily to the other side. As all the men came from Soledad, there followed a night of the usual festivities of drinking and leave-taking; but with the rising sun, the wind still holding out, we started on the real voyage up the great river.

Fortunately, the wind was favorable and continued to blow intermittently all day long; by ten o'clock at night we had covered about thirty miles and cast anchor at a point called *Boca la Brea*. The width of the river averaged about one mile and a half, and the entire bed is strewn with huge boulders, rendering navigation at night impossible.

Next day, a favorable wind did not reach us until late in the morning, and we had our first glimpse of wild life. The crew, a piratical-appearing band with unshaven faces, wearing short breeches only, and red and blue handkerchiefs around their heads, landed a number of large striped catfish; but their tackle was too light and others of greater weight broke the lines and escaped. Numbers of *caimans*, or crocodiles, floated lazily downstream with only the eyes and saw-like tails showing above the water; and a school of fresh-water porpoises jumped and raced around the boat.

On the days that followed, the wind either died down entirely or blew with terrific violence, so that slow progress was made. The *chubascos*, or squalls, not uncommon on tropical rivers, come up suddenly and without warning; a faint, funnel-shaped mass appears on the horizon, followed by a low bank of black clouds, and fitful little sandspouts that spring into existence on the vast *playas*. There is never time to seek the leeward banks, and not a minute is lost in lowering sails and placing every available object below to prevent its being washed overboard. While Captain Solano shouted hoarse orders and the crew worked like mad (the only time they really did work), we donned our oilskins and awaited the coming of the storm. To go down into the hatch was impossible, both on account of the lack of space and the stifling heat. The wait was never very long; with a roar the hurricane burst upon the quiet river, and in a few minutes everything was obliterated in the dense fog and wall of falling water. The wind tore through the rigging with agonized wails, and angry white-capped waves sprang suddenly into existence, sweeping over the boat and dashing it about like a cork in a millrace. There was nothing to be done but wait until the storm subsided and hope that no obstructing boulder, or the bank, would put an end to the madly careening craft in the semidarkness. This lasted from fifteen minutes to an hour; then the wind died down, the rain ceased, and the fog lifted. A changed river presented itself. Monstrous waves, capped with foam, dashed and tore at the high, crumbling banks, undermining them so that large sections tumbled into the water, carrying with them tall trees and massed vegetation. The agitated surface was littered with *débris* which bore good evidence of the violence of the storm.

After this there followed several days of calm; there was not enough wind to fill the sails, and all the "whistling for a breeze" of the sailors did exactly as much good as one would expect it to. Finally, in desperation, a long rope was tied to the mast, and two men going ahead in a canoe made the other end fast to a tree, a few hundred feet ahead. The remaining members of the crew then hauled on the rope, slowly drawing the boat forward. Progress was slow, of course, but on the 22d we reached the *Puerto del Infierno*, the best possible name for the narrow, rocky gorge through which the river rushes with uncontrolled fury. A large mass of granite covered with low vegetation divides the river into two narrow channels, one of them so protected by high, rocky banks that no wind ever reaches the water, consequently making it impossible for boats to sail up the passage. The other is a narrow, rock-strewn gorge, down which the water thunders in a series of cascades. On the right bank, perched high on the rocks, are a few mud huts called *Pueblo de las Piedras*. We spent the greater part of a day waiting for wind, and then made straight for the seething passage. Fortunately our pilot was a good one; his method was to steer directly for some great boulder, below which the water was quiet, and just as the ship seemed about to strike he swung the tiller, and the boat painfully nosed her way up the cataract that dashed down the sides of the rock. If the breeze slackened for a moment the ship drifted back with the strong current, which was extremely dangerous, as there was no way of regulating her course; but always, just in the nick of time, the sails filled and after an hour's struggle we left the rapids and sailed into the quiet water above.

Not far above the *Infierno* is the village of Mapire, a neat collection of perhaps fifty huts on a high bluff overlooking the river. In back of the town are vast *llanos*, or grassy plains, which are capable of supporting numerous herds of cattle. On the opposite side of the river, and some little distance up, is the mouth of the Caura, at one time believed to be the home of a tribe of headless people; but the old superstition has been overthrown, and during the first month of each year many adventurous parties ascend the river for a considerable distance in search of the *serrapia* or tonca-bean. The tree (*Dipteryx odorata*) upon which the fruit grows resembles a mango, with spreading branches and deep-green, dense leaves. The fruit also is very similar to the mango, though green, with tough, fibrous flesh and a large seed. While the fruit is still green great quantities of it are destroyed by macaws and parrots, which take a bite or two, then drop the rest on the ground. Upon ripening, the fruit falls, when it is gathered into heaps and dried; the seeds are later cracked open and the strong-smelling kernel extracted to be carefully preserved and sent to Ciudad Bolivar, where it is treated in casks of rum and then exported. It is used in making perfumes and flavoring extracts.

The water of the Caura is of a clear dark-red color, and for a great distance after entering the Orinoco the two waters flow side by side without mingling in the slightest degree.

The Orinoco widens into a majestic stream above this point, and we estimated that the distance from bank to bank must in some places be from three to five miles; also, vast sand-banks stretch along both sides for a distance of many miles.

Caicara, the only town of importance on the Orinoco besides Ciudad Bolivar, consisted at the time of our visit of about one hundred and fifty houses, but on account of a rubber and *serrapia* boom on the Cuchivero many of the inhabitants were leaving for the latter place. The next day we passed the mouth of the Apure, and just beyond the mouth of the Arichuma; a great low, sandy island rises out of the centre of the Orinoco at this point, on which thousands of terns, skimmers, gulls, and other water-fowl were apparently nesting. All day long and even at night the air was filled with darting, screaming birds that made such a terrific din that it was impossible to sleep. High waves prevented our landing on the island, but the natives visit it regularly, taking away canoe-loads of eggs; for this reason the island has been named *Playa de Manteca*, meaning in this case land of plenty.

The next settlement is called Urbana, and is on the south bank of the river, almost opposite the mouth of the Arauca. It consists of about a score of hovels. The Arauca is a river of considerable size, and is said to be bordered by vast marshes and swamps, the home of countless egrets and other water-birds. Hunting-parties ascend during the nesting-season and kill great numbers of the birds; the plumes are taken to Ciudad Bolivar and disposed of to the export dealers.

Leaving Urbana on the 29th, we entered one of the most difficult stretches of the river to navigate. The fish-hook bend of the Orinoco turns southward, and the eastern bank is dotted with a range of low granite hills which are, in fact, a chain of giant, blackened, dome-shaped boulders. The wind from the east, roaring through each cleft and opening, strikes the river from several directions and with cyclonic violence. One moment there is scarcely enough to make headway against the current; the next a gust strikes the sails and sends the ship wallowing on her beam until the boom drags in the water and it is an even bet if she will gradually right herself or go over. At such times of peril as well as on starting each morning it is the custom of the sailors to pray. Of course they were all Catholics. The captain or whoever steers said, "*Vamos con Dios*" (let us go with God), and the others answered in chorus: "*Y con la Virgen*" (and with the Virgin). Occasionally the person whose duty it was to lead was so occupied rolling a cigarette or slapping at flies that he neglected his duty; then someone was sure to remind him with a sarcastic "*Aha! Hoy vamos como los Protestantes*" (Aha! To-day we are starting like the Protestants). It often happened that the crew was remiss. The captain repeated his lead several times without being heard; finally, his patience exhausted, he shouted at the top of his voice, "*Vamos con Dios, caramba*," and the crew immediately yelled back at the top of their voices: "*Y con la Virgen, caramba*."

Added to the danger of the shifting gales is a rapids named San Jorge. There was just enough water to cover the rocks which dot the river-bed, causing a series of cross-currents and whirlpools which only a Venezuelan boatman, trusting mainly to luck, can hope to pass through. The rigging of the *Hilo de Oro* was old and rotten, and ropes were constantly snapping and sails splitting. No matter how obvious a defect was, it was never remedied until an accident had occurred. The boom had been threatening to break as each sudden gust of wind struck the mainsail, but a few boards nailed across the weakened place it was hoped would give sufficient strength for any emergency. An hour after leaving San Jorge, however, the boom parted with a loud report and dropped into the water, nearly upsetting the boat. Then, while the craft wallowed on her side with the deck awash there ensued a good deal of mingled praying, swearing, and frantic work until the heavy boom was fished out of the water. We tied up at the bank, cut down a tree, and worked the greater part of the night replacing the broken member.

One of the curious granite battlements rears its head out of the water to a height of several hundred feet, and is somewhat suggestive of a small edition of the famous Sugar-Loaf Rock at the entrance to the harbor of Rio de Janeiro. This is called Treasure Rock, and no Venezuelan ever passes the spot without casting envious glances to the top. In the days when the old Spaniards were still wandering over the newly discovered lands in search of El Dorado, so the story goes, they penetrated far into the Cerro Sipapo and found rich treasures in gold and precious stones. The Guajibo Indians, in whose domain they had penetrated and whom they had robbed, finally tired of their unwelcome guests and chased them down the river. In desperation the Spaniards formed a stronghold on this island rock, driving iron spikes into its sides as a means of reaching the top; for many weeks they resisted a siege by the savage hordes, but with the coming of the rainy season the Indians withdrew to their mountain fastness. Finally the Spaniards came down, cutting off the spikes as they descended; they feared pursuit, so left the treasure on the rock, hoping to come for it when reinforcements had been secured; they never returned, and to this day the fabulous wealth of the Guajibos lies entombed on the top of the impregnable boulder.

The Meta is a mighty river coming from the immense prairie region of eastern Colombia. It is navigable for the greater part of its course, and should be the means of opening up illimitable grazing areas when the Orinoco is thrown open to free navigation. Where the Meta joins the Orinoco, the latter is fully two miles wide; near its mouth the country is covered with a dense scrub growth. As we neared the mouth of the great river several large canoes filled with Indians, of the Guajibo tribe, shot from an invisible hiding-place near the bank and made for the centre of the stream. They have an unsavory reputation among the river-men, and Captain Solano added little gayety to the occasion when he prophesied an attack and armed his men. On they came, swiftly and silently, the dusky, naked bodies bending in perfect unison, and the great muscles of the arms and shoulders rippling in the sunlight as they drove the short, pointed paddles deep into the water with vigorous strokes; but our suspicions proved to be unfounded. They passed rapidly on some secret mission of their own without even condescending to glance in our direction. This utter indifference to strangers, I found later, is a characteristic common to all Indians of the Upper Orinoco. A man might be drowning or stranded on a rock, but they would pass him quietly in their canoes without apparently seeing him; they would pay not the slightest attention to his cries for help. Their ill treatment at the hands of strangers has been so great that they have lost all confidence in any one unknown to them, and so they retaliate by feigning indifference to him, even in his direst need.

The nights were usually spent aboard ship. If there was no wind it was safe to tie up to some tree; or if darkness overtook us near a playa the anchor was carried ashore and buried in the sand. While the cook prepared supper on the brazier or over a fire built on the bank, hammocks were strung in the rigging, and then we fished until time to retire.

Fish were always abundant and of many varieties. One kind that was taken frequently and that was excellent eating was a catfish, weighing up to twenty-eight pounds, of a deep brownish color with wavy bluish-gray lines running along its sides, called *vagre tigre*; another species of catfish, frequently of a weight of seventy-five pounds or more, and of a deep slate color, was not uncommon; there was also a third about eighteen inches long, with a large, narrow head and "feelers" as long as the body, that was always sure to be among the catch; but neither of the two last named was ever eaten, as the flesh was said to be poisonous. The crew was always careful to clean all fish immediately and place them under cover; if left exposed to the moonlight overnight they were unfit for food.

The hoarse cough of jaguars was heard almost nightly; it was the season when great numbers of turtles left the river at nightfall to deposit their eggs in the sand-banks, and the jaguars left the forest at dark to dig up and feed on these eggs. One night, just as the boat had drawn up to the high sand-bank preparatory to tying up, one of the huge cats was discovered sitting ten feet above us quietly surveying the scene on deck; there was a rush for the guns, but when they were secured the jaguar had disappeared. A clear sweep of loose sand with a low bush here and there stretched back a mile from the river to the heavy forest, and in the brilliant moonlight it was easy to trace the animal's tracks as it started toward cover. Several times its shadowy form was visible, slinking from one bush to another a few rods away, but always out of range; after half an hour the tracks were lost in the edge of the forest. We returned to the ship. Before replacing the guns in the hatch someone casually broke his, which action led to the discovery that it contained no shells; neither were the others loaded. One of the men while cleaning them that afternoon had removed the cartridges and failed to reload them. Fortunately, the jaguar is not quite as savage as he is usually pictured, or there might have been a lively scene on the *playa*.

There is but one other rapid of importance in the Orinoco before reaching the cataracts of Atures, and that is San Borja, not far above the mouth of the Meta. Just above this narrow stretch of seething water we met another boat about the size of the *Hilo de Oro*, which was cruising back and forth near the bank, her crew directing loud shouts toward the forest at frequent intervals. Upon inquiry we found that one of the men had gone into the woods to cut a pole; the other members of the crew had heard him chopping, as he had not entered the matted vegetation more than fifty feet; suddenly the chopping ceased, but the man did not come out; although they had searched far and near, no trace of him had been found, and this was the fourth day after his disappearance. The supposition was that he had been killed and carried away by Indians.

Perrico was formerly the port of call for sailing craft below Atures. At the time of our arrival there was nothing whatever there, not even a single hut. We continued up the river half a mile to a place called Vagre; here we found the remains of two palm-leaf huts, long since fallen down and overgrown with vegetation. In the small clearing a few cotton-stalks, beans, pawpaws, and castor-bean bushes still struggled for existence with the invading hosts of creepers and second-growth sprouts; the forest was rapidly reclaiming its own. On the sandy river-bank were the tracks of jaguars and caimans. At this point the river is divided into a number of branches by islands, and the one on which Vagre was situated is not over five hundred feet wide. Beyond this point a boat of any size cannot proceed; it is the foot of the series of cataracts, six miles long, known as the rapids of Atures. We sent a man overland to Zamuro for a *falca*, which is a canoe with the sides heightened with boards; and while our luggage was being rowed up the swift stream, we walked near the bank.

The aneroid, read at water-level, gave an elevation of three hundred and fifty feet; perhaps this is somewhat too high. Between Vagre and Zamuro a row of rounded, black rocks rise to a height of two hundred and fifty feet above the river, on the eastern side. Many boulders of enormous proportions lie sprinkled about in the most irregular manner, as far as we could see, and in spots there are outcroppings of ledges of quartz. The tops of the rounded granite hills are hard and glazed, so that they glisten in the sunlight as if covered with a coating of ice. There are but a few stunted trees, and where any vegetation can get a foothold tough, wiry grass grows; this is the home of many rabbits and rattlesnakes.

Zamuro we found to consist of three grass huts newly built and occupied by sick, miserable Venezuelan families. The heat is terrific, and mosquitoes and sand-flies first begin to make their presence known in considerable numbers. The river scenery is really magnificent; huge boulders of fantastic shape strew the river-bed, and rear their heads high above the seething torrent; against them the water dashes ceaselessly, surging and swirling in mad endeavors to destroy them, only to be baffled by the immovable sentinels and hurled back again to collide with their brethren equally unrelenting and equally impervious to the roaring onslaught. The scene is awe-inspiring.

The next step was to secure ox-carts to carry the *impedimenta* to the Rio Catañapo, three miles away, and this we crossed in a canoe, landing practically at Atures. The governor of the Upper Orinoco, General Roberto Pulido, made Catañapo his home. He was ordinarily supposed to reside in San Fernando de Atabapo, but on account of his arbitrary methods of government he was so greatly disliked that he decided it was "healthier" to live elsewhere.

The Catañapo is a turbulent stream of clear, cold water that dashes down from the near-by Cerro Sipapo. Not far above its mouth is a good-sized village of Piaroas, who come down occasionally with plantains, pawpaws, and other fruits which they exchange for cloth and sugar at Atures. When the Indians come down they apparently bring with them numbers of freshly killed monkeys, the flesh of which is greatly esteemed as food. We saw several heaps of the charred bones near frequently used camping-sites, here as well as at Zamuro.

The clear water of the Catañapo abounds in fish which may be seen twenty-five feet or more beneath the surface. Some were fully two feet long and resembled giant black bass; they refused to be tempted with meat bait, but rushed greedily for bright-colored objects such as fruit and flowers; they would take half an orange at a gulp.

Atures, consisting of six or eight mud and grass huts, owes its existence to the fact that the governor lives on the Catañapo and all the residents are his employees. Formerly the town was larger and there were thirty ox-carts plying back and forth across the portage; but the governor promptly selected the few he wanted and then discouraged competition in such a manner that he was shortly left alone in the field. To us he was most cordial, and immediately placed his carts at our disposal; nor did he examine our luggage, which was his self-imposed duty, and extract anything that suited his fancy.

The two miles from Atures to Salvajito, the port of embarkation above Atures Rapids, were covered in ox-carts which lumbered slowly along over the uneven semiarid country. Salvajito was only a small cleared space in the forest fringing the river.

The next step of the journey was to traverse the forty miles of river between Atures and the second great cataract at Maipures. Only a small canoe was available, so leaving my assistant and a number of the men to guard the left-over luggage, I started with three paddlers. The canoe was only eighteen feet long, with about two inches of freeboard, but fortune favored us and after two days we reached the mouth of the Tuparo. The first night out had been spent on a *laja*, or shelf of rock which extends over the water; the men set the dry vegetation in back of the camp afire in order to keep away jaguars, and built a fence of brands along the outer edge of the rock to frighten off the crocodiles. The second night was spent on a large sand-bank just below the rapid of Guajibo. In approaching this site the canoe had been caught in a sudden hurricane and swamped before land could be reached; but fortunately we had gained shallow water, so nothing was lost. On this sand-bar lived three species of terns, one of very small size that came in immense flocks after nightfall and, dropping on the sand, immediately disappeared from view; also numbers of yellow-legs and a few gulls. The wind blew steadily all night, so that by morning everything and everyone was half buried in the loose sand.

The rapid of Guajibo is one of the most treacherous in the whole Orinoco. Each year the rubber-gatherers pay heavy toll in lives while traversing this notorious spot. A great horseshoe-shaped ledge of rock extends across practically the entire river, and over this the water rushes at great speed; below is a series of scattered rocks extending for a quarter of a mile, and forming a raging, roaring gorge. We portaged around the spot, although the country is very difficult, owing to the many high rocks and the deep crevices between them. An acquaintance who had just passed attempted to have his men drag their boat through, with the result that they lost the canoe and three men. Shortly after a large *piragua* coming from up-river attempted to run the rapids to save time; seven of the crew, as well as the owner of the outfit, paid for their folly with their lives, and the entire cargo of rubber, together with the boat, was lost. A few days later another party wrecked their canoe and lost two men. These are all cases which came under our notice, and I was told of many others.

The port of Maipures is on the Rio Tuparo, about half a mile above its mouth. This river, some two hundred yards wide, comes rushing out of the interior of Colombia down a rocky river-bed. Where the landing was effected we found only the parched plain, a trail leading away from the river to the settlement of Maipures, a good three miles away. We pitched camp near the water, and the canoe and two men were immediately sent back for another load of the equipment. There was not much life along this part of the river. Numerous iguanas spent the hot hours burrowing in the sand, and if disturbed either ran away in the brush or plunged into the water. Both green and blue kingfishers clattered noisily on the opposite side, and a few large gray herons flapped up and down over the centre of the stream. We could constantly hear the loud roar of the Maipures Rapids, and the water rushing down the course of the main river was covered with foam.

Five days after our arrival the second load, in charge of my assistant, arrived. They had met with a mishap in the rapid of Guajibo, and one man and the canoe were lost. For nearly two days they had been stranded on an island and besieged by a party of Indians from the Sipapo; the occupants of a passing canoe, seeing their plight, came to the rescue, and brought them on to the Tuparo. While the borrowed canoe returned for the remaining members of the party, we busied ourselves transferring camp to Maipures, above the head of the rapids. The intervening country is level and covered with a sparse growth of clumps of wiry grass and patches of low woods; near the watercourse the trees are taller and the vegetation more dense. The town, consisting of six adobe houses with thatched roofs, nestles in a little grove of mango and tonca-bean trees, and from a short distance away is very picturesque; but like all the rest of the plain it is insufferably hot and the myriads of sand-flies quivering like heat-waves in the air make life almost unbearable.

Figure 29. The village of Maipures.

Figure 30. The *Hilo de Oro* at the end of the voyage.

While waiting for a boat of ample size to take us up the river to San Fernando de Atabapo, we had time to explore the surrounding country and to visit the rapids, three in number, which obstruct the river. The woods are wonderful beyond description; most of the trees are gnarled and low, as if grown under the guiding hand of a skilful Japanese gardener, and have the appearance of being hundreds of years old. Stunted spiny palms rear their crowns here and there, and an occasional tangle of red-flowered creepers forms an umbrella-like mass on the tip of some slender, dead stub. The ground is sprinkled with rocks of fantastic shapes, and some are of enormous size, rising in needle-like, fluted columns, or as the crumbling tiers of massive walls amid the curiously distorted vegetation. Along the river are other masses of rock, but of an entirely different formation; we saw caves and grottos, and ledges honeycombed with hundreds of pot-holes exposed by the low water.

Beyond the woods are large areas of cacti, pineapples, and low, thorny bushes, springing from crevices in the granite ledges. Bird life is abundant and varied. Quail and red-breasted meadow-larks occupy the open country, as well as a species of the much-sought tinamou; but a bird that proved to be the most interesting was a small, obscure individual called nunlet or swallow-wing. All day long the little creature, about the size of a king-bird, black above and gray below, with a saffron band across the throat, sits on the top of some dead tree, seemingly asleep; but let a fly or an insect of almost any kind pass along and the bird immediately becomes charged with activity and darts into the air in hot pursuit, catches its victim, and returns to its perch with graceful flits of the wings. It remains on the same twigs for hours, and usually returns day after day. If a stick is thrown at it the little creature flies away and comes back again and again. But stupid as the bird appears to be, it is nevertheless a skilful architect. I have seen them dig perfectly round holes deep into a bank of sand so loose that the whole mass would crumble at my touch; while one bird digs with much scratching and working of wings, the mate sits on a branch nearby and gives a twitter of alarm upon the approach of danger. Some members of the family build a huge pile of twigs on the entrance to their burrow to hide it. At the end of the tunnel, a foot or two back, the snow-white eggs are laid upon a thin layer of straw and feathers.

The highest falls in the river are known as Carretia, and are supposed to be about thirty feet high; they block the eastern channel of the river, here divided into two branches by the immense Isla de Raton. In the western arm the Raudal del Conejo and Raudal Saltinero effectively block this watercourse to navigation. It is said that the Spaniards built a road from Atures to the foot of the Cerro Sipapo above the falls of Carretia, and that the Indians still follow this route occasionally. If true, this was doubtless a great convenience, as it did away with the necessity of navigating some fifty-odd miles of the most difficult and dangerous waterway of the entire river.

A large boat called *piragua* was obtained at Maipures, and in this the expedition travelled to San Fernando de Atabapo in six days' time. The river is dotted with a number of islands, the largest being the great Isla de Raton, all heavily forested; the current is frequently so strong that no headway could be made either by rowing or poling the heavy boat. At such times a thick cable of the braided fibre of a palm called *chiquechique* had to be requisitioned, and everybody walked on the bank, dragging the boat slowly along. The very first day the man in the lead ran into a bushmaster fully eight feet long, and narrowly escaped the vicious thrust of the deadly reptile; a charge of shot soon put an end to the creature's menacing career, but the men jumped into the boat and did not want us to take along the dead snake, or they said its mate would be sure to follow and inflict a terrible revenge for the loss of its companion; this kind of superstition is very common among the natives on the Orinoco. Few of them would dare shoot a jaguar, as they firmly believe that for every one slain a member of their own family would be carried away by one of the huge spotted cats.

The country on the Colombian side, from below Atures onward, is level llano, covered with a good growth of grass, and with an abundance of water. Someday, no doubt, and in the near future, numerous herds of cattle will graze in the rich pasturage awaiting them, and another source will be added to the world's limited supply of meat. A fringe of trees grows along the river; among them are the valuable "cachicamo" and "cedro," the trunks of which are frequently fashioned into canoes by the natives.

The Vichada, at this season, had dwindled down until at its mouth it was not more than a hundred yards wide. We could see a range of hills far to the west, dimly outlined against the sky and finally fading into obscurity in the haze; in this direction the river has its origin. Several Piaroa families had settled near the junction of the two rivers and built a large hut of palm-leaves and grass. The men lounged in their hammocks all day long, drinking rum and fighting the clouds of sand-flies which feasted on their half-naked bodies; at night they crossed to one of the numerous sand-banks and collected basketsful of turtle eggs and also as many turtles as their canoes would hold. Some of their canoes were mere shells, so small that we could never learn how to negotiate them; no matter how quietly we sat they upset as soon as pushed out into the current, but an Indian or even two would calmly squat down in the bottom, take up their paddles, and glide away without the least concern.

The women were making cassava bread; after the tubers (*Manihot utilissima*) are ground and the juice has been extracted a thin layer of the coarse meal is spread on the bottom of a shallow pan about three feet in diameter; the heat causes the particles to adhere, forming a tough, round wafer which can be turned without breaking; it is thoroughly baked on both sides. When cold it hardens, and the huge slabs are then done up in bundles of twenty to forty each, tied up in plantain leaves, and in this way it can be kept indefinitely. This is the bread of the Orinoco, and is always carried as the main article of provision by Indians and travellers alike; when needed pieces are broken off, dipped in the river to soak a few minutes and then eaten. While not particularly appetizing, the slightly acid flavor is not unpleasant, and if there is time to freshly toast it just before using it is really quite palatable. Another article commonly prepared by the Piaroas is the bark of a certain tree, called "tabari." Long, narrow strips are cut from the trees and alternately soaked in water and beaten between rocks until the thin layers separate into tissue-like sheets; these are used in rolling cigarettes.

One of the granite ledges flanking the river just above the Piaroa dwelling bears on its surface a number of curious figures, carved in the face of the rock; unfortunately the water was so low that we passed far beneath them, and I was unable to make out just what they were; but the canoemen who had seen them a number of times said they were figures of men and date back to prehistoric times.

The country now rapidly grows wilder; tall forest replaces llanos or scattered growth, and the camps of rubber-collectors dot the river-banks. One afternoon, as we poled quietly along, we came upon a huge anaconda coiled up on a sand-bank; all about were iguanas three or four feet long, digging nesting burrows in the loose sand. The snake had just caught one of the big lizards and was crushing it into a limp mass, but the others paid not the slightest attention to the tragedy which was being enacted in their midst, and ran about or worked but a few feet away. When we approached to within twenty feet the anaconda dropped its victim and flung itself into the water; some of the iguanas followed it, and others scampered away over the sand.

That night we reached the low, sandy island of Tanaja and, ascending one of the branches of the river, made camp on the rocky mainland. The water was sluggish and shallow, so that we could easily see the muddy bottom six or eight feet below. As the boat moved slowly along we became aware of masses of black, flitting shadows underneath, and soon made out vast shoals of fish of various sizes that literally covered the bottom. There were rays, electric eels, catfish, and *piranhas* by the thousands, besides many others which we could not identify; the reason for their congregating in this shallow place is hard to guess.

The boulders on the bank were dotted with what we at first took to be lichens; but examination showed them to be night-hawks (*Chordeiles rupestris*) of a light gray color, which clung to the rounded tops silent and immovable, as if carved out of stone. When we paddled across to the island a short while after, we found scores of others, but these were the females squatting on one or two fragile speckled eggs which had been laid in shallow hollows scooped out of the warm sand. They were very tame and permitted me to walk up to within a few feet of them; then they took wing and with noiseless, graceful flaps flew a short distance away and dropped back on the sand.

Flocks of red-and-blue macaws flew screaming across the river in quest of some favorite tree in which to spend the night, far in the depths of the forest; after them trailed parrots of various sizes and colors, always flying two by two. Herons flapped lazily up-stream, and snake-birds perched on snags looked down at the masses of fish below, apparently regretting their limited capacity for eating. Exciting as this naturally must be to a field-naturalist, it was but a foretaste of what we were to find each day farther up the river.

As the morning of January 24 sped by, the water of the Orinoco began to assume a dark color, and by four o'clock that afternoon we had reached the mouth of the Atabapo; an hour and a half later we had ascended the clear red water of that river for a distance of three miles, and tied the *piragua* to the ledge below San Fernando.

San Fernando de Atabapo is the last settlement on the Orinoco and was the base from which we hoped to make our dash to the unexplored regions about Mount Duida.

Chapter 11

THE MAQUIRITARES' LAND AND THE UPPER ORINOCO

San Fernando, on the Atabapo, consists of about fifty adobe huts of the usual type, and at the time of our arrival was all but deserted. Almost the entire population had gone up-river to the scattered rubber-camps, as this was the season for collecting the valuable latex.

The town is situated on the Atabapo, where this river and the Guaviare unite, and its elevation above sea-level is three hundred and seventy feet. The mean temperature is about 80° F., although in the sun the mercury ascends to 112° F. or more, but the place is not particularly unhealthful.

The water of the Guaviare is muddy, while that of the Atabapo is of a clear red color and unfit for drinking. There are few fish, no crocodiles or sand-flies, and practically no mosquitoes, all of which is attributed to the discolored water. Two small springs near the town furnish an abundant supply of potable water, and when during the rainy season these are covered with the overflow from the river it is necessary to paddle across and fill the water-jars from the Guaviare.

To secure a crew of men for our trip up the river was a difficult undertaking and required a great deal of time. This gave us an opportunity of exploring the surrounding country.

In the immediate vicinity of San Fernando the forest has been cut down and tall second-growth sprouts form dense thickets; this is a favorite resort of many small birds, and several species of night-hawks make it a daytime rendezvous. The basic granite crops out in many places, the strata occasionally standing on end, and it is often streaked with narrow seams of quartz. There is no cultivation of any kind; the inhabitants lack all initiative for work and eat tinned foods and mandioc received in exchange for trinkets from the Indians.

When we returned a few months later a changed town confronted us. The rubber-collectors had returned from their several months isolation in the interior, and were spending the fruits of their labor as rapidly as possible. Dance-halls, gaming-dens, and almost every conceivable device for relieving men of their money had sprung up like mushrooms, and there was drinking and merrymaking day and night. Then suddenly, and without presage, a tragedy occurred; it will never be forgotten by the few who survived.

Governor Pulido, so it was rumored, had imposed a new tax on all rubber collected in the district, and had come to San Fernando to personally collect the extortion. Naturally, there was a good deal of dissatisfaction, and one night, just after we had been provided with a canoe and secretly advised to leave as soon as possible, the storm broke. A band of men, said to be under the leadership of one Colonel Funes, an Indian and the most notorious man in the district, attacked the town, killed the governor, and practically the entire male population, and rifled the shops and dwellings. If one may believe the tales of the few who escaped the brutalities committed that night, the deeds rival those of the most barbaric ages.

Perhaps some of those who perished deserved their fate, others assuredly did not; but it is a fact that government offices had been conducted abominably. In the post-office, for example, stamps were sold for twice their face value, and if one did not purchase them there and place them on the letters in full view of the postmaster, the mail was destroyed. A physician who chanced to be there, named La Page, and who apparently belonged to the military organization as he wore the regulation uniform, tried to collect over four hundred dollars gold for a few injections of quinine; and so the robbery went on until the whole band was exterminated.

Having engaged a captain with some experience on the Upper Orinoco, and a crew, we on February 3 loaded the low *batelão* and started on our mission, reaching a point called *Puerto Ti Ti* that night; from this spot a wide trail leads through the magnificent forest to the clearing wherein stands San Fernando.

For six days we made slow but steady progress up the river, and then entered the formidable *Raudal de Santa Barbara*, which extends across the entire delta of the Ventuari.

The Orinoco is wide but with few exceptions so shallow that we pushed along with long poles. Where the water was deep and the current swift, long-handled hooks were used to catch the overhanging vegetation and pull the boat along. This latter mode of travel was always slow and dangerous and the swarms of wasps and other insects living among the leaves, and shaken down, were far from being agreeable travelling companions. The banks were covered with dense, virgin forest; but there were extensive sand-banks and flat ledges of rock at convenient intervals, and one of these was always chosen for a night's camping-site. If we chanced to be on a *playa*, the early hours of the evening were spent in fishing. Armed with *machetes*, a bag, and acetylene-lamps, we waded out in the shallow water and "shined" the shoals of fish much in the manner that frogs are caught in parts of this country. At night the fish swam near the surface, and by directing the rays of the strong white light upon them one could approach to within a short distance and then strike with the knife: in this manner large numbers were taken. Occasionally a stingray, electric eel, or crocodile was suddenly encountered and then there ensued a hurried scramble in the other direction; this gave the pastime a decided element of sport. We also became more familiar with the dreaded *caribe* or cannibal fish, known as the *piranha* in Brazil, with which the water teemed. In the Orinoco they attain a weight exceeding three pounds and are formidable indeed. The natives will not go in bathing except in very shallow water, and I know of two instances where men were attacked and severely bitten before they could escape. The fish somewhat resembles a bass in shape, although the mouth is smaller; the jaws are armed with triangular, razor-edged teeth; and as they travel in immense shoals they are capable of easily devouring a man or large animal if caught in deep water. Floundering or splashing in the water attracts them, but they seldom attack unless their appetite has been whetted by a taste of blood; and then woe to the unfortunate creature which falls into their power. To catch them, we used a large hook secured to a long wire leader and baited with any kind of raw meat, and they always put up a good fight. Without a wire, a line would be bitten in two every time a fish struck. When taken from the water they are first killed by a blow on the head with the *machete*, and then removed from the hook.

At night there was always a heavy dew, and it rained intermittently each day. On dark nights, and often after a shower, the banks of the river where there was forest glowed with twinkling phosphorescence. Examination showed that the decaying vegetation was filled with myriads of small, wriggling insects, greatly resembling our well-known cellar-bug (*Isopod*), and one day we paddled for many hours through a mass of flying ants which had come to grief in the river. The water was covered with them and the waves had tossed them up on the banks to a depth of several inches. Another thing that attracted our attention was the large number of bats. On one occasion we heard a dull rumbling among the granite ledges near camp, and not long after a stream of bats began to emerge from the cracks; from a distance they resembled a cloud of smoke. There must have been many thousands, for the black masses continued to rise until darkness obscured them from our view. Spruce records that on one occasion he saw not less than a million under similar circumstances. This brings up an interesting problem. The individual range of these bats is probably not very great, the result of which is that immense numbers of them are distributed over a comparatively small area. Now, if the struggle for existence is as keen as is often supposed, how can the female, encumbered with her offspring fully three-fourths as large as herself, compete successfully with the unhampered males, and secure enough food not only for herself but also for her young? The fruit-eating varieties might not suffer seriously from this handicap, but it does seem as if the agility of the insectivorous kinds catching their food on the wing would be greatly affected.

There are numbers of curious formations along the river which cannot fail to attract the interest of the traveller, no matter what his particular mission might be. One of these is the Cerro Yapacana, a square block of granite not over one thousand five hundred feet high; it is a very conspicuous landmark as it towers above the forest like a giant monument, and can be seen many miles away. We did not come abreast of it until eight days after first sighting it.

There are few rubber-camps along this part of the river, but several Indian families had come to spend some weeks collecting turtles and eggs on the sand-banks. At night absolute quiet reigned on the *playas* so long as the moon shone; but no sooner had the brilliant orb disappeared below the horizon than the water was broken with ripples as numbers of turtles emerged to deposit their eggs in the loose, warm sand, and jaguars came from the dark forest to feast on the defenseless creatures and rend the still night air with ugly coughs and grunts.

In returning from fishing excursions we usually cut across the several miles of sandy waste toward camp, guided by the bright fire which the cook was required to keep burning, and in this way learned a good deal about the turtle's habits. After leaving the water the creature wends its way toward the highest point on the island or *playa*, and with a few powerful strokes of the flippers excavates a deep hole; the eggs, twenty to a hundred in number, are then deposited, after which the sand is scooped back into place and patted down so carefully that it takes a very experienced eye to locate the spot. The turtle then hurries back to the water, where it apparently remains until the following year. When the eggs, warmed by the sun's rays, finally hatch, the *playas* swarm with small turtles which are eagerly collected by the natives, boiled entire and eaten. The egg contains a great deal of oil, and although cooked a long time always remains soft. Iguana eggs are taken, also, and boiled and eaten, even when about to hatch.

Besides the turtles there were many other signs of life on the sand-banks. Water-birds, squatting low in some cup-shaped hollow, looked stupidly at the dazzling light of the gas-lamps, and could be approached to within a few feet; downy young birds waited quietly until nearly touched with the hand and then ran away into the darkness, like puffballs rolling before a breeze.

The *Raudal de Santa Barbara* is a wicked stretch of water. The Ventuari, coming from the neighborhood of the Brazilian border, forms an extensive delta near its mouth. There are many islands, some of great size, and all heavily forested. The Orinoco is very wide, and hundreds of sharp, tall rocks protrude above the water, causing a series of rapids which are hard to ascend. It took us three days of the most trying kind of work to traverse this stretch of agitated water, and finally to haul the boat up the falls, which come as a sort of climax at the end. A strong wind blows from the north almost constantly, whipping the water into a choppy sea. On the bank stands a good-sized rubber-camp, and extra hands can usually be secured to help pull the boat through the rapids. The men from this place had just returned from a hunt in the forest, bringing two jaguars and an armadillo weighing sixty-five pounds. One of the jaguars was black. All of these animals were eaten, and of the two species the flesh of the jaguars was the better. One night, not long after, one of these animals invaded our camp. As the sand-bank we had selected was a narrow one, the crew chose to sleep on the forest side; they greatly feared the crocodiles in the river. Early in the morning I was awakened by a jaguar's roaring mingled with frightened wails, and upon investigation discovered that the men had come to our part of the camp near the water, leaving the captain's wife in their former location. They had reasoned that she was the least useful member of the party and had compelled her to remain as "bait." Maria was sent back to San Fernando in the next canoe we met bound down the river.

The abundance of the big, spotted cats and their harmlessness under ordinary circumstances is astonishing, although at times they will attack human beings. At one of the rubber-camps we were shown the skin of a recently killed animal which had stalked a two-year-old child at play not far from the hut; the mother, a negress, seeing the animal in time, attacked it with a *machete* and killed it.

The next river of any importance to be encountered was the Rio Lao, reached February 17. Up to this time the strong north wind had continued to blow without interruption, and the course of the river was dotted with islands. Rubber-camps were situated on the river-bank, and we had our first glimpses of the Maquiritare Indians. Owing to the frequent rains, the year had been a bad one for the *patrones*, or managers of the camps; also, a kind of malady had broken out among the *peons* and Indians which killed many and frightened others away. Nevertheless, those who remained seemed quite contented, and if we chanced to spend the night at a camp or *barraca*, our men always joined them in their pastime of drinking, playing the guitar, and singing songs about one another, far into the night. Some of the men were clever at improvising songs apropos of the occasion. At one place, for instance, they heard of the jaguar's visit to the sand-bank, and that the captain's wife had been sent back to San Fernando. Without hesitation one of the peons sang:

> Qué tristeza en nuestro campamento, Pobrecito Ildefonso está llorando, A caramba, nadie está alegre, Será porqué Maria fué á San Fernando.

The largest *barraca* by far which we saw was owned by an old Turk named Parraquete. He received us cordially, shook our hands, and embraced us, apologetically explaining that a slight fever prevented his rising from the hammock; later we found out that he was a leper in the last stages of the disease. He had fifteen Maquiritares in his employ, each of whom collected the latex from several hundred rubber-trees every morning; in the afternoon the milk was smoked, one hundred pounds of the liquid yielding about forty or fifty pounds of crude rubber. A species of heavy, deep-red wood called *mazarandul* was used to produce the dense smoke necessary to coagulate the latex.

Hevea only was gathered here, although *balata* was also collected farther down the river and on the Guaviare. The governor of the district told me that about fifteen million trees of the *balata* had been cut down along the latter river during the last ten years, as the method used to secure this class of rubber necessitates felling the trees.

The proprietors of rubber-camps use the same system of keeping their employees that the commission merchants in Ciudad Bolivar, who are the purchasers of the crude product, employ in dealing with themselves; namely, they keep them constantly in debt by advancing quantities of merchandise at exorbitant prices. It is not unusual for one *patrón* to sell some of his men to another for the amount of their indebtedness, or more, if he can get it, and sometimes an unsatisfactory *peon* is turned loose in the wilderness to shift for himself; we picked up one who had been abandoned on a sand-bank, in a half-starved condition.

The Cerro Carriche is another granite mass similar to Yapacana, but not quite so high. It stands on the south bank of the river, between the mouths of two small rivers called Carriche and Trocoapure.

Early on February 21 we had the first distinct view of the Cerro Duida, looming, faintly outlined, in the distance. From afar it resembled a high, level plain, but as the vapor clinging to the huge, dark mass slowly dissolved itself, a well-defined short range appeared with twin peaks showing high above the rest of the mountains.

The Orinoco steadily decreases in width until the distance across is not more than half a mile; in many places the banks are high and composed of pink and white clay streaked with layers of dark-blue clay. On both sides the jungle presents an unbroken wall of tangled verdure; occasionally a slender palm rears its delicate head high above the riotous mass, as if gasping for one more breath of air before being strangled by the figs and creepers slowly entwining its stem with their death-dealing tentacles. Among the lower growth are vast areas of palms, known as *coco del mono*, with long fronds resembling those of the *Seaforthia*, and bearing small, hard nuts; the leaves are used in thatching huts and the *carroza* or covering of the boats. Another palm, tall and thorny, resembles the well-known *chonta* of Colombia; it bears large clusters of red nuts, which are very palatable when thoroughly boiled.

The heat was always intense and most oppressive; even the cool nights brought no respite, and in the early morning a thick vapor slowly rose from the water, to be later wafted above the tree-tops and disappear.

Flocks of hoatzins, or lizard-birds (*Opisthocomus cristatus*), were seen almost daily. They fussed and fluttered among the dense vegetation, but could not be induced to leave their dark retreat. There were also nesting-trees of the black and yellow orioles, better known as caciques, which are about the size of a blue jay; sometimes a single tree contained thirty or more nests placed close together, and also invariably a number of large wasp-nests were placed in the same tree. The nests of these birds differed from those of the giant orioles (*Ostinops*) in being smaller and having the opening at the top instead of at the side of the swinging bag.

Figure 31. A rubber-camp on the Upper Orinoco.

At the end of the twentieth day we reached the mouth of the Cunucunuma, and camped upon its sandy banks for the night. This river is approximately five hundred feet wide at its mouth, shallow, with dark, clear water, and flows southward, joining the Orinoco at right angles, as the course of the latter river at this point is eastward; a low, forested hill called *Ventana* rises to the north. One may cover the distance from the mouth of the Cunucunuma to the Cassiquiare in a day, and reach the plains on which was located Esmeraldas by continuing his journey an additional day.

Not far above the mouth of the river is the dry bed of a stream, said to have been the former course of the Cunucunuma; short, soft grass now covers the ancient, sandy route and the lines of trees on each side present such clean-cut edges as to suggest well-kept hedgerows. Tapirs and capybaras have worn many paths through the luxuriant sward; apparently these animals come out into the open at night to feed.

The current of the river is so strong that we could not average more than four or five miles a day. Through the clear water we could see shoals of fish and numbers of large sting-rays darting about over the bottom. One fish, resembling a beautifully spotted trout, rose eagerly to a trailing hook baited with a strip of white cloth; it weighed about a pound, and was called *pabón* by the natives; on two occasions members of this species leaped clear of the water and into the boat as we poled along after nightfall. Another kind greatly resembled a flying-fish, and leaving the water singly or in pairs, skimmed over the surface for a distance of twenty yards or more, and then dropped with a splash; when "flying" it left a train of ripples in its wake, as if long appendages were trailing after it.

Figure 32. Unloading for the portage, Raudal del Muerto.

There now followed a series of low, disconnected mountains which might be called the foot-hills of Duida. The first of these is the Cerro Piapoco, one thousand three hundred feet high; parts of it are covered with low scrub growth, and the river winds around three sides of it. Next comes the Cerro Tapicure, a rounded granite mass approximately one thousand four hundred feet high. At the base of the latter is a Maquiritare plantation of *yucas* (Manihot), pineapples, and plantains, on the edge of which stood the communal house, conical in shape and one hundred feet in diameter. The place was temporarily deserted, as the Indians were down-river gathering the rubber harvest. Near by also grew a palm new to us, the *Tamiche*; it is thirty feet high, with erect, undivided leaves, and the crown resembles a huge, green, opening tulip.

While tramping in the forest across the river from the Indian plantation we came suddenly upon a Maquiritare woman and her four small children, squatting around a small fire built under a rough lean-to. She was roasting a curassow and tearing off pieces for her young brood, which was devouring them with the voracity of wolves. The frightened glances of these wild people and their gnawing at the half-cooked flesh was quite in keeping with their surroundings, and stamped them immediately as a perfect part of the virgin wilderness.

Rapids are not wanting in the Cunucunuma. The first is the *Raudal del Muerto*, formed by a wide ledge of rock which extends across the river, and over which the water rushes with a deafening roar. Next comes the *Raudal del Sina*, which is longer but not so difficult to navigate. Just above this we entered the Sina, a small stream which comes from the direction of Duida, and ascended to its highest navigable point; this, however, was only a few miles above its mouth.

Figure 33. The Cerro Duida.

The Cunucunuma, it may be stated, rises in the vicinity of the little-known Cerro Cuachamacari, and may be ascended to the foot of the Cerro Maravaca. On most maps its course has been marked east of Duida, while in reality it is on the western side. Its tributaries from the east are the Tabarí, Sina, Cua, and Rio Negro; and from the west the Yacaré and Cumichi. There are numerous rapids. Besides the two mentioned, the Indians named the San Ramón, Rayao, Chacherito, Vaquiro, Mapaco, Chipirima, Picure, and Culebra, all of which must be passed before reaching Maravaca.

When rocks and low water barred a farther ascent we made camp on the high bank and began the arduous work of cutting a trail to Duida, about six miles distant. We had secured the services of a number of Maquiritares; two men, two women, and a boy, and these, together with the members of our crew, were immediately put to work on the trail. While this was in progress we devoted ourselves to the exploration of the forest and its inhabitants.

Apparently the Indians, who in common with many South American tribes seek the smaller streams for their habitations, and who live in small groups all along the

Cunucunuma, rarely visited this locality. Game was so abundant and so tame that it was impossible that the animals had been persecuted to any considerable extent. We also visited the house of the chief of the tribe, named Antonio Yaracuma, whose *cunuco* (clearing) was on the Cunucunuma, a few miles above the mouth of the Sina (Sina is a Maquiritare word meaning wolf). This place he chose to call Yacaré. Surrounding the great, conical house was a small patch of *yucas* and pineapple-plants, walled in on all sides by the interminable forest. The edge of the roof came down to within five feet of the ground, and there were eight fireplaces, equal distances apart, showing that eight families occupied the dwelling. A perfect network of poles and beams supported the ragged grass and palm-leaf canopy, and from these various articles were suspended: Drums, made of sections of hollow tree-trunks and covered with the skin of a red howling monkey on one side and of a peccary on the other; long tubular baskets of wickerwork used to express the poisonous juice of the *yuca* root in making *mandioca*; blow-guns ten feet long, hammocks, and fishing-tackle. Everything was immaculately clean and well arranged. On one side two small rooms had been built of adobe, one for the chief, and the other for storing baskets of *mandioca*, each of which held about a hundred pounds.

A walk around the edge of the clearing disclosed an obscure trail which zigzagged and wound through the forest about a mile and then opened into an immense plantation, which we estimated contained not less than a hundred acres. The trees had been cut down and burned, and *yucas* neatly planted in hills stretched to the very edge of the clearing. Through the centre ran lanes of plantain and banana plants, bordered by rows of pineapples, sugar-cane, and cashews. The ground was carefully cultivated, and there were no weeds; the stalks of uprooted plants had been piled around the edge of the field forming a thick fence. The reason for maintaining such large plantations is that the women make a good deal of *mandioca* to sell to the traders for cloth, matches, perfume, and trinkets. The men clear the ground; the women plant and care for the crops. From the juice of the *yuca* a very intoxicating drink called *casire* is made, and of this great quantities are consumed during the wild orgies which take place at frequent intervals. Boiling and fermentation destroy the poisonous effect of the fresh juice.

We found the forest around camp to be all but impenetrable on account of the underbrush and creepers. Also, there were a number of windfalls where cyclones had cut wide, clean swaths through the forest, leaving an upheaved barrier that could not be crossed without the liberal use of axe and *machete*. Small birds were abundant and travelled in mixed flocks. Of the larger birds there was an unfailing supply; guans and curassows strutted unconcernedly about, or flew into the lower branches of the trees to look at us with surprise or resentment; large tinamou ran about in pairs like chickens and were slow to take wing. Occasionally we ran into a flock of trumpeters (*Psophia*), which stared at us in curiosity for a few moments and then flew into a tree, and raised an unearthly din, cackling and screaming until dispersed by a few shots.

The Indians told a curious story about a trumpeter and a curassow. In the very beginning of things two of these birds decided upon a matrimonial alliance, but domestic troubles soon broke out and there was no possibility of a reconciliation; it was thereupon decided to lay the case before the gods who live on the summit of Mount Duida. The wise gods ordered them to fight it out; in the course of the combat that followed, the curassow pushed the trumpeter into the fire, burning off the feathers of the latter's tail; the trumpeter promptly retaliated by pushing her mate into the fire, singeing his crest. Thereupon the gods decreed that they should remain in this humiliating plight for the rest of their days, and so even to this day the curassow wears a curled crest and the trumpeter has a very short tail.

No matter how far we chanced to go during the morning's hunt it was always easy to determine the exact location of our camp. A colony of caciques had built their nests in the top of a tree near the tent, and quarrelled and chirped so noisily all day long that we could not get out of hearing of them.

After the trail had been completed for a distance of several miles, hunting was rendered much easier. It was a delight to wander noiselessly along the clean path and watch the wild things pursuing their daily activities. Tapirs moved quietly across the narrow lane, like shadows; but if disturbed crashed through the brush and thundered away like frightened horses. Large red squirrels frisked in the trees or fed in the nut-bearing palms. Monkeys were always about; there were red howlers, cebus, and small black woolly monkeys with gold-colored hands; the latter travelled in small troops and raced through the tree-tops at great speed, making long jumps from branch to branch; at frequent intervals during the morning and evening they raised their voices in shrill little cries of distress, resembling a series of quickly repeated ohs.

The river was teeming with fish. At night, after their work had been completed, the Indians, who camped on the water's edge, threw in their lines and never failed to catch a goodly supply. While in our presence the men always wore blue cotton trousers and the women loose dresses of the same color, but when alone they threw aside all clothing.

Occasionally a light canoe containing women and children passed our camping-site, but they always remained as near as possible to the opposite bank and paid no attention to us whatever if we chanced to call to them; in fact, they could not even be induced to look in our direction.

The nights were always sultry and it rained frequently. If the weather permitted, a huge fire was built; into this a steady stream of fireflies or click-beetles winged their way to destruction. Late one night we heard a queer pattering on the top of the tent-fly; back and forth scurried the little feet, and up and down the sloping roof. Our acetylene-lamps revealed a family of opossums which had discovered an ideal playground. Often, too, we heard cautious footsteps nearby, and the suddenly flashed light disclosed the glowing eyes of a deer, tapir, or jaguar, which gazed stupidly a moment into the dazzling brilliance and then darted away.

On account of the dampness mould formed so rapidly that cameras and all leather goods had to be cleaned daily, and there was great difficulty in drying specimens.

We had frequent views of Duida. Each morning at about ten the mist drifted from the summit and revealed the jagged, rocky peaks; our calculations placed the altitude of the mountain at approximately five thousand five hundred feet. Toward the Orinoco the mass presents a bold front, the sheer walls of granite rising to a height of several thousand feet. The western slope is gradual and any attempt to ascend the mountain should be made from that side.

Cutting the trail required more time than we had anticipated. It was our intention to remove the equipment to the very base of Duida, and this was impossible until a suitable way had been prepared. The intervening country is rolling and the hollows are filled with a network of deep, water-filled canyons; across these trees had to be felled to provide a means of crossing. Also, neither the Maquiritares or the Venezuelans proved to be very industrious, and were about as poor a class of assistants as can be found. However, work progressed steadily, and there came the day when the last bridge had been placed across the winding river, and we were able to proceed to the foot of Duida.

Near the mountain the forest assumes a different aspect. Instead of the tall trees there are vast groves of palms which form such a dense canopy that the sunlight never penetrates to the ground; for this reason there is no undergrowth, but the earth is covered with a soft carpet of dry leaves. Some of the plants attain such giant proportions, with fronds thirty or forty feet long and fifteen feet wide, they form great tent-like shelters.

As we neared the mountain the Indians became restive and finally refused to go any farther. They firmly believe that it is the abode of spirits who will be quick to resent any intrusion into their sacred precincts. Besides, the rainy season was fast approaching, and at night blinding flashes of lightning played among the crags, and the dull boom of distant thunder pierced the sultry blackness. Wind swept through the forest in fitful blasts, and it rained frequently. Sometimes the blasts attained the velocity of a cyclone and sent tall trees crashing down on all sides. The Indians could endure the strain no longer, so one night they quietly disappeared, taking the boat with them. At first this loss seemed anything but pleasant, but a raft was soon constructed, and two of the men were sent down to the nearest rubber-camp on the Orinoco for another craft. We never saw our Indians again, but one afternoon two men of the tribe visited our camp. They emerged silently from the forest, having concealed their canoe somewhere above or below, laden with baskets of plantains, sweet potatoes, and bananas, and several cakes of cassava bread, also a large, freshly killed curassow—enough provisions to keep two men a week. I thought they wanted to stop with us for the night, and showed them the fireplace. They paid no heed to my implied invitation, but dropped their burdens at our feet, reluctantly accepted a few fish-hooks which were offered to them, and then departed as mysteriously as they had come. Perhaps they had been sent by our erstwhile companions, who may have been conscientious enough to make some reparation for the theft of the canoe.

The rainy season advanced with such rapid strides that further work was impossible. Vapor hung over the forest like a pall for days at a time, and the river, rising with each passing hour, was quickly inundating the lowlands. The sight of the new canoe coming up the river was therefore a welcome one, and it did not require many days to pack our collections and outfit, stow them aboard, and steer a course downward with the rapid current. It required only nine days to reach San Fernando de Atabapo.

The results of the expedition are surprising and interesting. Duida is not the isolated "mountain island" it was commonly supposed to be, but is connected with the mountains of the Ventuari and Parima by a series of hills, some of which reach a height of over a thousand feet. Its elevation is comparatively low, being less than that of the Maravaca. To attempt its ascent from the Orinoco side seems hopeless on account of the frowning precipices facing the plains near Esmeraldas. The proper placing of the Cunucunuma and an elaboration of the map of the region are other results.

It should be remembered that the dry season is much shorter on the Upper Orinoco than on the lower river, and work must be pushed with the utmost speed. The tributaries of the Orinoco, as well as the main river, leave their banks soon after the beginning of the steady downpours, and the whole country is flooded many miles inland; in places the river is then one hundred and twenty miles wide; all the rubber-camps we had seen on the upward trip were totally deserted when we passed them going down, and of some of the huts the roofs only showed above the water; others had vanished with the yellow flood.

The collections of birds and mammals were large and interesting; they yielded a number of species and one genus new to science.

And finally, a word about assistants; under no circumstances should Venezuelans or Indians be depended upon. It is possible to secure experienced river-men in Trinidad, and with proper treatment they make faithful and efficient companions.

Chapter 12

LIFE IN THE GUIANA WILDS

A naturalist might spend many years in Venezuela and still exhaust but a very small fraction of the possibilities offered to the field-observer—so vast are the resources of that zoological wonderland. Exigencies beyond our control, however, recalled us to Trinidad, and after a brief rest we turned our eyes toward British Guiana.

The distance between the island and the low Guiana mainland is not great; it required just two days of uneventful sailing for the *Sarstoon* of the Quebec Line to plough through the deep water and schools of flying-fish, and finally nose her way carefully through the mud to Georgetown.

The city is built on the low coastal land, and a great stone wall prevents the sea from reclaiming its own at high tide. The streets are wide and bordered with trees. No more suitable style of architecture could be desired for a tropical country than that employed in constructing the houses of the better class of inhabitants; they are practically all doors and windows, giving admittance to every passing breeze. The wide verandas are carefully screened.

Numerous canals, spanned by picturesque little wooden bridges, divide the city into sections. At low tide the locks in the sea-wall are opened to permit the excess of water to escape; at high tide the locks are closed to keep the lowlands from being flooded. Growing in the water are masses of *Victoria regia* lilies with white or pink flowers; the giant leaves, with upturned edges, and often several feet across, resemble huge pies; but the plant is lovely from a distance only, as the veins and midribs are covered with long, sharp spines that effectively prevent any intimate advances on the part of an overenthusiastic admirer.

Mosquitoes are not lacking, but they appear at night only, when one can easily evade them by remaining indoors; and through the hours of darkness the twanging and peeping of myriads of frogs fill the air with a not unmusical din.

The population is the most cosmopolitan imaginable. It ranges from dignified, helmeted British officers down to the meanest Chinese or Hindu coolie living in a dilapidated shamble on the border of a marshy rice-field.

Our first care was to secure the admission of our equipment by the customs officials. This was accomplished without an undue amount of difficulty; and within a short time we had also obtained a permit to pursue our scientific work, for in British Guiana birds are wisely protected. We also opened negotiations with Sproston's, Ltd., who operate many large lumber, rubber, and mining enterprises in the interior of the country. This step is a most essential one, as the concern, through its agents, can be of the greatest assistance to the traveller.

On July 7, we boarded a comfortable little steamer and started up the Demerara. Rain fell in torrents throughout the day so that it was impossible to see anything but the fleeting, yellow water against which the straining craft battled vigorously, and the long rows of trees faintly outlined in a world of blue-gray mist. Wismar was reached that night and passengers and luggage were hurried aboard the waiting train, which soon covered the eighteen miles of intervening country to Rockstone, on the Essiquibo River. A delightful bungalow hotel is maintained by Sproston's at the latter place, and every need of the visitor is superabundantly supplied.

A launch of considerable size, towing a house-boat provided for first-class passengers, left Rockstone early the following morning. The Essiquibo is truly a very great river, and the height and magnificence of the forest covering its banks is not exceeded in any part of South America. In some instances, the trees are one hundred and seventy-five feet high; cottonwood, greenheart, and wallaba mingled their leafy crowns far above the mere rabble of palms and lower growth, shutting out the light and effectively killing their competitors until—after hundreds of years of successful fighting—the strain begins to tell and the monarchs are compelled to bow before the inevitable onslaught of old age. At the first signs of weakness enemies spring up on every side. The struggle for life is constant and in deadly earnest. Of the hundreds, perhaps thousands, of saplings which appear as the light and air gradually penetrate through the opening made by the dying giant, only one can eventually survive. Naturally, the strongest and fittest possesses every advantage in the mad fight for existence, and as it quickly outstrips its weaker rivals they wither and die.

The launch called at a number of rubber-plantations and lumber-camps. Great quantities of greenheart (*Nectandra rodiœi*) are cut and exported; the wood is very hard and durable, and resists decay when under water, for which reason it is used largely for submerged work such as wharfs and piles. Next in importance is crab-wood (*Carapa guianensis*) employed in building houses; third in value are several varieties of wallaba (*Eperua*); this wood has a coarse but even grain and is very resinous, being suitable for the manufacture of shingles and vat-staves.

Figure 34. Wismar on the Demerara River.

Figure 35. Tumatumari on the Potaro River.

The rubber industry ranks third in importance in British Guiana. By far the greater part of this product is balata, collected from indigenous trees that are tapped under careful government supervision. *Sapium* yields the better quality of rubber, but exists in limited quantities only, and the majority of the trees on plantations have not yet reached the productive age.

Our first headquarters were made at Tumatumari, a short distance above the mouth of the Potaro River. The river is at this point encumbered with a series of rapids ending in a fall of twenty or thirty feet.

Tumatumari is a small negro settlement, and owes its existence to the gold-mines scattered throughout the surrounding country. We made headquarters in a comfortable bungalow provided by Sproston's. A good trail leads through the tall forest, a distance of many miles, with numerous side trails branching off in various directions. Along the latter we immediately began to prosecute our work. On our very first excursions we heard the enchanting song of the Guiana solitaire, or quadrille-bird as it is locally known. From the depths of the dark forest there arose a low, mournful note, so liquid and melancholy that the music of no instrument made by the hand of man could equal it in ethereal beauty; gradually it swelled louder and louder, but always preserving the same exquisite quality until the eight notes had been uttered and the song died with a wistful sob. To hear this song is to experience one of the most enchanting of earthly delights, the memory of which will remain as long as life itself and gild the other reminiscences of sweltering days spent in tropical lowlands, of plagues of insects, of fever, and even the hard-fought battles against odds that seemed overwhelming. The bird is a shy little creature, and is obscurely colored; among the deep shadows where it spends its lonely existence the brown and gray of its modest attire blend so well with its surroundings that it is rare indeed to have even a fleeting glimpse of the captivating songster.

Perched in the dead tops of some of the tallest trees, we found a bird which, seen from below, resembled a giant long-tailed hummer; but a short time spent in observation soon disclosed the fact that it belonged to another family, so different were its habits. It sat motionless many minutes at a time, and darted off a short distance, presumably in pursuit of an insect, at infrequent intervals, only to quickly return to its perch. This was the paradise jacamar, a resplendent bird with a metallic green back and black underparts. The jacamars form a peculiar family, and have been frequently called "forest-kingfishers" because of their superficial resemblance to the *Alcyones*. The greater number of species are gorgeously colored and inhabit the lower branches of forest-trees, feeding on insects. The nest is placed in a hole in the bank of some wild ravine or river.

The abundance of bird-life, and also the variety, found in the lowland forest of British Guiana is bewildering, even to the seasoned field-observer; and nowhere in all South America are the feathered folk clothed in more brilliant and gorgeous colors. Evolution, it seems, has run riot in almost every conceivable direction in an effort to provide each species with some special color or characteristic that might enhance its beauty or better suit it to compete successfully with its hosts of neighbors. Thus we find the king-tody, a species of small flycatcher that preys upon insects. The body of the bird is inconspicuously colored, but the head is adorned with a crest of the most vivid scarlet feathers. As the bird sits quietly upon some low perch, the crest is depressed and invisible; then suddenly the flaming crown is erected and spread in fan-shaped formation, when it resembles a brilliant flower newly burst into bloom. Is it not possible that this flashing bit of color may attract some passing insect, which instead of finding nectar meets destruction?

However, I do not believe that the survival of every species is dependent upon some one particular patch of color or exotic appendage which it may possess. It does not seem to me possible, for instance, that one species of humming-bird owes its existence to a green throat-patch, or another of similar size and habits to a red or blue one; nor that one bird of paradise persists because it has curious, long appendages on its head or shoulders while a second one may have similar ones in its tail; but rather does it show that evolution tries many experiments. Each animate thing is full of latent buds, it would seem, any one of which might break out at any time, prompted by an impulse or conditions of which we know nothing. If the result of such newly acquired variation is beneficial, the species would naturally persist; if injurious, it would result in extermination; if indifferent (neither harmful nor of value) it would have no effect one way or the other, and might still be retained. This latter, I believe, has occurred in a number of instances.

Our visit to Tumatumari was supposedly at the end of the wet season; notwithstanding this, it rained copiously nearly every day, and invariably each night. We spent the evenings on the wide veranda of our habitation, preparing specimens or writing notes. Myriads of insects, attracted by our bright lamps, fluttered in and out of the darkness and settled on the white walls. Our two colored assistants, whom we had brought from Georgetown, were trained and enthusiastic entomologists, having been employed by Doctor Rodway of the Georgetown Museum, and spent several hours each night with net and cyanide bottle. Frequently they caught several hundred specimens in a short time. They also prepared cages of fine wire netting in which caterpillars were imprisoned and carefully fed, and glass boxes, or "incubators" for cocoons; in this work they were most successful, and a number of moths of rare and desirable species were reared to a state of perfection. Sometimes the downpour was so heavy that it disturbed small birds in their sleep in the bushes; on several occasions finches (*Sicalis*) fluttered up to the lamp in a dazed or bewildered manner, when we caught them easily and placed them in a cage, liberating them the next morning.

Numbers of Indians of the Patamona tribe live in the surrounding forest. They are a friendly though primitive people, and some of them speak or understand a few words of English. We accompanied the Protector of Indians, a British official living at Tumatumari, to one of the Indian dwellings one day. It seems that a negro had promised to marry a Patamona woman, then ran away, when she promptly married a man of her own tribe. Learning of this, the former suitor had written a letter to the officer demanding either his bride or damages. The official spent a very bad hour trying to explain the situation to the woman in the limited Patamona vocabulary at his command, while she sat stolidly in a hammock. When he had finished, she calmly remarked, "Well, you tell him I think he is a damn fool," in perfect English!

This tribe of Indians has a curious custom of torturing themselves in various ways, which performance is called "beena." It is supposed to insure success in any undertaking. A favorite method is to insert tough, pliable creepers into the nostrils and draw them out through the mouth. Another consists of slashing the breast, arms, and legs, and rubbing into the wounds the acrid juice of a plant. The official to whom I have previously referred had an Indian in his employ whose duty it was to supply the table with fresh meat. He hunted daily in the forest, bringing in deer, peccaries, agoutis, and other game in abundance; but on one occasion fortune conspired against him. Thereupon he tried his favorite beena, but it failed to bring him luck; every other means of mutilation known to the man was then resorted to in rapid succession, but still his long tramp and careful stalking yielded no meat. He became greatly discouraged and told his employer that he would make one more attempt at hunting, and should he fail in this would use his weapon upon himself. The officer thought it unwise to permit the discouraged man to return to the forest on the day following this declaration, so ordered him to cut weeds in his back yard. This the Indian reluctantly consented to do, but scarcely had he begun when he cut down a bush containing a wasps' nest and was severely stung. He immediately took his gun and hurried away, saying that a new "beena" had been sent to him, and that at last the evil spell was broken. Strange to relate, that night he returned laden with game.

A daily launch service is maintained from above the falls at Tumatumari to Potaro Landing, a day's journey up-stream. The boat's crew are all negroes, and are ordinarily a careless, slovenly lot. A short time before, they had failed to make proper allowance for the strength of the current when approaching the landing, and the launch, together with its thirty or more occupants, was swept over the falls and lost. Accidents such as this have caused the government to make wise and stringent rules regulating navigation on all streams, and applying to all craft, even canoes, containing passengers other than the owner; as a result accidents are now of rare occurrence.

One day's time is required to reach Potaro Landing, the end of launch navigation, from Tumatumari. Tourists who visit the justly famous Kaieteur Falls proceed overland from this point, a distance of seven miles, and then embark in canoes manned by full-blooded Patamona Indians. There are other but shorter portages farther up the river, though as a whole the journey is not difficult and well worth making.

The appeal of Potaro Landing was irresistible to us, so we decided to remain a week or two. Unfortunately, Sproston's maintains no rest-house here, as touring-parties continue to Kangaruma, at the other end of the portage, to spend the night. However, we found a good-natured Chinaman, who operates a store in the one lonely building at the landing, and he permitted us to use half of his barn; he had to remove his horses in order to supply even these limited quarters.

A good cart-road leads through the forest a distance of eighteen miles to the mining country on Minnehaha Creek, and many negro miners passed along this way each day; the greater part of them are what is locally known as "pork-knockers," because they live largely on salt pork and knock about from one place to another. They secure a small stake from the government with which to buy a pick and shovel, and then go into the interior to prospect. If, as frequently occurs, they strike a rich pocket, or find a nugget of considerable size, they immediately drop their implements and rush back to Georgetown to spend their newly acquired wealth. Carriages are engaged by the day, servants employed, and clothes of a bright and flashy nature are purchased in quantities. For a short time they revel in luxury and live in contempt of their erstwhile companions. Quite naturally their wealth soon disappears, and the tawdry finery is pawned to provide money for more necessary things; but there is an end even to this resource. Soon they again seek the stake of a few dollars and hie themselves back to the wilderness to once more try their luck as ordinary pork-knockers. To strangers the negroes are courteous and obedient, but among themselves they are quarrelsome, unfeeling, and even cruel. I heard of an instance where a number of them had been commanded to take a very sick companion down the river in search of medical treatment. As they paddled along the pilot frequently called to the man nearest the sufferer: "Ain't dead yet?" The person addressed roughly turned the sick man over with his paddle to inspect him, and then answered with a curt "No." "My! dat man dead hard," replied the pilot. They were most eagerly awaiting his death because it would save them a long trip, and they had planned to divide among themselves his possessions the moment life departed.

We met an American at the landing, who had experienced several unpleasant encounters with the negroes. He was engaged in searching for diamonds and had many of the colored folk in his employ. So far all the stones discovered had been of small size, but one day two of his men found a gem of good proportions. They immediately entered into an argument as to whether or not it was a real diamond, and to settle the dispute placed it on an anvil and hit it repeatedly with a sledge-hammer. "If it a diamond, it can't broke," was the gist of their theory. However, it was a real diamond, and it also broke; their outraged boss found the worthless particles a short time later. On another occasion this same man was confined in a hospital at Georgetown with a severe attack of fever. One night the colored head nurse swept in majestically, gave him a short, condescending look, and then directed his private nurse as follows: "Look through Mr. M.'s drawer to see if he's got a white shirt to bury him in!"

At frequent intervals throughout the day we heard a deep, powerful note coming from the forest. It was a long-drawn *Wow* that lasted eight or ten seconds, and exactly resembled the sound made by a circular saw cutting its way through a log. This we found was made by the bald-headed cotinga (*Gymnocephalus*), a bird the size of a crow, and of a dark-brown color; the head is entirely devoid of feathers, like a vulture's. Invariably several of these curious creatures were together, fluttering about among the lower branches and making the woods ring with their queer, outlandish cries. Another species of cotinga (*Xipholena*) was very rare; it was of smaller size and of the deepest wine color, with long, graceful wing-coverts and white primaries. When several were together in some tall tree-top they kept up a continuous quacking like a flock of ducks. If a skin of this bird is exposed to heat the color rapidly fades to a sickly bluish-gray.

One day an Indian hunter brought in a very small red howler monkey, and as I was aware that the species had not been known to live in captivity more than a few weeks, I was very eager to see if I could rear it. On account of its small size it had, of course, to be fed on milk (condensed), which it soon learned to take from the point of a fountain-pen filler. While it thrived and grew rapidly, it was always a sad little fellow and made no attempt to play or show signs of great friendliness. The only advance it ever made was to come up to me occasionally when I spoke to it, and feel of my face with its little black hands. After a time it was given full liberty about the camp, when it would spend hours sitting quietly beside a basin of water gazing at its reflection. After two months, and just as I was congratulating myself on having raised it past the danger-point of its existence, it climbed to a high shelf and ate a quantity of the arsenic compound used in preparing specimens.

Learning of our presence at Potaro Landing, a Mr. McKenzie, manager of the Minnehaha Development Company, very kindly invited us to his bungalow, eighteen miles away, and later sent a carriage for our transportation. The distance was covered in half a day, and lay mainly through the heavy forest, although there was occasionally an area of considerable extent covered with tall, rank grass and bushes. The company was operating one small dredge in Minnehaha Creek, and notwithstanding the fact that the entire region had been gone over before, quantities of gold were being recovered from the bed of the stream. As there had been no "clean-up" for two weeks, one was arranged for our benefit. The gold, which was in very fine particles, was brought up from the dredge in tin cans, and then placed in an iron retort and heated to a very high temperature; this freed the mercury with which the yellow metal had been collected from the mud and water in passing over the sluiceway of the dredge. Later it was again placed in the retort, together with pulverized glass and borax, to gather up the impurities, and melted; then it was poured into moulds. Four bars, weighing one hundred and twenty-five ounces each, were recovered. It was then inspected and passed by an official, who also made a note of the amount of tax due the government. A coolie servant was despatched to take it to Georgetown to the company's headquarters, and although he would be on the way a number of days and be compelled to mingle with all sorts of people, he carried no weapon of any kind with which to protect his precious burden. This speaks well for the law and order maintained throughout the colony.

The country along Minnehaha Creek is rolling and covered with a good stand of timber. Numerous small streams flow through ravines between the hills, and while the current is strong the streams are not deep. A footpath continues to a point seven miles beyond, on the Konamaruck, and from this a network of short, narrow trails branch out in all directions. The rainfall is very great in the entire region; during the month of August (1913) it was twenty-seven inches, while only nineteen inches fell at Tumatumari in the same period of time. One result of the great amount of moisture is that there is an increase in density of the lower growth, and the branches are covered with hanging moss.

As one moves quietly along the narrow lanes, enclosed on both sides by walls of trees, the lofty tops of which form a leafy vault overhead, he cannot fail to be impressed with the great breathless silence of the forest. The gloomy solitude seems pregnant with mysterious forces that draw the thoughts of the lonely wayfarer to far-off regions of blissful oblivion. Then, suddenly, a low, wailing cry of anguish rising in tremulous crescendo, but with liquid smoothness, smites the wanderer's revery and brings him back to earth with palpitating heart and throbbing pulses; the whinny rapidly decreases in volume and dies with a few short sighs. "Something, perhaps the combination of all these, makes one feel as if he had been caught with his soul naked in his hands; when, in the midst of subdued and chastened revery, this spirit voice takes the words from his tongue and expresses so perfectly all the mystery, romance, and tragedy that the struggling, parasite-ridden forest diffuses through the damp shade." It is the voice of the forest tinamou.

The notes of several species of ant-thrush (*Grallaria* and *Chamæza*) are remarkable for their quality and even beauty. One of them has a peculiar call resembling the words *compra pan* (buy bread), and by this name it is known among the natives of Colombia. Another gives a very good imitation of a moon whistle, the song lasting fifty seconds at times, without the slightest intermission. These birds are very long-legged, almost tailless, and obscurely colored above; the breast is frequently streaked. They spend their entire lives in the damp gloom of the forest floor, and although the song may come from but a few feet away, it is impossible to get even the briefest glimpse of the bird in ninety-five per cent of the cases where it is heard.

If we stopped to rest on the buttressed roots of some great cottonwood, we saw a few of the minor creatures whose existence is hardly suspected by the casual observer. What at first appeared to be a maze of cobwebs filling the entrance to a dark cavern under the roots, resolved into a moving, living mass. A closer inspection, and small, black specks could be distinguished in the madly weaving and revolving haze, and also long, threadlike legs dangling so idly that one wonders why they do not become hopelessly entangled with those of their neighbors. This peculiar, wavering flight of the crane-fly seems to form the delicate, spidery creature's chief occupation, for I rarely found them at rest. Presently, other little insects, encouraged by the silence, make their appearance. First among them may be a small *Gastaracantha* spider, slowly letting itself down from an overhead twig on a thread of finest gossamer. At first glance one may easily mistake the insect for a minute crab that has fallen from the leafage into a silken snare, but when, at the watcher's first movement, it either runs nimbly up the dangling thread, or drops to the ground with a rapid slacking of line, one is convinced that it must be a spider. The hard shell, or back, is fringed with sharp, upturned spines and is of an orange color marked with a number of small black dots.

After a shower, mosquitoes were numerous and attacked with the utmost persistency. This irresistible thirst for blood is very extraordinary; it does not seem possible that more than a very small proportion of the countless millions of these insects living in a given area ever have an opportunity for satiating their appetite for blood during their entire lifetime; yet the instinct remains, and they attack on sight ferociously and without hesitation any living thing whose skin their beaks can penetrate. It is also a well-known fact that malarial fever, so prevalent in the tropical lowlands, is transmitted by a genus of mosquito, *Anophiles*. The germ of this fever, however, passes only one period of its existence within the insect's body, and the spores must be secured from some living creature, and after development transmitted to another to complete the life cycle. Some of the areas in which malaria abounds are practically uninhabited by human beings, so this agent in the propagation of the disease is of course lacking—at least to a considerable extent. It naturally follows, therefore, that some other creature or creatures, may be preyed upon and inoculated by *Anophiles*. I have on several occasions observed pet cebus and woolly monkeys (*Lagothrix*) that showed decided symptoms of suffering from malaria, and to me it seems highly possible that monkeys may be at least one of the animals that serve to keep the infection alive.

While at Minnehaha Creek I received the information that Colonel Theodore Roosevelt was shortly to embark on a voyage to South America; and also, much to my pleasant surprise, that I had been selected as a member of his expedition. The time remaining at our disposal was very limited, so we rather reluctantly gave up our intended visit to Kaieteur Falls and Mount Roraima, and returned to Wismar for our last work in British Guiana. A strip of land several miles wide on either side of the railroad connecting Rockstone and Wismar, is owned by Sproston's, and the greater part of it has been cleared of forest. Instead of the dense growth of tall trees there are now impenetrable thickets of high slender sprouts and bushes. These jungles harbor almost every bird and animal found in the region, and while it is impossible to enter them for any great distance, we had not the slightest difficulty in making large and varied collections along the borders. One evening the superintendent of the line was returning from a tour of inspection, and as the motor-car in which he was riding slowly rounded a curve, a jaguar suddenly appeared on one side of the track; he promptly killed it with a shotgun as it was only a few yards distant.

We returned to Georgetown, from which place Mr. Igleseder, who had been my assistant, started for New York, while I sailed for Barbados, where I planned to await the arrival of Colonel Roosevelt and join him on his expedition into the wilderness of Brazil.

Chapter 13

First Weeks with the Roosevelt South American Expedition

The S. S. *Van Dyck* of the Lamport and Holt Line, with Colonel Roosevelt and his party on board, arrived at Barbados on the morning of October 10, and late that afternoon pointed her nose southward toward Bahia. The plans of the expedition, with which I was immediately made acquainted, called for a rather short and not too difficult trip up the Paraguay River and down the Tapajos, having for its prime object the study of the fauna and collection of zoological specimens in the region traversed; but all this was changed within a very few days as we shall subsequently learn. Besides Colonel Roosevelt, the expedition consisted of Geo. K. Cherrie, Anthony Fiala, Jacob Sigg, Father Zahm, and myself.

Bahia was reached on the 18th; Kermit Roosevelt joined the expedition at this place. The *Van Dyck* remained at anchor the entire day, thus allowing sufficient time for a casual inspection of the city. Two days after, we arrived in Rio de Janeiro. The paucity of the English language does not permit of an adequate description of the natural beauties of the harbor and the city. All steamers entering the bay must sail through the narrow passage between the famous *Pão de Azucar* and the mainland on the opposite side. The great loaf-shaped rock rises to a height of twelve hundred feet above the water; if one craves excitement, it is possible to ascend to the top in a small car travelling on steel cables.

In few cities is there such a display of great wealth. The main street, the Avenida Central, is wide and beautiful, and the sidewalks are of coarse mosaic. There are numerous palatial buildings, though some of them are too ornate to appeal to North American taste, and gold-leaf and carved marble have been used lavishly in their decoration. The public squares, filled with the finest of tropical trees and plants, give a park-like appearance to at least parts of the city.

Of interest to the tourist, perhaps, are the numerous curio-shops filled with a varied assortment of almost everything ranging from minute, brilliantly hued beetles, to feather flowers and the skins of anacondas. Brazil is of course popularly believed to be the land of huge snakes; one dealer calmly told us that he frequently had skins forty *metres* long, but the longest he happened to possess measured less than twenty feet in length. The number of stories in common circulation concerning serpents of monstrous proportions in South American countries, is astonishing; and it was interesting to note that the farther south we went, the longer the reptiles grew.

Thus, in Barranquilla, near the Caribbean coast of Colombia, I was told that specimens thirty feet long were to be had frequently; this did not seem quite probable. In Venezuela thirty-five feet was not considered unusual, and I was sorry that none were to be obtained during my visit. In British Guiana, snakes forty feet long were said to be fairly common, although I could find no one who had actually seen one of that size. The climax was reached in Rio de Janeiro, when a curio-dealer told about the forty-metre snakes. I frankly expressed my doubts, and he proceeded to tell of how a man standing beside a snake of this size, that was coiled up, could not look over the top of it—it was such a great heap.

It must be admitted that Brazil with its great Amazon basin produces many strange and unusual creatures; but when it comes to one-hundred-foot snakes, it can only be said that there is absolutely no proof of their existence. No dealer I ever visited, and there were many, could ever produce a skin over twenty feet long.

The traveller into the interior hears many stories of great serpents and their doings; for instance, the story of the horned snake is famous all over South America, and while the details may vary, the main features are always the same. It is the tale of a person (usually the one telling the story) who came suddenly upon an enormous snake with a long horn on either side of the head. Of course, the reptile was immediately killed, sometimes with a rifle or revolver, or occasionally with a knife, after a desperate struggle. As the slain monster writhed its last, the heroic hunter made a startling discovery; the snake did not actually have horns; it had merely swallowed an ox, which feat it performed without difficulty until the head was reached; this refused to go down on account of the spreading horns lodging crosswise in the corners of the snake's mouth. Hence the old, old story of the horned snake.

Another favorite anecdote which I have heard repeated a number of times is that of the man who with his wagon, to which two oxen were hitched, attempted to ford a stream; suddenly an anaconda of enormous size emerged from the water and, enveloping both animals in its coils, crushed them to death. I never encouraged those telling this story to continue, because I was afraid that they might say the snake had swallowed both oxen and perhaps even the wagon at the same time!

Not many years ago a South American explorer brought back photographs of the "trail" made by a huge snake in crawling along the sand. It would be easy to manufacture such a trail by dragging a bag full of sand along the ground, and while it is impossible to say that this was really done, such a photograph would be of no value, anyway, as it would be impossible to determine the size of the reptile from such a picture.

In this way the evidence of the existence of gigantic snakes gradually dwindles away, and we are compelled to look for material on which we can lay our hands, whereon to base our knowledge. That is, the stories of the average traveller and native as well must not be taken too seriously; and only the skins or living specimens known to exist can be taken into consideration.

The longest South American snake of which I could obtain any definite information is in a Brazilian museum, and was said to be about twenty-five feet long. A skin of this size may be stretched several feet during preparation, so the snake may have been somewhat shorter in life. In the Bronx Zoological Gardens, New York, there is a living anaconda fourteen feet long; the largest boa-constrictor is eleven feet in length.

No visit to Rio de Janeiro is complete without an inspection of the botanical gardens, which cannot fail to appeal to all lovers of the beautiful. Immediately upon entering, one is confronted by avenues of stately royal palms, ninety to one hundred feet high. The "mother of the palms," towering above all the others, is pointed out with pride by the gardeners. It is said that this was the first of the species to be planted, and that all of the others were grown from seed taken from this one plant. There are also attractive little lagoons filled with flowering pond-lilies and fishes, and bordered with graceful travellers' palms introduced from Madagascar. Rows of bamboo form sheltered lanes where the visitor may seek relief on comfortable benches from the midday sun.

The palace Guanabara, dating back to the time of Dom Pedro, was opened for the use of Colonel Roosevelt. Its location is in the most attractive spot imaginable. Sitting at the table in the immense dining-room, one may look down a palm-lined avenue to the blue water of the bay, a half-mile distant; it was through this lane of tall, beautiful trees that Isabella, daughter of the King, drove to her daily bath in the surf.

Acting upon the invitation of officials of the Brazilian Government, Colonel Roosevelt abandoned the plans he had made previously and changed the character of the expedition from a zoological to a geographical one. Colonel Rondon, who had been engaged some years in making a survey through Matto Grosso for a telegraph-line, had discovered the headwaters of an unmapped river. This he had called the *Rio da Duvida*, or River of Doubt, for no one knew whither it went. The invitation to explore and map this stream was tendered to Colonel Roosevelt, and he accepted it.

We left the colonel at Rio de Janeiro, after making arrangements to meet subsequently, and continued on to Buenos Aires, spending a day *en route* in Santos, and one in Montevideo, the capital of Uruguay.

Although we had read and heard a great deal about the city of Buenos Aires, we were hardly prepared for the pleasant surprise that awaited us. The population of this metropolis of the south is more than a million and a half, and the city presents a clean, dignified appearance. In many respects it is as modern as New York City. There are numbers of tall edifices patterned after our own skyscrapers, large hotels, and theatres. An electric subway was just being opened, and the crowds in the Calle Florida in the late afternoon rival those of Broadway. The climate is cool and agreeable. One of the things that particularly attracted our attention was the presence one day of swarms of dragon-flies flying in a steady stream high above the city; they were blown in by violent winds, or pamperos, which sweep across the level plains country, and gave one the impression of a raging snow-storm.

As Mr. Cherrie and I were eager to devote every available moment to zoological work, we left Fiala and Sigg, whose duty it was to look after the rather appalling amount of luggage, and secured passage on the Argentine Northwestern Railroad, which had just inaugurated through service to Asuncion, Paraguay. We took only the small amount of equipment necessary for few weeks' work, as the two others were to come up with the remainder of the baggage on the first available freight-boat. Our train was the second to make the through trip and was scheduled to run biweekly. It was composed of seven Pullmans, two baggage, and a dining car; the service was good. Leaving Buenos Aires on the afternoon of Sunday, November 2, we reached Rosario at dark. Here the train was run onto a steel boat and carried up-river a distance of sixty miles, after which it continued the journey on the east bank of the Paraná. The next night we recrossed the river on a ferry, and were landed at Encarnacion, Paraguay. Asuncion was reached late on the afternoon of Tuesday.

The railway journey had been through level plains, interspersed at long intervals with small clumps and strips of low woods; but it is essentially a grazing country and we passed numerous herds of cattle feeding on the vast, fence-enclosed ranges. Stalking about unconcernedly among the herds were small bands of semi-domesticated rheas, but they were not abundant; I doubt if we saw a hundred during the entire trip. Caracaras, or carrion-hawks, glossy ibises, our old friends the jacanas, which resembled huge grasshoppers when on the wing, rails, and spur-wing plovers, or lapwings, were plentiful. Frequently we saw the domed mud-nests of oven-birds perched upon fence or telegraph-poles, or on the lower branches of trees. Villages are few and far between, and the natives, a motley crowd of dark-skinned individuals, usually left their shambling, grass-thatched huts and came down *en masse* to see the train.

Asuncion is a quaint old town, plainly showing the marks of violence that have been left by frequent revolutions. Mr. Ferris, the American consul, who met us at the station and rendered us every assistance possible, had witnessed five revolutions in as many years; there had been seven presidents in the same period of time. The streets of the city are narrow and paved with cobblestones; the buildings are low, constructed of adobe, and have red-tile roofs. There are one or two banks, a college, several churches, a public market and good hotels, as well as fair electric car and light service; there is also the inevitable lottery. We noticed little business activity. An air of depression seemed to hang like a pall over the people, and this may be readily accounted for when one recalls the tragic history of their country. Many of the women were in deep mourning. One authority estimated that the proportion of women to men was eleven to one, although this is probably an exaggeration.

One of the most interesting places in Asuncion is the market. Paraguayan lace is offered for sale in quantities. It is made in intricate and dainty designs, and many of the pieces consist of numerous small "wheels" or squares that are made separately and then united to form collars, handkerchiefs, or covers. One is astonished at the quantity of fruit displayed; oranges are brought from the surrounding country in cars and barges, and shovelled, like coal, into piles or carts. Some of them are of large size, delicate texture, and excellent flavor. The choicest of these are exported and may be purchased in Buenos Aires at rather high prices.

Figure 36. Camp on the Rio Negro in the Gran Chaco of the Paraguay.

After spending a few days at Asuncion we were invited to the home of one Professor Fiebrig, who lives at Trinidad, a few miles from the city. Professor Fiebrig is a scientist of more than local note, an instructor in the University of Paraguay, and curator of the museum. While journeying to his place we entered into conversation with two Paraguayans, apparently men of the upper class, who were travelling in the same car. When they learned our identity they shook their heads in a pitying and condescending manner. "How sad," said one of them; "you North Americans do nothing but pursue the almighty dollar. Now, in Paraguay we live for art, literature, and science." We had visited the natural history museum in Asuncion a few days before, and had taken note of the bullet-holes in the walls, the rents made in the stuffed animals by bayonet thrusts, and other marks decidedly not of an artistic or scientific nature.

Figure 37. Selling oranges in the market at Asuncion.

Our first zoological work was done in the country near Trinidad. All about were tracts of land of considerable size, covered with low forest, patches of brush country, grassy fields, and cultivated plots. Birds were plentiful, and as practically all of them were new to us, work in this region was doubly interesting. We here formed our first intimate acquaintance with the white ani (Guira), member of a subfamily of cuckoos, large flocks of which sat like rows of beads on the fronds of palm-trees. They are slender birds, about fifteen inches long, and are striped with brown, black, and white; a row of long, narrow feathers forms a high crest. They remained soberly on their perches, awkwardly jerked their tails from side to side, and mewed dolefully. The birds seemed utterly out of place among the vivacious tanagers, creepers, and finches, and seemed to belong more properly to some remote and unrecorded past. Their flight is slow and uncertain, the birds flapping their wings and sailing alternately; when alighting they strike a most ludicrous pose and barely avert falling over frontward before finally securing their balance. The long tail helps the bird to keep its equilibrium, although adding to the awkwardness of its appearance. The bird always gives one the impression of being exceedingly miserable, and particularly so during cold, rainy weather. Then all the members of the flock will crowd close together for warmth and protection, often placing their wings over one another in an affectionate manner, and even standing perhaps on the backs of their companions. On account of its scanty covering of feathers,

Guira guira is not well suited to resist cold weather. When the breeding-season arrives a huge nest is built in a cactus or low bush, usually at no great height from the ground; but the mass of sticks is not conspicuous, despite its bulky size. Occasionally a number of birds occupy the same nest, when many eggs are laid; the adults keep up a constant wailing and shrieking if their domicile is approached.

The eggs are among the most beautiful laid by any bird. They are elliptical in form and of a deep turquoise color, covered with a lace-work deposit of calcareous material. As incubation advances the shell becomes stained and the white, decorative layer wears away where the eggs rub together. Then the heretofore lovely egg bursts, and from it emerges the ugliest creature imaginable. Apparently the natives can think of no homelier object, for when they wish to call attention to the fact that one of their neighbor's children is of a superlative degree of ugliness, they call it *Pichón de Urraca* (young urraca).

Mammalian life was scarce, but considering the short time available, a comparatively representative collection was made, including specimens of a small gray wolf (*Cerdocyon*), which roamed singly and in pairs in the country bordering the Paraguay River. A few rabbits and opossums visited the mandioc-fields at night to feed upon the succulent tubers. We had abundant opportunities to observe the rural populace in the vicinity of Trinidad. They are of a rather unambitious type, and seemed contented only when taking their noonday nap or *siesta*, or while drinking *maté*. The general language of Paraguay is Guaraní, although Spanish is used by the upper classes.

"Yerba maté" is the modern name for the *caá guazú* of the Guaraní. It is applied to the dried leaves of a species of South American holly (*Ilex*) growing abundantly in parts of Brazil, Argentina, and Paraguay. The tree is very bushy and beautiful, and remains green the year around; the leaves are small, and those of a light-green color make the best quality of tea. Several methods are employed in gathering the leaves; one is to cut down the branches, pile them into huge stacks, and apply heat for about twenty-four hours, when they are dry and ready for the next stage of manufacture, consisting of pulverization. The heating and drying process is known as torrefaction. In preparing the beverage a quantity of the powdered leaves, and sometimes sugar, also, are placed in a small, hollowed gourd, and the container is then filled with boiling water. The liquor is taken through a metal tube called *bombilla*, with a hollow, spoon-shaped expansion filled with small holes on the end that is placed in the gourd. It is customary to refill the container with water many times before recharging it with leaves, and to pass it around among all the members of a family and any guests who chance to be present. Everybody drinks in turn from the same maté and tube. A kettle of boiling water is kept on a charcoal brazier near at hand. Some of the containers or matés are very elaborate affairs, made of pure silver and elegantly carved or chased.

The amount of *yerba maté* consumed annually is enormous. It is estimated that no less than ten millions of persons in South America indulge in the habit. In Chile the annual consumption per capita is about one hundred and twelve pounds; in Paraguay thirty-four pounds, and in the Argentine twenty pounds. Quantities of it are also exported, principally to Holland. Some years the supply falls short of the demand, but plantations have added very materially to the available wild growth.

Yerba maté has much in common with both tea and coffee, but does not contain as much tannin as either; of caffein it contains about as much as coffee, and this imparts to it the sustaining virtues. In many parts of the maté-drinking belt the beverage only is taken for breakfast, and I have seen a man in western Argentina take thirty-two matéfuls in rapid succession. The flavor is very agreeable and not unlike that of rather strong tea.

After spending a few days at Trinidad we returned to Asuncion. A launch was placed at our disposal, through the courtesy of the President of the republic, and on November 11 we started on a short voyage up the Rio Pilcomayo, into the Gran Chaco of Paraguay. Several men had been sent with us to look after the luggage, which was carried in a separate boat towed behind the launch, and three local naturalists, representing the museum of Asuncion, went along to collect specimens for their institution.

The Pilcomayo is a river of great size, coming from the northeast and emptying into the Paraguay a short distance above Asuncion. The greater part of its course is in the Gran Chaco, a wild, uninviting region inhabited by savage Indian tribes, and of the interior of which practically nothing is known. We proceeded up the river but a comparatively short distance to the little settlement of Porto Gallileo, the headquarters of a concern engaged in extracting tannin from quebracho-logs. A comfortable home had been erected for the management, and their attention and courtesy were most touching. They were a polyglot community, consisting of a Frenchman, a Brazilian, a Swede, an Argentinian, a Paraguayan, and a German. However, they lived on the friendliest possible terms, and all co-operated for the general good of the company. We came unexpectedly, so no preparations had been made for our accommodation; but each man had a private store of treasured articles from home hidden away somewhere, and before long one brought sheets, another blankets, a third monogrammed towels, etc., until we were as comfortably provided for as anyone could wish. The men were very fond of a pet jaguar which they had taken when a cub, but as the animal grew older its temper became uncertain, so it was necessary to confine it in a barred cage. Its wild brethren came from the forest at night to pay it a short visit occasionally, as attested by the footprints left in the soft ground near the cage.

The factory at Porto Gallileo for the manufacture of tannin was of considerable size. Upon arrival from the forest the trees were stripped of bark, ground, and boiled in huge vats. The extract was boiled down to a concentrate and pressed into small cakes; it is very valuable in tanning hides, and its use shortens the time usually required for the process. A number of valuable by-products are also obtained, including dyestuffs.

A narrow-gauge railway line was being built farther and farther into the interior as the land was cleared; this had been completed a distance of fifteen kilometers, and the road-bed was in course of construction for forty additional kilometres. The morning after our arrival at Porto Gallileo we proceeded to the end of the line on the daily work-train, and pitched camp on the bank of a small stream, the Rio Negro.

Figure 38. A street in Buenos Aires.

Figure 39. Porto Gallileo on the Rio Pilcomayo.

Our camp was merely a rough shed built of sheets of corrugated iron supported on poles driven into the ground. The river-water was salt and unfit for use, so each morning several large jugs of fresh drinking-water were sent in from Porto Gallileo, together with a supply of provisions. All about lay marshes, swamps, and large grass-covered areas, the latter type of country predominating.

The Rio Negro teemed with a species of *piranha*. They are deep-bodied and blunt-nosed, and the jaws are armed with sharp, triangular teeth. Although they grow to a length of eighteen inches in the Orinoco and some of the other large South American rivers, those we found in the Rio Negro did not exceed eight inches in length; but they travelled in enormous schools, and made up in numbers what they lacked in size. During the hours of late afternoon, when our day's work was over, I tried many experiments with the *piranhas*. They have a bad reputation and are known to attack animals much larger than themselves, and even human beings who enter the water. Usually they are slow to attack unless their appetite has been whetted by a taste of blood from a wound; then, however, their work is done with lightning-like quickness, and unless the luckless victim succeeds in reaching the shore immediately nothing but the skeleton will remain within a very short time. If I fished with a hook and line baited with any kind of raw meat the fish would scarcely wait for the bait to sink below the surface of the water. The number caught depended entirely upon the amount of time spent in fishing. The bodies of large mammals, such as monkeys, after we had skinned them, were thrown into the stream; instantly the ravenous hordes charged the spot and tore greedily at the bloody flesh; so great were their numbers that they threw one another out of the water in their mad struggles to reach the gory repast. On several occasions I threw dead or stunned individuals of their species into the midst of the frenzied mob, but, strange to relate, they floated on the surface of the water untouched. Unplucked birds were not molested, either. A struggle in the water seems to attract the fish, but I must admit that their behavior is very erratic. While washing my hands in the edge of the stream one day a *piranha* snapped a piece out of a finger; a few days later a man in passing over the river on a bridge dropped his purse into the water in almost the exact spot where I fished, and where the *piranhas* were most abundant; he stripped, waded out very slowly and cautiously so as not to create a disturbance, and felt about with his toes for the lost article; although the water was over four feet deep and he remained in it fully fifteen minutes, he remained untouched.

It is in the dark swamps dotting the chaco like low, glossy islands that the precious quebracho-trees grow. It was also from these same swamps that clouds of ravenous mosquitoes issued with the first signs of failing daylight, and drove us to the refuge of our net-covered hammocks. There we sweltered through the long hours of the night, listening to the angry buzzing of our outwitted assailants, which was not unlike the sound produced by a swarm of enraged bees. I could distinguish a number of different pitches and qualities in the music, blending harmoniously in one general chorus. The varying size of the insects, which ranged from individuals nearly an inch long to the small infection-bearing *Anopheles*, doubtless accounts for the different tones produced by the vibrations of the wings. Brockets were seen occasionally; they left the forest morning and night to feed. In the tall pampas-grass cavies abounded. They came out into the opening beside the railroad just before sunrise and ran about, or sat motionless, when they resembled clods of earth or shadows. Ocelots had worn well-defined paths through the fields in their nightly raids on the cavy community. In the trees we found black howlers, night-monkeys, and giant weasels (*Tayra*); opossums and various species of small rodents held sway on the ground.

Figure 40. Fort of Coimbra on the Rio Paraguay.

While there was no scarcity of birds, they were largely species already known to us, and one day one of the men brought in an anaconda ten feet long, that he found basking on the river-bank.

After spending a week on the Rio Negro we returned to Asuncion, where we were joined by the commissaries who had just arrived with the equipment. Two days later we boarded the comfortable little steamer *Asuncion* and started for Corumbá.

The four and a half days' trip up the Paraguay was most interesting, although the heat and insects at times were troublesome. We had entered the great pantanal country, and the vast marshes teemed with bird-life. As the *Asuncion* fought the strong current and moved slowly onward countless thousands of cormorants and anhingas took wing; lining the pools and dotting the marshes were hordes of wood and scarlet ibises, together with a sprinkling of herons and spoonbills; egrets covered the small clumps of trees as with a mantle of snowy white, and long rows of jabiru storks patrolled both shores. Scarcely a moment passed in which we did not see hundreds of birds. Some of the passengers were armed with rifles and revolvers, with which they kept up more or less of a fusillade on the feathered folk; but fortunately their aim was poor so that little injury was inflicted.

The day before reaching Corumbá we passed an interesting old landmark. It is the fort of Coimbra, built on a rocky hillside with a cluster of thatch-roofed huts nestling against the base. As Coimbra is near the Bolivian border, the fort figured prominently in several of the bloody controversies of bygone years between the neighboring republics.

Chapter 14

HUNTING EXCURSIONS ALONG THE UPPER PARAGUAY

Corumbá is a very hot, dusty town built on a high, rocky elevation on the west bank of the Paraguay. The settlement bears the unenviable reputation of being the rendezvous for fugitives from justice from many climates, but we saw nothing of the lawlessness and disorder said to prevail, and the treatment we received was all that could be desired. The heat at midday was great, but frequently a breeze came up at night. Rows of low, spreading mimosa-trees lined some of the streets and cast a welcome shade; their branches were covered with clumps of gorgeous scarlet flowers.

A railroad in course of construction will soon connect Corumbá with Rio de Janeiro. There is also a cart trail leading through the heart of the chaco to Santa Cruz, Bolivia; to travel over it is a difficult undertaking, the ox-drawn carts requiring a minimum of thirty days for the trip. During the rainy season a large part of the country is inundated, when the caravans must, of course, suspend their activity. I met two men who had made this journey but a short time before. One night a party of Indians attacked and killed all the members of a caravan, stopping only a half-mile distant from the spot where one of these men and his family had made their camp. The tribes along this route are the Penoquies, Guaranokas, and Potoreras, and they are said to be of a treacherous, hostile disposition.

As there was little zoological work to be done in the immediate vicinity of Corumbá, we moved to a place called Urucúm, about nine miles away. The road lay through scrub growth and forest, and was all but impassable on account of the deep mud and rocks. Numbers of native cabins are scattered along the wayside; some of the occupants conduct dairy-farms, and the cows carry bells tied to the tips of their horns.

Urucúm proved to be a garden spot of clear, cold springs, shady groves, and plantations of tropical fruits and vegetables. In the centre of all stood comfortable cottages with large, well-ventilated rooms and delightful shower-baths. Fields and forested hillsides, marshes, and lagoons were easy of access; in them dwelt an abundant and varied fauna. A grove of magnificent mango-trees grew near the house that had been assigned for our use; hundreds of bats came to the trees each morning just as dawn was breaking, to seek their diurnal sleeping-quarters among the dense foliage. They arrived in unbroken streams and spent a great deal of time whirring through the branches, squeaking and making queer little noises that sounded as if they were grating their teeth. Then they finally settled in clusters of from six to a dozen individuals in some particularly thick clump of leaves and, suspended by the claws of their hind feet, began their daytime slumbers. On windy or rainy days they lost little time in becoming settled, and did not seek the swaying branches, but clung to the tree-trunks or on the under-side of the thick limbs. This species of bat (*Vampyrops lineatus*) has a leaf-shaped appendage on top of the nose which may be of some use to the animal, but is probably of little consequence. This "leaf," the nose and face, including the tips of the ears, were tinged with delicate green. As the bats hung head downward, the green-tinted extremity naturally pointed toward the earth; but if the color was intended as a protection it was of little or no avail, as it could not be seen unless the animal was examined at close range. Other individuals of the same species were collected in a dark cave in the near-by mountains. They, however, showed only a very faint or no trace at all of the green coloring on the face. I am convinced that this color is not a vegetable stain, but that the pigment exists in the skin; it fades soon after death.

A footpath leading through the forest a distance of several miles ended at a manganese-mine which penetrated into the mountainside about three hundred feet. Although the mine had been by no means exhausted, it was no longer worked, owing to the great expense of transporting the ore. The dark, deserted tunnel was an ideal resort for bats of not less than four species; one of them (*Mimon bennetti*) was of considerable size. We entered the mine with a lighted candle, but the bats invariably soon put out the light with their wings. Each kind, it seemed, occupied a different part of the tunnel. At first they were slow to leave their places of concealment in the crevices between the rocks, but after a few days' persecution numbers of them rushed from the mine and disappeared over the top of the mountain at the mere appearance of the lighted candle in the entrance. The men who accompanied me on these excursions refused to enter the dark opening in the mountainside, as they said it was infested with poisonous snakes; but, although we explored it thoroughly on several occasions, not a single reptile was ever seen.

In walking through the forest we always saw animals that were of more than passing interest. One day I surprised a tiger-cat in the trail; it ran a few yards and then started up a tree, rapidly climbing about twenty-five feet, and then clung to the rough bark; it remained perfectly motionless and permitted me to walk up to within a short distance of the base of the tree. A short time later I came upon two cebus monkeys feeding in the branches above the trail. I shot at one of them, wounding it. The other was fully ten yards away, but rushed to the rescue, and taking up the wounded animal started off with it at a rapid pace. Most South American monkeys will promptly desert a comrade in danger or trouble, but in this instance it was a female with her two-thirds-grown offspring, and the mother-love was so much stronger than her fear that she exposed herself to danger without hesitation, in saving her distressed young.

One of the most surprising animals encountered in the forest was a large, red, hairy armadillo (*Euphractus*). It sprang up suddenly, almost beneath one's feet, and bounded away with such great speed that it always reminded me of a boulder hurtling down a hillside. Within a few moments it was lost from view among the undergrowth, but the bumping noise as it struck the earth at each jump could be heard for some time after the animal had disappeared. At night these armadillos came out into the clearings and did a great deal of damage in the fields newly planted in corn. We desired to trap some of the creatures, so, following the advice of the natives, we cleared a path one thousand metres long and one metre wide on the edge of the field, and next to the forest. Four salt-barrels were sunk in this cleared lane, their tops flush with the earth; then we covered the openings with a thin layer of dried grass. Grains of corn were strewn all along the cleared stretch, and a liberal amount was sprinkled on the grass covering the pits. The armadillos, in their nocturnal excursions from and to the forest, were attracted by the line of corn and followed it, eating the kernels as they went; when they arrived at one of the barrels they plunged into it and were unable to clamber out. We caught several in this manner. One of them was despatched to the Bronx Zoological Park, but it died *en route*. It is a remarkable fact that after the armadillos fell into the barrels, which contained no wooden bottoms, they made no attempt to burrow out. Their long claws and strong limbs enable them to dig with ease and rapidity. When cornered they fight viciously with the claws and teeth and are capable of inflicting dangerous wounds.

One of the owners of Urucúm stated that at one time he owned a pet jaguar that subsisted entirely on armadillos caught in the manner described above. The flesh is esteemed by the people, also.

On several occasions we saw the gaping entrance to the tunnel of a *Tatu canasto*, or giant armadillo, but at no time did we have a glimpse of its occupant. This is one of the curious, archaic creatures persisting, together with the giant ant-bear, sloth, and hoatzin, long after the star of their age has passed its zenith. Apparently they were not at all uncommon, for we saw scores of the enormous carapaces, looking like casques of armor, in the curio-shops at Asuncion. The animal is fully four feet long, and weighs upward of sixty pounds. A single claw that I found on the Upper Orinoco was seven inches long.

Another visitor to the plantations was a kind of small, red forest-deer or brocket (*Mazama*) with single-spike horns. They spent the days in the heavy timber or dense, low thickets and wild banana-brakes. They were particularly fond of growing beans and destroyed quantities of the legumes in a single night. The natives' way of ridding themselves of the plunderer is to erect a high platform on poles in the centre of the field, commanding a view on all sides, and then shoot the animal as it emerges from its hiding-place.

We also secured a good specimen of one of the rarest animals found in South America. It is the red wolf (*Chrysocyon*), or guaraguasú, of the Brazilians. However, very little is known of the animal's habits even by the Indians and natives who are usually so prolific with stories about the wild creatures coming under their observation. My own experience is limited to two fleeting glances of the huge red forms dashing away at breakneck speed several hundred yards distant, and to hearing the weird, strange wail at night. It equals or exceeds in size the gray wolf of our north woods. It is said to live singly, frequenting the *chapadão* and papyrus marshes, and to travel great distances in quest of rabbits, cavies, and other small mammals that form its principal items of food.

There were also peccaries, black howler monkeys and marmosettes, and among the smaller mammals living in the deep forest was a curious little woolly opossum (*Metachirus*) that ventured out only after dark in search of fruits, insects, birds, or almost anything of an edible nature. It is essentially an animal of the deep shadows; if taken out into the brilliant sunshine it dies within a very short time. Frequently our traps were sprung by black lizards three or four feet long (*Dracæna*); they fought fiercely and clung tenaciously to a stick or other object within their reach. Their teeth are so strong that they scratched the steel barrel of a shotgun. Rattlesnakes were not rare in the open country, but they were of small size; I saw none more than three feet long.

Among the hosts of birds parrakeets were by far the most abundant. They came to the mango-trees by hundreds and were so noisy that they became a decided nuisance. In feeding they frequently took a bite or two out of a fruit and then, letting it fall, proceeded to another. In this way a great amount was wasted, but the people were good-natured over the matter and doubtless realized that there was fruit enough for all, as they never molested the parrakeets. Many of the birds were nesting. A red-breasted thrush (*Planesticus*), not unlike the robin, had its mud and grass nest in the low crotch of a tree on the edge of the forest, but the three eggs were heavily speckled with rusty brown instead of being of a plain-blue color. There were cunning little pigmy owls in the brush-patches, but in spite of their small size they are very bold and ferocious and kill birds nearly as large as themselves. In turn they are preyed upon by members of their own family. Some of the larger owls habitually catch small owls whenever possible. One day my attention was attracted by a commotion in a clump of dense bushes and, as I neared the spot, an owl of moderate size (*Ciccaba*) made a number of attempts to fly up from the ground, but apparently it was carrying something too heavy to permit it to fly. Finally it deserted the object and flew to a branch a few yards away. Going to the spot, I found a screech owl with a portion of its head eaten away. Pigmy owls are eagerly sought for by the natives. They become very tame and are supposed to bring good luck to their owners. We had brought a small owl of another species with us which had been named "Moses." When we found him in the market at Asuncion he was a forlorn and hungry little creature, but showed such a friendly disposition that he was promptly purchased and soon became the very popular mascot of the expedition. At Urucúm Moses was given his liberty among the rafters of our home; he walked about gravely overhead and came down only when hungry or when the half-filled wash-basin lured him to the delights of a cool bath. Sometimes I put him out in a tree for an airing, but carnivorous ants were abundant and nearly always discovered him before very long; then he danced about, clattered with his bill, and made queer little cooing noises until I rescued him.

We spent nearly three weeks at Urucúm. They passed very quickly, for Urucúm is one of those delightful places found all too rarely in South America. Word reached us that Colonel Roosevelt and his Brazilian escort had reached Corumbá, so we hastened back to town; there we met the entire party and made the acquaintance of Colonel Rondon and the other members of the Brazilian Commission.

A hunting-trip on the Taquary had been planned to secure some of the large game that is found in the region. December 16, therefore, found the hunting-party aboard the *Nyoac*. This boat, which was a river steamer of considerable size, had been placed at the disposal of the expedition by the government, and served as our "home" during the weeks that followed, until we reached Porto Campo. Besides Colonel Roosevelt, there were on board Colonel Candido Mariano da Silva Rondon, Kermit Roosevelt, Captain Amilcar de Magalhães, a photographer, physician, taxidermist, and myself. Mr. Cherrie had returned to Urucúm to finish the work in that locality, and Mr. Fiala remained in Corumbá to complete the examination of the enormous amount of impedimenta which he had so ably brought together.

Figure 41. S. S. *Nyoac* on the Paraguay River.

Figure 42. Corumbá.

The *Nyoac* steamed up the Paraguay a few hours, and then turned into the mouth of the Taquary. The water of the latter river being pretty low, a steam-launch was towed along as a precaution; should the steamer become stranded it would have been possible to proceed on the launch. We had been travelling but a short time, when cries from members of the crew drew our attention to the water; and there, where the launch had been but a moment before, were only a few sticks of fire-wood floating on the water. A man had been placed aboard the smaller craft to operate the steering-gear; he had fallen asleep at his post, and in rounding a sharp bend the launch had capsized and sunk. We spent several hours trying to drag the submerged boat to the bank, but the task had to be abandoned, and the launch was left—a total loss.

There were scores of caimans along the Taquary. As these reptiles are justly classed as vermin they may be destroyed on sight. Frequently rows of them dotted the edges of the sand-banks, lying with wide-open mouths. A shot in the head was instantly fatal, and the only movement perceptible was the sudden closing of the mouth as the bullet went home. Crocodiles frequently enter the forest to quite some distance from the water; I know of no more repulsive sight than to come suddenly upon one of the huge saurians lying quietly in wait among the shadows; the evil, grinning expression; the leering green eyes and the glistening, scaly body of the creature suggest treachery and cruelty combined with agility and cunning. One of the reptiles that we saw had cornered a school of fish in a small inlet, blocking the entrance with its body. As the frantic fish tried to escape by jumping out of the water and over the obstruction, the crocodile caught them in mid-air and swallowed them.

Late in the afternoon we saw a giant ant-eater galloping across a grassy field. The steamer was brought to the bank instantly and a hunting-party with dogs landed. Soon the animal was brought to bay and shot. When it was brought aboard darkness had set in, so no photograph could be made of it, and as game spoils within a few hours in the damp, hot climate, the animal could not be left until morning. We took the necessary measurements, skinned the creature, and then spread the hide out on the upper deck. Later we found that the tamanduá bandado, as it is called, was not at all rare in the *pantanales*. This occasioned some surprise, as a great deal of this country is marshy and there are consequently few termites, on which it was thought to feed exclusively. Recent observations by Mr. Cherrie, however, explain why this animal can exist in the *pantanal* type of country. He found it *climbing trees* and devouring the soft part of nestling birds, both of which acts are about the last things one would expect of such a highly specialized animal.

The tamanduá bandado stands about two feet high and is six feet long. The body is covered with long, coarse hair. The color is gray. A broad black band, bordered with white, begins on the chest and passes obliquely over the shoulder, ending in a point as it approaches the loins. This marking gives the animal a peculiar, "cut-up" appearance. The nose is greatly elongated, and the mouth is a mere slit through which the pensile tongue is thrust in licking up ants. As it gallops clumsily along, for the enormous back-turned claws of the front feet impede its progress, the flattened tail is thrown up and seems to aid in balancing the animal. When pursued by either men or dogs, it runs until closely pressed, and then rears up and makes short dashes at its assailants. It is easily capable of inflicting fatal wounds with its claws. P. Lydekker (Royal Natural History) states that its habits are nocturnal and that it has "usually a regular lair ... generally situated among tall grass, where it spends the day in slumber...." In the same paragraph he speaks of the animal tearing open the hillocks of termites with the powerful claws of the forefeet; and "as soon as the light of day is let into their domicile the ants or termites rush to the surface...." Without commenting on this inconsistency, I believe that the giant ant-eater is at least partially diurnal. The stomachs of the specimens shot by Colonel Roosevelt and his son Kermit contained ants and termites, a quantity of earth, and bits of dry and green leaves. The colonel expressed the opinion that the earth and leaves had been picked up with the ants. The walls of the stomach are thick and muscular, like the gizzard of a fowl. In captivity they thrive on finely chopped meat.

We spent the night aboard the *Nyoac*, which had been made fast at a landing where there was only a dilapidated thatched-roof shed. Early the next morning horses were brought up and saddled and we started on a five hours' ride to the ranch-house that was to serve as camp.

Before us stretched vast marshes, dotted here and there with little islands of pastureland and groves of trees or thorny bushes. It was typical pantanal country. Parrots, parrakeets, and macaws flashed by with raucous shrieks, and *kis-ka-dee* flycatchers calmly surveyed the cavalcade from the uppermost branches. Sometimes we flushed a small flock of beautiful Brazilian teals, and in the distance we saw ibises and jabiru storks standing in the long grass, like foam-flecks on a sea of green. For the greater part of the distance we rode through water knee-deep to the horses, but in spots the marshes were drying. In the little pools that were all that remained of what had formerly, perhaps, been an immense lagoon, myriads of imprisoned fish wriggled and churned the water into thin mud. They formed an almost solid mass, and at the borders numbers were constantly leaping out; the ground was strewn with the dead and dying by thousands, and of many species. The stench from the decomposing fish was almost overpowering. Numerous animals coming out of their hiding-places at night to gorge on the bountiful repast left their foot-prints in the soft mud. Apparently opossums, coatis, tiger-cats, and even jaguars haunted these places. In the daytime the countless numbers of water-birds exacted their share of the spoil.

The *fazenda*, or ranch-house, called Palmiras', was reached at noon. It was an interesting place; the long, low, rambling buildings formed a square with an open court in the centre, in which trees and flowers grew and pigs and chickens roamed at will. All about lay marshes, papyrus swamp, fields, and forests. Herds of half-wild cattle grazed on the vast range, and marsh-deer stalked among them or along the borders of the thick papyrus growths. The main object of this excursion was to obtain the lordly jaguar. Men were sent out to locate fresh spoor of the animals, and after a several days' hunt were successful. Then a motley cavalcade, headed by the colonel, set out to find the big, spotted cat. Some of the party rode horses or mules, and a number of natives were mounted on steers. A pack of dogs, used to tree the quarry, trotted excitedly beside the riders. After many hours the faint call of a bugle far away announced the return of the hunting-party. Other bugles took up the signal, and by the time camp was reached all of the natives were lined up and eager to inspect the trophies. Within a week two jaguars, a second ant-eater, and a few deer had been secured. There was not sufficient time to undertake a systematic study of the bird life, but the species found in the immediate vicinity of the house were of ample interest to occupy the attention of a naturalist for many months. Foremost among them was the hyacinthine macaw, largest of the entire parrot family. The dazzling blue creature is more than a yard long, and the beak is so powerful that it can gnaw through the tough hull of the *castanha*, or Brazil nut, a feat unequalled, perhaps, by any other bird. It is a powerful flyer and usually there were only two or four together; but some of the flocks we saw numbered ten or twelve birds. But as a whole, the bird is rare, and as it inhabits the wildest *pantanales* and jungles, its graceful flight and loud screams are one of the rare rewards of those only who venture far beyond the beaten route of travel. The closet naturalist may inspect the stuffed skin, but it can no more convey to him an impression of the gorgeous, living bird, than the dry, shrunken bush at midwinter suggests the flowering rose.

Small colonies of blackbirds dwelt in the papyrus swamp. Their heads were of a fiery red color, and as they sat on the swaying reeds they, from a distance, resembled brilliant blossoms. However, these birds were not abundant.

The preparation of the skins of large mammals was a difficult undertaking. No provision had been made for this branch of the work, as the object of the expedition was not zoological but geographical. However, none of the large game was thrown away; it was skinned and preserved in the best manner possible under the circumstances.

Returning to Corumbá on the evening of December 21, we were joined by the other members of the expedition and immediately proceeded on the up-river voyage toward São Luis de Caceres. A short side-trip was made up the Rio São Lourenço, with brief stops at various points where there were evidences of game, but very little was added to the collections.

On January 1, early in the morning, we halted at a place where there were fresh jaguar tracks on the river-bank. Colonels Roosevelt and Rondon, and Kermit, accompanied by a number of *camaradas* and the dogs, immediately took up the trail and disappeared among the trees. We spent a part of the day on board the steamer, and the remainder collecting in the immediate vicinity. One of the men ran into a nest of maribundi wasps; one of the enraged insects stung him on the head and for several hours the poor fellow was in great agony. His head was swollen to an enormous size, and his companions bathed it constantly with water to relieve the pain; they feared he would die. I have very good reasons for remembering these wasps. While on the Chaparé River, in Bolivia, one of them crawled under the mosquito-net covering my cot; when I retired at night I put my arm on the insect and was stung four times before it could be captured. The effect of the poison was as rapid as it was remarkable. It produced a kind of paralysis within about five minutes, which the prompt action alone of my companion prevented from ending fatally; but more extraordinary still, the same symptoms returned regularly at six months' intervals during the following two years. Each attack lasted from a week to ten days.

The day gradually drew to a close, and finally darkness settled over the landscape, but there was no sign of the hunting-party. The captain, therefore, began to cruise up and down the river, giving frequent blasts of the ship's whistle, for it was feared that the hunters might have become lost. After an hour or so we suddenly rounded a sharp bend and heard a loud voice singing cheerfully somewhere on the bank. A boat was sent in the direction whence the sound came, and after a short time it returned with Colonels Roosevelt and Rondon. They had been pursuing the jaguar through forest and swamp for twelve hours on foot, and without food or drink. Their clothing was torn and covered with mud; it had been necessary to swim frequently, in their clothes, holding their rifles above their heads; the lagoons were infested with *piranhas* and crocodiles. In running through the vegetation fire-ants and wasps had been swept from the leaves and branches, and the insects had been quick to retaliate with bites and stings. But Colonel Roosevelt had enjoyed the experience thoroughly and at once sat down to a hearty dinner, during the course of which we heard the story of the hunt. Kermit returned some hours later. Most of the *camaradas* were so tired they spent the night in the forest and did not come in until late the next morning.

We always passed the nights ashore; the temperature in our cabin aboard the *Nyoac* was 118° F., so we much preferred to sling our hammocks among the trees, where it was cooler. One morning upon awakening I was surprised to see the gently waving palm-leaves overhead. It seemed queer that I should have forgotten to adjust the mosquito net the night before; but an investigation showed that the greater part of the netting had been carried away during the night by the *carregador* ants. In my several experiences with these insects I have never known them to carry away woollen clothing, but all articles of cotton to which they had access were destroyed.

The jabiru storks were nesting on the São Lourenço; we saw several of their great platform nests of sticks perched in the crotches of giant trees. The young storks, two in number and fully feathered, were continually exercising their limbs by running back and forth in the nest, flapping their wings all the while, preparatory to launching forth into the big world. If we tossed short sticks up to them they caught them in their bills, held on for a few moments, then dropped them. Caimans were particularly plentiful on the upper Paraguay. Scores of the evil-looking reptiles lay on the sand-banks, with wide-open mouths and staring, glassy eyes. A fringe of trees flanked the water; through them we could see the boundless wastes of *pantanales* beyond. Troops of black howler monkeys ambled leisurely away as the boat drew near; the males only were black, the females being of a straw-color. There were immense flocks of a species of gray-throated, green parrakeets; some of them were building enormous nests of sticks in the branches. When a single tree contained three or four of the huge structures, its strength was strained to the breaking-point, for some of the nests were five or six feet across and contained hundreds of pounds of material; but not all of them were of this size; some were composed of no more than an armful of sticks and were occupied by a single pair of birds. The larger ones harbor dozens of birds. The nesting cavities had been in the under-side of the structures; entrance to them was gained through tubular openings underneath.

The number of water-birds in the *pantanales* bordering the upper Paraguay is almost unbelievably large. There were such countless thousands of cormorants and anhingas that they confused the eye. Colonel Roosevelt never permitted useless slaughter, and when one day, one of the *camaradas* forgot himself and shot a bird, he was compelled to go for it in a rowboat; then the bird was skinned and preserved. After that no one ventured to shoot at the winged hosts. Egrets were present in such vast numbers that the trees were white with them; and when they flew the twinkling wings filled the air like snowflakes. They were not molested in this locality for the reason that their habitat is impenetrable. I later learned in another region that thousands of these birds are killed for their plumes, in a most atrocious manner. About the time the egret's feathers are at their best, which is also the time when the nests are filled with young birds, the annual floods have begun to recede, leaving small lakes and marshes teeming with imprisoned fish, such as we had seen *en route* to Rancho Palmiras. This is the season of harvest for the water-birds, and they repair daily to some favorite resort to gorge on the luckless fish. The plume-hunters, taking advantage of this combination of circumstances, collect quantities of fish, poison them, and then scatter them over the birds' feeding-grounds. Occasionally poisoned shrimp are used if the inundations extend beyond the usual time. This method is, of course, cheaper than shooting; the birds are not frightened away as they are by the loud reports of guns, and the success of such relentless persecution must be obvious. A whole colony could be exterminated in its feeding-grounds even if the rookery is impregnable.

São Luis de Caceres was reached January 5, and at noon the next day the *Nyoac* weighed anchor again and started up-stream. A short stop was made at a small landing called Porto Campo, where a few days' hunt produced tapirs, deer, and white-lipped peccaries. January 13 found the expedition aboard a launch, struggling against the swift current of the Sepotuba. A heavy house-boat full of provisions and luggage was towed alongside, and we made not over a mile an hour. The end of the river journey came on January 16. We had reached Tapirapoan, the farthest outpost on the frontier, and immediately preparations were begun for our long dash across the *chapadão* of Matto Grosso.

Chapter 15

A Forty Days' Ride Through Wildest Matto Grosso

Tapirapoan presented a scene of festive gayety upon the arrival of the expedition. The large, open square around which clustered the low, mud-walled huts was decorated with lines of pennants, while the American and Brazilian flags fluttered from tall poles in the centre. Flag raising and lowering were always impressive ceremonies; everybody lined up and stood at attention while the banners were elevated or taken down, as the case might be, to the strains of martial music. However, if Tapirapoan bore a festive outward appearance, it acted merely as a mask to cover up the general confusion that even a casual inspection could not fail to disclose. Numbers of horses, mules, and oxen had been gathered from the surrounding country; an array of natives or *camaradas* who were to have charge of the animals and the impedimenta, had assembled, and several warerooms were filled with provisions and equipment. To organize properly a cavalcade of such vast proportions required some little time—in fact just six days. We did not particularly regret the delay, for it gave us an opportunity of making daily excursions into the near-by country. This was mostly of an open character and yielded no big game, but it teemed with interesting little creatures. Several small tracts of land were fenced in and planted in maize, and it was wonderful to note how these restricted areas had been discovered by small rodents which apparently came from the surrounding wilds, found an abundance of food and favorable conditions, and multiplied so rapidly that within a short time they were so abundant as to be decidedly harmful. One would almost expect their natural enemies to increase in the same proportion, but such was obviously not the case. Wherever there was a patch of ground under cultivation, rats and mice teemed, particularly the latter, belonging to the genus *Oryzomys*; they are several times the size of a house mouse, have rather short tails, and are of a very deep brown color. The small burrows in which they live are made at the bases of weed-stalks, bushes, and under fences and logs; or, lacking these protective agents, they dig down into the ground almost anywhere. If one sits still for a few minutes, preferably at dusk, they may see the beady-eyed little animals steal forth, whiskers twitching nervously, and ears alert to catch any sound which might apprise them of danger. I have never seen them go very far from the protection of their underground runways; and even while nibbling hurriedly at some tempting bit of food, they frequently dash away suddenly, then stop short, look around, and come back—all apparently without the slightest provocation.

Some of the men had caught a huge tortoise known in various parts of South America as the *morrocoy*, farther down the river. This became a sort of general pet, and while it was at first intended to use "Lizzie"—for that was the name that had been given to the friendly, inactive creature—for food, it was later decided that the animal was worthy of better treatment. It was therefore agreed upon that Lizzie should go to the Bronx Zoo. A comfortable crate was constructed, and just before loading it on the launch bound down-stream, we gathered around the box and dropped an abundant supply of sliced melon and other succulent food through the bars. Then we learned an interesting bit of natural history. One of the *camaradas* had stood by until he thought enough perfectly good food had been wasted on the tortoise. "Don't give her all that," he advised, "a turtle is just like the camel and the elephant; it can go six months without eating." We were glad to learn later that Lizzie survived the trip to New York, and proved to be the largest of her species in the Zoo collection.

Order was finally restored out of chaos, and each member of the party was given a mule and a complete saddle outfit. The pack-animals were divided into squads, each in charge of a chief mule-man and his assistants; then the impedimenta were sorted out and arranged for easy and quick packing on the mules.

At noon, January 21, the first detachment of the expedition started. This included all of the Americans and several of the Brazilians to whose number Lieutenants João Lyra and Joaquin de Melho Filho had been added. Captain Amilcar was to follow the next day with the remainder of the caravan. This division of the party was absolutely necessary as, on account of the great number of men and animals required, the expedition would have been unwieldy if it had attempted to move in one body.

The first day's ride was a short one. Early in the morning the men started to load the pack-animals, many of which were apparently fresh from the ranges and had never been broken to work of any kind; as a result of this there was a good deal of confusion at first. The corrals reminded one of a wild-west show. Guachos, wearing fringed leather aprons, and wicked, keen-edged knives in their belts, and who swore fluently in two or three different languages, lassoed the panicky animals, blindfolded them, and adjusted the packs. When the covering was removed from the animals' eyes they frequently gave a few sharp snorts, and then started through the corral in a series of rabbit-like leaps, eventually sending the packs, saddles, and all flying in every direction. After freeing themselves of their burden, they gave a few extra high kicks of exultation, and then ran into the huddled mass of their fellows for concealment. Gradually, however, the men became more adept at their work, the mules and oxen quieted down and little groups left the corrals, wound up the trail, and disappeared in a cloud of dust.

Our mounts were good, strong animals. We cantered up the trail at a brisk gait while Mr. Fiala, who had gone on a few miles in advance, took a motion-picture of the entire outfit, beginning the long journey through wildest Brazil that would end—we knew not where. Unfortunately Mr. Fiala was not present to take a film of the expedition when it emerged at Manaos; the two pictures side by side would have told an interesting story.

Figure 43. Colonel Roosevelt in the Brazilian *chapadão*.

A few hours' ride through forest and brush-covered country brought us to the Rio Sepotuba again, but quite some distance above Tapirapoan, and we crossed the stream on a pontoon made by laying a platform of boards across three dugout canoes. There were a number of new palm-leaf houses on the river-bank, so these were used for the night's camp instead of erecting the tents.

Next day we were in the saddle by nine, riding through tall virgin forest with occasional stretches of sandy soil in which an expanse of low bushes only grew. It was evident as we penetrated farther into the interior that the forest zone was fast disappearing, to be replaced by the vast *chapadão*; this latter type of country is high, nearly level, and covered with widely scattered, stunted trees. The heat was intense; there was no rain, but troublesome insects were lacking. At three o'clock we entered an old clearing. Formerly rice, plantains, mandioca, and corn had been cultivated here, but now the place was deserted and overgrown with weeds. Kilometre 52, as the place was called, had been an important camp of the telegraph commission while work was being prosecuted in that region, but had long since been abandoned.

Figure 44. A camp in the *chapadão*.

On January 23 a 32-kilometre ride took us to the site of an old Indian village known as Aldeia Queimada, meaning burnt settlement. A single hut was all that remained, and in this lived two Indian women, each of whom had two husbands and a number of small children. We were adhering closely to the telegraph-line, following the wide swath that had been cleared to protect the wires from falling trees and branches, except when a short détour was desirable to find a better crossing for some small stream. The country was of a gently undulatory character, covered with wiry grass and a very sparse growth of scrubby, gnarled trees. This vegetation is typical of a great part of Matto Grosso. With the exception of a few small deer and a limited number of wood-hewers and jays, there were no evidences of animal life. A clear, cold spring rippled over a pebbly bottom near our night's camp. It was the last stream we should see that discharged its water (via the Sepotuba) into the Rio de la Plata system.

Colonel Rondon had employed a number of motor-trucks in constructing this section of the telegraph-line; several of them were still in serviceable condition. It was therefore thought advisable to send a portion of the baggage ahead on the cars as far as the trail permitted, and as it would take several days for the rest of the expedition to catch up, Mr. Cherrie and I went along to devote to collecting the time thus gained. Father Zahm and Mr. Sigg also went in this party. We started from a point called Rio Mandioca, two days beyond Aldeia Queimada. There were three trucks, great, well-built machines of foreign make, laden to their fullest capacity with the heaviest and most cumbersome pieces of equipment. It was a strange sight to see them racing across the uninhabited *chapadão* at a speed of thirty miles an hour. It rained frequently, but the powerful cars charged through the blinding sheets of falling water, and sent streams of mud flying from the inundated trail. Each car was provided with two wide belts of heavy slats; one of them was fitted over the wheels on each side of the car, so they formed a sort of endless trail and gave greater traction in the uneven roadway. Surely this was exploring *de luxe*; but we were to reach the other extreme before long.

The car in which we travelled had a full-blooded Indian mechanician, who seemed to be fully initiated into the mysteries of handling an automobile, from gathering up branches and stones with which to fill up the roadway when the cars mired deep in the loose sand, to repairing the engine on the rare occasions when such a procedure was necessary.

On the afternoon of the third day we reached a point called Macacos. A few decaying huts marked the spot, and in them lived a number of Parecís Indians, the first we had seen. They were a wild-looking lot; some of them wore breech-cloths, others loose, long, shirt-like garments, and all had a thick mop of tousled black hair. A few of the children were nearly covered with ropes of black beads cut from sections of thin rattan or bamboo. They rubbed their stomachs with their hands and said "*fome*," meaning hungry; so we gave them half of a deer that had been killed a short time before, and they rushed into the huts to feast. We continued on a distance of four leagues. This brought us to the Rio Sacre—the end of the wide road. The river is here broken by a fall of one hundred and fifty feet. As elsewhere in South America, we were impressed with the appalling lack of animal life. So far we had seen only a few rheas, a seriema or two, and several small deer.

On the morning of the 29th we crossed the Sacre on a pontoon boat and, using a number of mules that had been previously sent there, rode the two leagues to the Parecís Indian village of Utiarity. From afar we could hear the deafening roar of water, and the Indians eagerly guided us to a spot just below the settlement, where the Papagayo rushes over the edge of a precipice and falls into the gorge below in one sheer drop of two hundred and eighty feet. The river is fully five hundred feet wide, and the quantity of clear, cold water it flows is enormous. The spectacle of the descending wall of snowy water streaked with various shades of green and blue, the idly floating mist-clouds, and the thunderous roar is awe-inspiring. When it is remembered that these falls are higher than Niagara, one can easily picture the wonderful sight that meets the eye of the traveller in this virgin country.

The Parecís are a small tribe of semicivilized Indians who live in substantial huts and cultivate fields of mandioca, corn, and sweet potatoes. Formerly they were far more numerous, but an incessant warfare with the neighboring tribes and contact with the outside world have thinned their ranks until they are well on the road to extinction. Some of them wore clothes, while many wore only a breech-cloth of their own weaving. They also make hammocks and various articles for useful or ornamental purposes.

In stature the Parecís is rather short, but he is well built and sturdy. His color is a light shade of brown. The youths of the tribe engaged in a curious game of head-ball, using for the purpose a hollow rubber sphere a foot in diameter, manufactured by themselves. They chose sides and batted the ball back and forth across a line with their heads. At no time were the hands or feet used to strike or kick the ball. They displayed remarkable dexterity and tireless energy at this form of amusement; if the ball came bounding along the ground they made headlong dives for it like a baseball-player sliding to home plate.

One evening just before sundown practically all of the men joined in a sacred dance. On this occasion they were clothed in gaudy red head-bands, bead neck-chains and belts, also anklets made of bunches of curious dry seeds that kept up a continuous rattling sound as the dancers stamped in rhythm with the low, sighing music of reed flutes. They stopped at short intervals to drink chicha, and during certain parts of the dance they sang the names of their dead warriors and mighty hunters, calling upon them for guidance and assistance. We had previously seen a blue-and-yellow macaw about the village; it had the run of the place and seemed to be a great favorite with everybody. While the dance was in progress the bird sat disconsolately on top of one of the huts. Then we discovered that the Indians had pulled out its tail-feathers and used them to decorate their head-dresses.

The women were not permitted to witness the first part of the performance, but later the dancers visited each home and exacted tribute from the squaws in the form of several gallons of chicha.

Utiarity was a profitable collecting-place. Many small rodents and a few larger mammals, including a soft-shelled armadillo collected by Colonel Roosevelt, were taken, and a number of birds besides. We spent five days in the village (Colonel Roosevelt arrived three days after we did), at the end of which time Father Zahm and Mr. Sigg started back home. A short time later Mr. Fiala left the expedition for his trip down the unexplored Papagayo. Mr. Fiala undertook this work voluntarily, well realizing the hazardous nature of the venture ahead of him. He carried the undertaking to a successful close, but barely escaped with his life.

The first telegraph-station along the line was located at Utiarity. It was conducted by a young Brazilian; his wife acted as schoolmistress and was doing a really noble work in educating the younger generation of Parecís along mental and moral lines. The second telegraph-station was on the banks of the Rio Juruena, approximately one hundred kilometres away; it required five days to reach this place.

Figure 45. The Falls of Salto Bello of the Rio Papagayo.

By this time the order of the expedition had settled down to a regular routine. We arose as the first sharp blasts of a bugle smote the silence of early morning. A short time later the faithful Juan, a colored man who was as big-hearted and obliging as he was tall and powerful, appeared with coffee. At about eight o'clock a bountiful breakfast was served. Then we mounted the riding animals which were brought and saddled by the *camaradas*, and started on the day's ride. Each person was advised in advance of the distance to be covered, and it was easy to locate the camping-site by watching the numbers on the telegraph-poles; there were eleven of these to the kilometre, and as they were numbered consecutively it was of course simple to arrive at the numeral that marked the end of the day's ride. Usually we made camp at about 4 P. M., but sometimes it was much later. The cook and his assistants immediately began to prepare supper, and always had it ready in a short time, much to our relief, as there was nothing to eat between meals. The other men cleared spaces, erected the tents, and cared for the luggage and animals. Cherrie and I occupied a fifteen by twenty-five foot balloon-silk fly that I had used on the Orinoco, and this was one of the few shelters that withstood the entire trip; it was used later on the Rio da Duvida by Colonel Roosevelt. If there was time we went on a short hunting-trip and usually secured at least a few interesting and little-known mammals and birds. Night in camp was invariably delightful; when the weather was favorable the peons gathered great heaps of wood and made a huge bonfire. Then we sat around it and listened to Colonel Roosevelt telling of his hunting adventures on the Western plains, in the north woods, or on the African veldt—all told in such a way that we were enthralled and could visualize pronghorn, cougar, bear, and lion, as well as their actions in their native wilderness. Should the weather be unfavorable, Cherrie and I repaired to the colonel's tent for a visit; or the colonel and Kermit came to see us. We discussed history, literature, and science, and sometimes, if the conditions were propitious, we were favored with tales of ghosts, the werewolf, and other supernatural beings. I always looked forward to these occasions; they are among the never-to-be-forgotten events incidental to our journey through the wilds with Colonel Roosevelt.

It required four days' time to reach the Juruena. We had been compelled to reduce the amount of baggage very materially shortly after leaving the Parecís village, as many of the cargo-animals had given out on the trail, and the others were weakening perceptibly. It was of course impossible to carry along grain for the animals; each night they were turned loose to shift for themselves, and while there was an abundance of grass, the long-continued strain began to tell. We abandoned most of the tents, and all superfluous clothing was left behind. The equipment for collecting and preserving specimens had to be reduced to a minimum also, on account of its weight; we retained only a few hundred cartridges and about a dozen traps with which to prosecute the natural-history work. This reduction of the impedimenta was unavoidable and affected every member of the party, either directly or indirectly. It was one of the several instances where individual interests had to be sacrificed for the good of the whole expedition.

At Juruena we made the acquaintance of a primitive tribe of Indians who probably represent the lowest type of savage to be found anywhere on the South American continent. They are known as the Nhambiquaras. As we drew up on the river-bank they gathered about and stared in curiosity at the party, but betrayed no hostile feelings. Colonel Rondon had but recently succeeded in establishing amicable relations with them. On his first visits into Matto Grosso, numbers of his men had been slain by the Nhambiquaras, and they had resented his every step into their stronghold. In the days that followed, Colonel Rondon related some of his experiences with these Indians. As accompanied by a few companions, among whom Lieutenant Lyra figured prominently, he made his way slowly and painfully through the *chapadão*, parties of the savages constantly followed his movements. On account of the open character of the country they remained in concealment during the daytime; but when night spread a protecting cloak of darkness over the land, the Indians became bolder and harassed the camp. It was impossible to build a fire, for that would have enabled the lurking fiends to see their victims and make easy targets of them. After trying many schemes for making friends with the savages, Colonel Rondon took a phonograph into the wilds with him, and played it at night. The Indians were unable to understand the music, and finally their curiosity prompted them to leave the sheltering blackness and come timidly into the Brazilians' camp in search of the sound.

Colonel Rondon has persistently treated the wild people with kindness. During all their persecution of himself and his men, he permitted no one to retaliate in any manner whatever. They have therefore learned to look upon him as a friend and some even appeared to be heartily glad to see him.

In stature the Nhambiquara is short, but well built and of a rather dark-brown color. It seems possible that some of them have a slight amount of negro blood in their veins, obtained from runaway slaves many years ago; a few of this class had a light growth of hair on the face. The others were beardless and their bodies also were entirely devoid of hair.

Clothes are entirely unknown to these Indians, and practically the only ornaments in their possession were strings of beads given to them by Colonel Rondon. The men had the septum of the nose and upper lip pierced, and wore quills or slender pieces of bamboo in these perforations. They had the unpleasant habit of coming close up to one and jabbering at a furious rate of speed; this caused the labrets to move uncomfortably near one's eyes, and it was necessary at times to retreat a short distance in order to get out of range of the menacing ornaments. This tribe builds curious round huts or *maloca* of grass or leaves, and cultivates small areas of mandioca; but forest fruits, game, and wild honey form important articles of diet. Ants, snakes, and almost any creature they can capture are eaten. One day I saw several children playing with a calabash of honey, when they accidentally upset it on the ground; this, however, caused them not the slightest concern; they gathered around the spot, and scooped up handfuls of the saturated sand, which they ate. When they had finished, a deep hole remained to mark the site of their banquet!

The weapons of the Nhambiquaras consist of bows six feet tall, made of tough black or red palm-wood, and long bamboo arrows. The points of the latter vary according to the purpose for which they are to be used, and some of them are poisoned. A bamboo cap is placed over the points that have been treated with curare, to prevent the owner's causing injury to himself, and also to keep the rain from washing the poison off. Hunting-parties take long tramps at frequent intervals, subsisting on the fruits of their prowess. At night a rude lean-to is built of branches; a fire, started by rubbing two sticks together, is placed in front, and the game is roasted and eaten; then they stretch themselves on the bare ground to sleep, like so many sheep or dogs.

Colonel Rondon was always kind to the Indians. He gave them beads, trinkets, and food. A herd of steers was driven along with the expedition, and one of the animals was killed whenever meat was required. The Indians always received an entire quarter of beef. They built a huge fire, tore off pieces of the meat and threw them into the embers, where they were left until charred; then they were raked out with a stick and eaten. This was continued far into the night, until not a morsel remained. Sometimes the Indians danced for us, and once we joined them. They clasped hands and stamped about in a circle singing in a loud, shrill voice, words that sounded like "*Nã-na-ha-ha-ha*." After a time we regretted having entered into their pastime, for they kept up the dancing for an hour or more and refused to permit us to drop out.

We remained a day at Juruena to rest and develop films. The pictures taken by an expedition always form one of its important records, and great care must be exercised in developing all exposed films promptly or they will spoil in the hot, damp climate.

When we were ready to continue our journey on the second morning, we discovered that the Nhambiquaras in departing had taken two of the dogs with them. Colonel Rondon spent some hours hunting for the Indians, but their start was too long and he could not come up with them. I regretted heartily that they had not taken all of the dogs, as they were a mongrel, worthless lot; they were of no assistance in hunting, nor did they watch camp. On the contrary, they brought fleas and ticks into the tents, insisted on eating and drinking out of our dishes, and consumed quantities of food that might have been used to better advantage later.

The country beyond the Juruena is somewhat rolling, but there is no appreciable change in the vegetation. We rode twenty kilometres the first day, camping on the banks of the Rio da Fomiga (February 10). Next day we travelled but twelve kilometres, reaching the Jurina, a shallow though rapid stream six hundred feet wide; the crossing was slow and laborious, as there was only one very small *balsa* or ferry. Camp was pitched on the banks of a small stream a league beyond. Nearby were several deserted thatched huts and the comparatively new graves where two Brazilian soldiers and one army officer had been buried. The Indians had killed them, and interred them in an upright position with the head and shoulders protruding out of the ground. The following night, on the Rio Primavera we saw two other graves; the men who were buried there had been slain while asleep in their hammocks. This was the most dangerous part of the whole Nhambiquara country.

When we reached a place called Mutúm Cavallo in the afternoon of the 15th, the mules Kermit and I had ridden were so tired that we decided to give them a day's rest; that meant walking to the next camping-site, and rather than undertake the long journey during the hot hours of the next day we planned to start immediately after supper. There was still some time to spend, however, so we went about our work as usual. An army of ants was foraging near the tents; they had discovered a large, hairy caterpillar, but the half-inch long "bristles" with which it was covered protected its body from the onslaught of the marauding host. The ants, however, were not to be deterred from their purpose; they made repeated rushes at the caterpillar, clipping off a bit of hair each time they struck. After continuing these tactics for twenty minutes, a small patch of the plump insect's body had been cleared of hair, and one ant got a good hold with its vise-like mandible. The caterpillar, upon feeling the pain, promptly began to wriggle, thus exposing its unprotected under-side, and the ants immediately rushed at that vulnerable part and soon succeeded in overwhelming their victim.

Nearby lay the dry, bleached skull of a steer. A fer-de-lance three feet long had apparently been struck with the possibilities as a safe hiding-place presented by the interior of the skull, and proceeded to crawl into it *via* the nasal openings. Then it discovered that this was not the proper entrance and tried to back out; but bits of sharp, splintered bone caught under the plates and scales of the reptile's body, holding it as securely as a trap, until it died, perhaps of starvation.

At 8.30 P. M. we started on our long walk. It was very dark at first, so that it was impossible to see the trail. We had taken one of the dogs with us, and this is the only time, so far as I know, that he was of the slightest use. He was of a light color, so we could make out his dimly outlined form in the darkness. He was therefore permitted to go in advance, and we followed in his footsteps; not once did he lose the trail.

Each of us carried a hammock and blanket, also a gun, as Colonel Rondon had warned us against bands of prowling Indians and jaguars. But to our disappointment we saw absolutely nothing, and did not hear so much as even the hoot of an owl. The only excitement was occasioned when streams blocked our way, and it was necessary to start across without knowing just exactly what was ahead. At midnight we saw a bright light in the distance, and soon after passed the sleeping sentinel and entered Captain Amilcar's quarters; he was camped on a grassy knoll called Campo Novo.

Formerly the third telegraph-station was located at this point, but it now stands on the Rio Nhambiquara, a league away. We were now on the border of the great Cerro do Norte, a vast tract of country composed of high, broken plateaus or mesas covered luxuriantly with grass. Many small streams flow through deep gorges, and near some of the watercourses tall, dense forest grows. The soil is fertile and would produce abundant crops of corn and vegetables. Countless herds of cattle could be reared on the extensive plains, and the climate is cool and healthful. There are few portions of South America so well suited to colonization by Europeans, but on account of the remote location and the lack of means of communication it will be several decades before this vast and fruitful region will become inhabited.

It required about a week's time to cross the extensive Cerro do Norte. The type of country gradually changes. The vegetation of the *chapadão* gives way to a taller growth, and the banks of the numerous streams are heavily forested. Occasionally all other vegetation is superseded by extensive areas of wild pineapples. Many square miles are covered with dense thickets of the plants; during the greater part of three days' ride we were seldom out of sight of them. The fruit was just ripening by countless millions; it was small but of delicious flavor. The Indians ate quantities and also made wine of it.

We added few specimens to the collection after leaving Utiarity. Animal life was not abundant, and the rapid pace at which the expedition was compelled to move left no time for collecting. The Nhambiquaras came to our camp almost daily. They usually approached unarmed, having concealed their bows and arrows some distance away; that was a sign of peaceful intentions. One day we passed one of their settlements; it contained a few low, round huts made of poles covered with grass; one small opening served as the doorway. We also encountered a number of them on the march. A solitary man walked first, carrying his bow and arrows only; about fifty yards behind came a woman, heavily burdened with baskets, calabashes, and children. Another man followed, and then a second woman, and so on until the whole band had passed. The reason for this formation is apparent. As the men are first and are the fighters, they must be on the alert and ready to face danger without an instant's delay; were they encumbered with the family impedimenta the delay occasioned in ridding themselves of it before being able to use their weapons might be fatal to the whole family.

We found an interesting little animal called cururú (*Ctenomys*) at a place named José Bonifacio, reached February 23. It is of gopher-like appearance and habits, and is said never to come out of the ground. It throws up mounds of earth at irregular intervals of from a few feet to ten yards apart, and some of them are very large—three feet across and eighteen inches high. We were very desirous of securing one of these animals, but as there were no traps available for the purpose, six Nhambiquaras were induced to dig open one of the burrows. At first the Indians, guided by the mounds and aided by a sharpened stick, followed the galleries, which were about a foot beneath the surface, and at intervals of ten yards blocked them by stamping down the earth into the hole. We returned a half-hour later and found that the plug between two of these sections had been opened, so knew just where the creature was bottled up.

The Indians now opened the entire section of the gallery and found a hole going almost straight down, which, they explained, led to the nest. A soldier was now called with a hoe, and the work of excavation was begun. In order that the hole might not be filled up, a long, pliable stick was inserted, and this served as a guide. The Indians worked with pointed sticks and threw out the loose earth with their hands. Frequently they relieved one another. When near the end of the work the animal could be felt with a stick; they became greatly excited and worked in feverish haste, as a fox-terrier might after a rat, and kept up a continuous yelling. They were covered with earth from head to foot; ears, eyes, nose, and hair were caked with sand and clay, and the naked bodies looked as if they had just emerged from a mud-wallow. Finally one threw away his stick, inserted his arm into the hole, and with a yell of triumph jumped up, holding aloft the kicking little creature by the tail. Then he flung it from him into the grass. The animal seemed bewildered above ground and could not run fast.

The hole, after leaving the upper gallery, descended eight feet, and then ran in a horizontal direction fifteen feet. At the end was a small cavity, but no nest. Small bunches of grass were found in the gallery which had been pulled down by the roots.

The excavation measured fifteen feet long, eight feet deep and three feet wide, and it required half a day for the Indians to complete the work.

The Indians are fond of the animal's flesh, and often dig them out to eat.

At a camp named Sete de Setembro the two divisions of the expedition were reunited. Captain Amilcar and his party had arrived a day or two before, and a halt was made to divide the equipment and provisions between what were to be the Duvida and Gy-Paraná parties. The Rio da Duvida was only ten kilometres away, and on February 27 we reached its banks. It is a silent, swift stream about sixty-five feet wide at this point, spanned by a substantial wooden bridge. A number of canoes, some of them old and water-logged from use, were tied at the landing. No time was lost in loading them and making ready for the start into the unknown.

Chapter 16

THE DESCENT OF THE RIO GY-PARANÁ

While we were on the Paraguay River it was decided that upon reaching the Rio da Duvida the expedition should be divided into two sections, as a large party always decreases the chances of success in an unexplored region. Colonel Roosevelt suggested that Cherrie and I draw lots, or to settle the matter in any way we chose as to which one should accompany him. After due consideration it seemed to me that, as Cherrie was the older and more experienced man, he was justly entitled to accompany the colonel on the journey down the new river; so I volunteered to join the party which planned to descend the Gy-Paraná.

The eventful day arrived at last, when the expedition must separate; we had looked forward in eager anticipation to the end of the long ride across the Brazilian *chapadão* and the beginning of river work, but now that the goal had been attained without serious mishap, thirty-seven days after leaving the Upper Paraguay, the division of the expedition seemed to have come all too soon.

To better organize the two different forces, our party had halted at a point called Sete de Setembro, ten kilometres this side of the Rio da Duvida, while the other division had pushed on to the point of embarkation. We reached their camp early February 27, 1914, just as the tents were being taken down and the canoes loaded, preparatory to the plunge into the unknown. A short time later everything was in readiness, and farewells were exchanged with Colonel Roosevelt and with the Brazilian officers. Then, with a parting "Good luck!" their dugouts swung into the current and were whisked away. For several minutes we stood upon the fragile structure that bridged the unexplored river and stared at the dark forest that shut our erstwhile leader and his Brazilian companions from view; and then, filled with misgivings as to whether or not we should ever see them again, we turned our thoughts to the task before us.

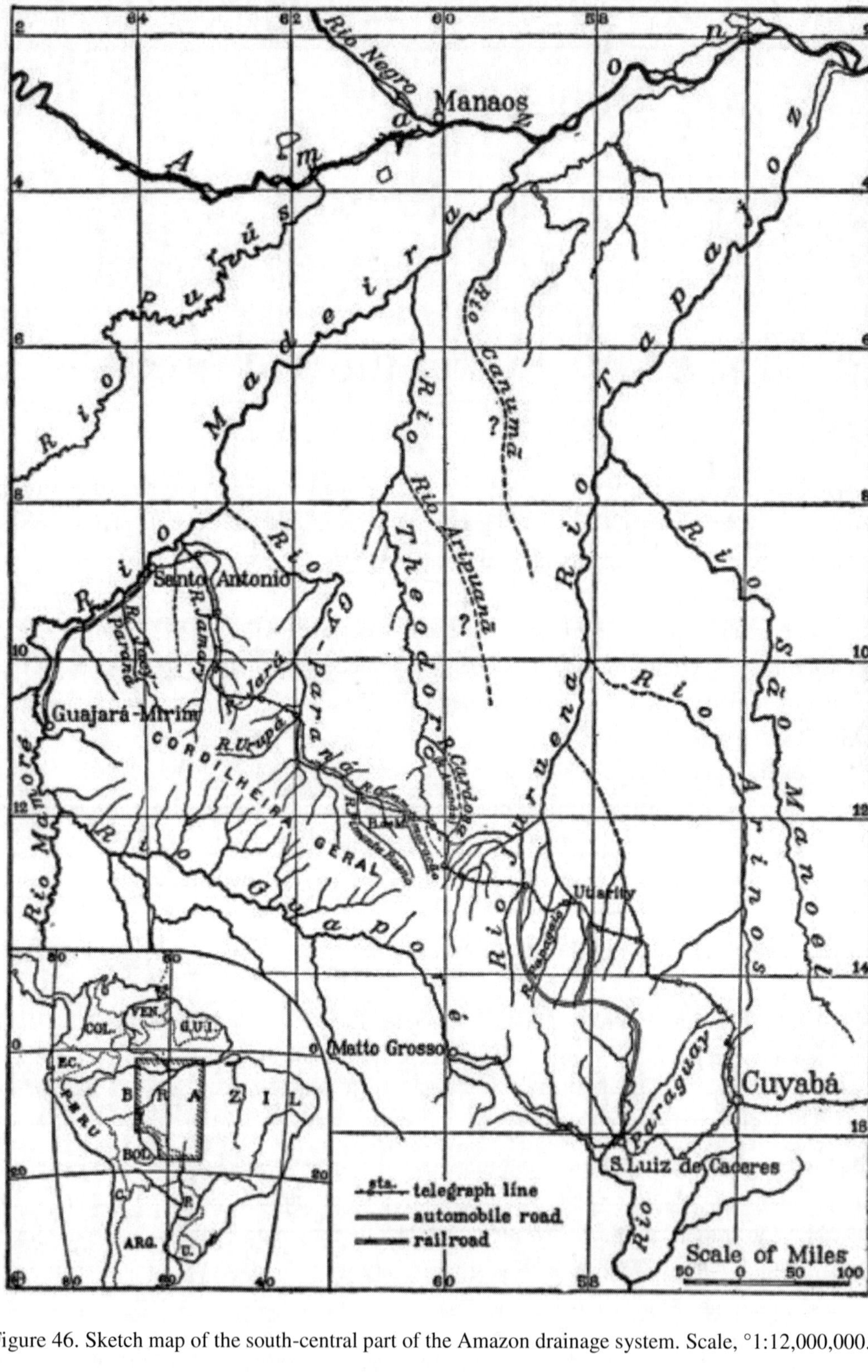

Figure 46. Sketch map of the south-central part of the Amazon drainage system. Scale, °1:12,000,000.

The party of which the writer was a member descended the Rio Commemoração and the Gy-Paraná to the Madeira. From here Manaos on the Amazon was reached by the regular steamer route down the Madeira.

The drainage between the Upper Paraguay and the Madeira is based on the surveys of the Brazilian Telegraph Commission, so far as available. Note the recently explored course of the Rio Ananás (*The Geographical Review*, January, 1916, p. 50, and February, pp. 143–144), and the completed telegraph-line from Cuyabá to Santo Antonio (*Bulletin American Geographical Society*, September, 1915, p. 693). The latter is taken from a map in the Rio de Janeiro newspaper *A Noite* for October 25, 1915.

Our party consisted of Captain Amilcar de Magalhães, a remarkably skilful and wholly tireless leader, and Lieutenant Joaquim de Melho, of the Brazilian Army, Doctor Euzebio Paulo de Oliveira, a geologist, and Señor Henrique Heinisch, a taxidermist, all of the Brazilian Telegraph Commission, besides myself; then there were some thirty-odd *camaradas*, or native assistants. We had a very large pack-train of mules and oxen, as that wing of the expedition, in charge of Captain Amilcar, which had hitherto travelled ahead of the main party, was to proceed with us from this point. Our plan was to continue overland to the headwaters of the Gy-Paraná and to descend that stream to the Madeira, taking observations as we went, for, in common with many of the rivers of the South American continent, the course of this stream has not been accurately mapped.[1] Zoologically speaking, we were in a most interesting and almost unknown country, and no opportunity could be lost to add to our already large and constantly growing collection of both mammals and birds.

We left the Duvida (now Rio Theodoro) shortly before noon; but it had rained nearly the entire day and the trail was indescribably bad; besides, the animals had completed their thirty-eighth day of travel without proper food or rest. That night we camped beside the trail on a site cleared for the purpose by the *camaradas*; we had taken only the canvas flies, as it had been found necessary to abandon the wall-tents some little distance back on account of their great weight. There was no feed for the animals, but the men had cut a quantity of palm-leaves growing abundantly in the forest, which the oxen refused to eat, however.

[1] The Gy-Paraná had been descended by two parties which Colonel Rondon detached for this purpose from his main expedition of 1909. The first, under the zoologist Alipio de Mirando Ribeiro, went down the Pimentá Bueno and the Gy-Paraná to the Madeira; the second, under Lieutenant Antonio Pyrineus de Souza, descended the Jarú and the Gy-Paraná (see map). It was the reconnoissance survey made by the first party that established the fact that the Gy-Paraná, instead of flowing northwest throughout its course, as until then supposed, turns abruptly north in 11½° south latitude, and flows in this direction for nearly three degrees, until, at another abrupt bend in 9° south, it turns west and empties into the Madeira.

Figure 47. Camp on the Rio da Duvida.

From this point the exploration of the river was undertaken by Col. Roosevelt.

The trail had now left the open *chapadão* and wound between high walls of dark forest; instead of the monotonous level of the plain, the country was broken and hilly, with numerous small streams trickling through the dividing ravines, and it rained almost continuously; if we had succeeded in evading the rainy season heretofore, it descended on us now with doubled vigor.

A very wide swath had been cut through the forest for the telegraph-line to protect the wires from falling trunks and branches; so recent had been the work that the shrivelled leaves still clung to the prostrate trees, and the thick second growth, which springs up immediately where the sunlight is permitted to reach the ground, was just sprouting. The ground was covered with fruits of many kinds, most of them insipid or of acrid flavor, but the herds of peccaries seemed to relish them; and the flocks of parrots and macaws quarrelled noisily overhead in their struggles to reach some particularly appetizing morsel. One of the things that especially attracted our attention was the great number of hard, cannon-ball-like shells that littered the trail; they were the empty casques of the *castanha*, or Brazil nut, which grew abundantly throughout the forest. The Indians had opened most of them, in what manner I am unable to say, as they are so hard the blows of a hammer fail to make any impression, and extracted the dozen or more triangular nuts from each. The trees upon which they grow are remarkable for their height and great thickness; not infrequently we saw one a hundred and fifty feet high and four feet in diameter without a single branch lower than sixty or seventy feet. Graves were numerous by the wayside; I counted fifteen, near one another, each newly made mound being marked by a rude cross without name or inscription; they indicated the burial sites of *camaradas*, victims to the dread beri-beri and malaria.

During our second night's camp we heard the gruff, cough-like roar of a jaguar not far away, and next morning the men reported that the creature had killed one of the oxen. I went to see the slain animal and found that it was badly bitten about the neck and that one of the thighs had been partly eaten; in its enfeebled state the ox had been an easy kill for the big spotted cat. We made no attempt to follow the jaguar, but shouldered our guns and started on the home-stretch of the long journey. Again it rained heavily, though intermittently, and frequently the mud was knee-deep: but the knowledge that the river and rest lay but thirteen kilometres away acted as a stimulus to the men, and even the weary animals responded to the ceaseless urging of their drivers and panted along as if they, too, understood that the end of their toil was at hand.

At about four o'clock that afternoon our destination was reached. From the top of a rather high hill we had an unobstructed view down the wide, newly cleared lane through the forest; a small cluster of mud-walled, palm-thatched huts nestled in the depression at the foot of the hill, with a patch of corn and rice growing to one side; a hundred yards beyond sparkled the river, and on all sides of the little clearing rose the Amazonian forest. The little building housing the telegraphic equipment was placed at our disposal, and tents were erected for the *camaradas*, who straggled in with the footsore pack-train until a late hour. The animals were given their liberty and bountiful feeds of corn and fodder, so that within a week many of them were in condition to start on the back trail, a comparatively easy trip, as there were no heavy loads to carry. Many of the natives were also sent back, while others were retained in the service of the expedition; one detachment was sent to the camp of the laborers who were working on the telegraph-line, which extended two kilometres beyond. This was the end of the survey, Barão de Melgaço being the name of the last station, and a force of about fifty men were engaged cutting an opening for the continuation of the line. At the rate they were working it was estimated that the line to Manaos would be completed in about two years.

We had expected to find a craft of some kind awaiting us so that we might immediately pursue our journey down the river, but in this we were disappointed, although, as it later developed, a boat was then on its way to us, sent by order of Colonel Rondon. There were only two small dugouts available, which were entirely inadequate for our purpose, so the men were put to work cutting down a tree of large size and hollowing out a canoe which would hold the party and the necessary luggage. This work we estimated would take some weeks, so in the meantime we busied ourselves exploring the country in the vicinity of Barão de Melgaço.

A short reconnoissance through the forest revealed a veritable zoological wonderland. I was consequently very glad that we were delayed, as this gave me an opportunity to study the fauna of a zoologically unknown region, and to work on some of the problems of nature with which we are constantly confronted, but of which so little is known. One of the facts that no field-naturalist can fail to have thrust upon his notice is the exact precision and nicety with which the balance in nature is preserved. Take the familiar example of the oyster. In its early stages of development it is subject to the raids of such a host of enemies and adverse conditions that out of a million eggs only a few bivalves reach maturity; to offset this wholesale destruction nature has provided that a single oyster may lay several millions of eggs, and thus the race is preserved. Birds, to a less extent, are subject to this same thoughtful provision; therefore we find that the species which are subject to many dangers during the nesting-period or which undertake long, perilous seasonal migrations, lay comparatively large sets of eggs; this is best evidenced by ducks and quail. Species which are subject to the natural dangers of migration only and are protected during the nesting-season, comparatively speaking, rear small broods of young; warblers, thrushes, and a number of our own native birds would come in this category; to further offset the loss, some of these latter may even rear two broods in a season. When we reach the tropics a marked change is noticeable; the extremes in climatic conditions are usually represented by the wet and dry seasons; there are few enemies and food is abundant, consequently the loss of life is comparatively small. If reproduction proceeded there at the same rate as in the northern lands, it must be obvious that the country would soon be overstocked; but again it has been decreed that the equity should be preserved, and the great majority of tropical birds nest but once a year, and then the full complement of eggs is but two. Of course there are a number of exceptions on each side, and on such matters it is difficult to generalize, but in the majority of cases this will be found to be true.

On one of my walks in the forest I came upon a troop of peculiar little monkeys of the saki family (*cacajão*) feeding in the top of a tall wild fig-tree. They differed from all other known members of the genus by being entirely black, with snow-white noses. While feeding they were quiet, and the only thing that betrayed their presence was the constant pattering of small particles of fruit upon the dry leaves carpeting the ground. Presently they took fright, and away they went in a series of leaps and bounds, so that the tree-tops were agitated as by a violent gust of wind; they uttered queer little whining squeaks as they ran and soon disappeared from view. A small one of the same species which I owned was a most amusing little pet and never failed to gain a place in the affections of any one who beheld it—even the most calloused *camarada*; it was of a most playful and friendly disposition and, if petted, made the most ridiculous faces and bubbled with laughter. Another monkey that was common in the forest was a species of *Ateles*, or spider-monkey, which is very appropriately named on account of its slender build and long, wiry arms and legs; it also is of a black color, and swings its way through the branches much after the order of a gibbon, although it lacks the latter's agility. The Indians are very fond of this species, both for food and as pets; but whatever epicurean merits may attach to the flesh, in appearance the creatures are most repulsive. The face is pinched and drawn, with a long-suffering expression about the eyes, while a tuft of long, stiff hair extending over the forehead like a ragged cap gives it a greater look of misery and grotesqueness. One specimen which I collected measured six feet two inches from the tips of the fingers to the tip of the tail.

Birds were not uncommon, but rather hard to observe on account of the density of the vegetation. Near the river stretched a wide band of bamboo, beautiful to look upon but impossible to penetrate without the aid of a machete. Just beyond, the trees grew tall and in close proximity, giant *castanhas*, *heveas*, and ironwoods intermingling their branches to form a canopy of deepest green, impervious to sunlight and through which rain filtered slowly; palms, ferns, and thorny shrubbery formed a dense undergrowth near the streams, so that progress at best was slow. From all sides came the clear, ringing "*hoo-whee-whee-hoo*" of the gold-bird, or whistling *cotinga*, often misnamed the bell-bird, and although the sound came from but a few feet overhead, it was usually impossible to locate the dull, slate-colored songster perched motionless on a well-screened branch. The smaller species of birds travelled in large flocks, doubtless deriving some mutual benefit from this mode of living; usually the band was preceded by a few scouting brown wood-hewers, some with slender bills four inches long bent in a half-circle, flitting silently from trunk to trunk, lighting low down and running up rapidly, while they searched the crevices in the bark for insects; then came the vast host of vireos, warblers, flycatchers, tanagers, and woodpeckers, completely investing the trees in their all-absorbing quest of a livelihood. Twigs snapped, seeds dropped, the woods seemed full of fluttering wings and chirping voices; but in a few moments the noise grew faint and stopped; the tireless army had gone its way, and the vanguard of trogons suddenly appeared, hovered in mid-air to snap off an enticing fruit, and then hurried away. Occasionally we were fortunate enough to shoot a curassow, a large turkey-like bird, and then our Brazilian chef prepared the national dish called *canja*; it consists of a fowl and rice boiled together and is delicious.

On account of its large size, work on the dugout progressed slowly; a section of the trunk, some thirty feet long, had been cut off where the tree had fallen, and this was being hollowed out with adzes, while short-handled axes were used in dressing down the exterior. After twelve days of continuous hewing the dugout began to assume the appearance of a seaworthy craft, and we figured that she would be ready to launch at the end of another two weeks; but the next day a *batelão* arrived. Her captain had been fighting his way up the Gy-Paraná over three months in his efforts to reach Barão de Melgaço, having been sent from the Madeira by order of Colonel Rondon.

We loaded our meagre outfit into the *batelão*, which was a good-sized craft built of boards nailed over heavy wooden ribs, and with a squared tree-trunk for a keel; an arched palm-leaf roof covered a section in the centre, under which we sat to avoid the rain or sun. This style of boat is in general use on the larger tropical rivers and corresponds with the *falca* of the Orinoco and the *champan* of the Magdalena. A crew of eighteen men was mustered, all of whom were more than willing to leave their pestilential environs, and we were soon shooting down-stream with the rapid current. Captain Amilcar had gone on ahead with the small canoes in order to survey the river. They carried a sighting-rod with red disks and a telemeter for measuring distances; a compass gave them the direction.

A quarter of an hour after starting we reached the camp of the telegraphic commission and made a short stop to take aboard a number of men who were suffering with fever and beri-beri; shattered wrecks of humanity whose only hope of life lay in flight. I saw a number of the *camaradas* who had come across Matto Grosso with us, and it was surprising to note the great change which only two weeks in the steaming, insect-infested forest had wrought; several of them were already suffering from violent attacks of malaria, and their faces were colorless and sallow; others who had been in the region longer stared at the *batelão* with sunken, lustreless eyes in which not even a vestige of interest in our visit or of hope was evident; a few had apparently reached the stage where the sight of the twelve newly made mounds on the hilltop no longer aroused feelings of dread or apprehension, but rather of indifference tempered with longing for a welcome release.

The Commemoração, the headwater branch of the Gy-Paraná, on which we were, is a deep river from three hundred to a thousand feet wide, with reddish water and a swift current. It was not necessary for the men to ply the oars except when rounding some sharp bend where steerage-way was required, and this was fortunate, as it rained so much of the time that the men were glad to seek the protection offered by the covered portion of the boat. In the intervals between the deluging showers the sun blazed down mercilessly; trees on both sides of the narrow lane of water sparkled as if bedecked with jewels. In places the forest rose from the river's edge in sheer walls of variegated green; tree-trunks, brush, and palms united into one solid battlement by mosses, climbing lilies, and ensnaring creepers. Again, clumps of graceful ita-palms leaned far out over the water and then rose in a series of stately, feather-crowned columns. At frequent intervals we had glimpses of the animal life that lurked within the impenetrable barrier of the forest fastness. Monkeys were especially plentiful, and within an hour after starting we had seen four distinct species, representing as many families; there were files of black howlers, the males jet-black, while the females are of a straw-color, moving leisurely through the branches; troops of dainty squirrel-monkeys, with deep-chestnut backs, grayish heads, and white faces, scampered over the tops of the lower trees. Black spider-monkeys sat in the highest crotches and gazed down at us in stupid perplexity, and once we startled a family of woolly little night-monkeys of a grayish color, which had selected a thick clump of overhanging vegetation as their diurnal sleeping-place. Large flocks of blue-and-yellow macaws, flying two by two, crossed the river high overhead, doubtless on their way to some choice feeding-ground. Kingfishers sped away in front of the hurrying *batelão*, and from the depths of the woods came the muffled sound of an ivory-bill's tapping on a hollow trunk.

That night we reached the junction of the rivers Commemoração and *Pimentá Bueno*, the latter a stream not less than a thousand yards wide, with a great volume of water. The river formed by the confluence of these two streams is known as the Gy-Paraná. We had covered a distance of eighty kilometres. In ascending, it had taken the *batelão* nineteen days to cover the same stretch of river that we had just descended in one day.

Of course, the surveying canoes could not travel at this rapid pace, so the two parties became separated. In the very beginning Captain Amilcar's party had suffered an accident which came near ending fatally for several of the men in his canoe. Their work necessitated frequent halts, and to bring the dugouts to a stop while racing down-stream was no easy task; so they had adopted the method of driving them into the vegetation and then holding on to the branches while a sight was taken with the telemeter. On one of these occasions a bushmaster fully seven feet long was shaken from the overhanging brush and fell into the canoe; the panic-stricken crew leaped into the water. Captain Amilcar retained his presence of mind and shot the snake, but in the meantime several of the men had been swept down-stream and were on the verge of drowning before he could reach them; the geologist had gone to the bottom, but was rescued and revived with some difficulty; thereafter he travelled with us in the *batelão*.

There were numbers of small alligators in the river, not over four feet long, called *jacaretinga*; later on we had the cook prepare one, as they were said to be good to eat. The flesh was of a white color when cooked, and tender, but it possessed an objectionable muddy flavor, so that we could eat but little of it; however, the natives liked it.

The next day we covered a distance of one hundred and eight kilometres. The current in the Gy-Paraná is not so strong as in the Commemoração, but, the weather being fair, the men pulled at the oars steadily during the twelve hours' travel, with only short periods for rest and refreshment. All meals were cooked aboard, on a fire built on a box of sand in the prow. Insects were not particularly troublesome, as we kept to the middle of the stream, which, receiving the water of numerous good-sized tributaries, was constantly growing wider. There were abundant signs of the close proximity of Indians on both sides of the river. We saw some palm-leaf lean-tos used for overnight stops, with the charred sticks of a camp-fire in front; where the water eddied slowly against a crumbling bank, bamboo stakes protruded above the muddy stream—remnants of an ancient fish-trap—and occasionally we passed a small cleared spot, now overgrown with rank weeds and second-growth sprouts, which marked the site of an old plantation.

Realizing the importance of obtaining the good-will of the wild folk of whose existence in the surrounding forest there was such abundant evidence, the Brazilian Government had erected a number of small bamboo and palm-leaf sheds various distances apart, near some of the more recently used trails that led from the water into the dark jungle. Under each rough shelter a bench, made of long poles laid across sticks driven into the ground, had been built. It was the custom of the officials in going up or down the river to stop at each of these stations and place beads, knives, and trinkets on the benches as a peace-offering to the Indians; but so reticent had been the latter that not one of the articles had hitherto been touched. Great was our surprise and joy to find that all the precious offerings had been removed, and that the Indians themselves had left a number of tokens of friendship in return. They were arrows six feet long, beautifully adorned with the bright-colored feathers of trogons, toucans, and other birds; parcels of Brazil nuts neatly done up in leaves; a few ears of maize, a feather head-dress, and a small pottery bowl. We collected all these treasures and left many more presents in exchange.

As we neared one of the last stations the sound of loud hallooing came from the forest on our right. We swung the great *batelão* toward the shore. We landed, but no sooner had we climbed to the top of the steep bank than we realized how cleverly had been arranged the plan by the Indians to effect a meeting with the mysterious strangers who were passing through their country. Following a wide path that led into the dense forest for a distance of twenty yards, we suddenly came upon a small, swift stream that sped through a dark tunnel-like opening under the dense canopy of leaves and branches. As we stared in blank amazement into the impenetrable tangle of vegetation on the other side of the stream, there emerged from the forest four nude, bronze figures, gesticulating wildly and chattering in a strange jargon which, of course, we could not understand; they were of good build, though inferior in physique to the Nhambiquaras we had seen on the *chapadão*, and not over five feet tall, with long, straight hair, and, remarkable though it is, the tangled hair of two of them was of a decided auburn color. Their bodies were plentifully besmeared with dark-bluish paint, applied in queer zigzag designs and giving a grotesque effect. No wilder scene can be imagined than the quartet of naked, trembling savages faintly outlined against the dim background of merging shadows and sombre green; somehow they seemed to fit into the picture and to complete the impression of primality conveyed by the vast wilderness of the Brazilian hinterland. Our captain held up bunches of bright-red beads and started to wade into the stream toward them, but they immediately withdrew into the thick cover, so he came back. A moment later they reappeared and again began talking and waving their arms; by signs we tried to induce them to come nearer and to assure them of our peaceful intentions. Finally, after a powwow with his companions, one of their number approached to the margin of the stream and held out his hands. He then pointed to one of our men and motioned for him to take off his clothes and come over with the presents, which was done; the Indian grabbed the trinkets from the native's outstretched hands, gave him a violent push back, and fled to his companions. This was repeated a number of times. Then we refused to permit our man to go farther than the centre of the stream—the water was nearly up to his chin—and after lengthy entreaties the Indian waded out and met him half-way. We laid out an attractive assortment of beads, knives, hatchets, and bright-colored trinkets on our side of the river and, retreating ten or fifteen feet with extended arms, invited the Indians to come over. Slowly they came, apparently with many misgivings. We approached them in a friendly manner; they made no attempt to flee, but cast meaning glances behind them where, obviously, an armed force was concealed near by to protect them in the event that our actions aroused suspicion. The chief was an intelligent fellow; his first deed was to enact before our eyes a drama that we shall never forget. Assuming a rigid pose, he pointed straight in front of him with one hand, as if taking aim; then with a sudden "pong" he clutched at his breast and fell upon his knees, gradually sinking to the ground, where he lay moaning. We understood the accusation; one of his tribe had been shot to death by our people, probably a rubber-collector farther down the river; that was the reason why they had mistrusted us. We showed them how to use the *machetes* and hatchets, and they seemed delighted; but when we demonstrated the use of matches their joy knew no bounds; they yelped and danced, made weird grimaces, and tried to set the trees and bushes afire, like so many monkeys. Finally, upon our urgent invitation, the chief shouted a guttural command, and three more savages appeared instantly and joined the group, making seven in all; the late arrivals were also treated in a generous manner, and then we withdrew to our boat. Before leaving, however, we promised to return and bring more *machetes* and matches, which they seemed particularly to appreciate, and they in turn promised to have the bench in the palm-leaf shed heaped with bows and arrows and other things of their making, promises which were religiously kept on both sides.

Figure 48. A rubber-camp on the Rio Gy-Paraná.

Figure 49. A rubber-camp on the Lower Gy-Paraná.

Our next halt was forty kilometres farther down-stream at a rubber-camp known as Urupá. There were several palm-leaf huts standing on a slight elevation, so we took our hammocks and mosquito-nets and spent the night ashore. Travelling eighty kilometres the next day, we reached another rubber-camp called La Pena. The surrounding forest appeared most attractive, and it was said that a footpath led far into the interior to the side of an old Indian village, so I decided to remain at this point a few days to collect. However, a short walk down the trail soon showed that this plan was not feasible; the whole country was inundated to a depth of several feet, and there were so many fallen trees and clumps of thorny undergrowth that hunting was out of the question.

The next day we reached Monte Christo, the depot of a large rubber concern which has its headquarters on the Madeira; about one hundred men had congregated here to await the coming of the dry season, when they would begin collecting rubber-latex from the *hevea*-trees which abound in the forest. Several long, thatched sheds housed the waiting crowd; hammocks were strung from every available post and rafter, giving the interior a cobwebby appearance, and around the edges of the huts, protected from the rain by the low, ragged roof of grass and leaves, numerous small fires smouldered, over which the men boiled their rations of beans or *farinha*. There were pure blacks, descendants of slaves who had been imported into Brazil from Africa many years before; also Indians, Portuguese, and men in whose veins flowed the blood of all three of these races. Many of them were ill with fever, and had large, vile-looking ulcers or "jungle" sores, which were said to result from the bite of a small fly. This was not surprising, as the place was entirely surrounded by pools of black, stagnant water in which clouds of mosquitoes hatched, and no sanitary precautions whatever were taken against infection.

The natives are very fond of pets, and numbers of animals taken from the forest while young were enjoying their full liberty, but never ventured far from the houses. There was a collared peccary, full grown and very amiable, which liked to be petted, and emitted short, low moans and grunts when any one was near it; three curassows, dignified but restless, spent much of their time preening their feathers on a half-submerged log. They were beautiful creatures of a deep blue-black color, with white under parts and a wonderful curled crest. A pair of trumpeters strutted about the camp; monkeys of the *Cebus* family and parrots of several species climbed about in the network of hammocks and added their chorus of screams and squawks to the general confusion.

We had to leave the *batelão* at Monte Christo on account of the cataract which obstructs the river at this point, and carry our luggage around for a distance of half a mile. Below the rapid we found another craft similar to the one we had just left—perhaps a trifle larger—and towed by a small wood-burning launch. On the 18th of March all our things, and the sick men, several of whom were in a serious condition, were carried aboard the waiting *batelão*, and the next morning again found us on our way. The Gy-Paraná was rapidly becoming a vast, muddy sea, comparing favorably in size to some of the larger affluents of the Orinoco, such as the Caura and the Ventuari. The character of the vegetation remained essentially the same, but some of the creepers that drooped from the tall trees and trailed in the water were covered with clusters of yellow, pink, and pale-blue flowers. We saw and heard little of the animal life, as we travelled too far from the banks. In the afternoon a violent wind-storm blew up the river, accompanied by a terrific downpour.

Soon after the storm cleared we reached São João, another rubber-camp, not unlike Monte Christo. The water was so high at this station that we had to use a canoe in going from one hut to another, and the whole place reeked with pestilence. It is infinitely more dangerous to traverse country of this kind than to pass through an entirely uninhabited region; the huts are fertile propagators and harborers of contagion of all kinds, to say nothing of the danger to which one is exposed on account of the more or less constant mingling with the natives. Just below São João the river is again broken by rapids; we rowed down to the beginning of the turbulent water in a canoe and then carried around to the foot of the falls. The distance is not great, but we had to cross a high, rocky hill, so that we were delayed a day in making this portage. The rapids are called São Feliz and are of a formidable character, as the bed of the river is dotted with huge granite boulders over and among which the water rushes with a roar that can be heard half a mile away. During the dry season these rocks are exposed by the receding water and left covered with a thin scum of mud impregnated with salt; it is said that parrots, parrakeets, and macaws then come in thousands to eat of the saline deposit, and that they become so tame great numbers of them are killed with sticks and eaten by the rubber-collectors. I saw two macaws nearly three feet in length, and of a blood-red color with blue-and-golden wings, that had been caught the previous year; they were beautiful creatures, but had the curious habit of spending the entire day squatting in a dark hole under the floor of their owner's hut, coming out only when hungry and at night, when they climbed to a perch above the door to sleep.

After dark our men indulged in a curious native dance which I had never seen before in South America; they collected a great heap of wood and soon after supper had a roaring bonfire going; then they formed a circle, with one man in the centre who began to sing in a high, strained voice, and after each line the whole chorus answered with a wail that sounded something like "*oh-tee-oh-tee-ah*." The centre man bowed and hopped about on one foot in a most ridiculous manner and made frequent sudden charges into the surrounding company, and if he succeeded in knocking one of them down that man took his place in the middle of the ring. The whole performance looked very much like an imitation of a cock-fight. Some of the onlookers had rattles made of small calabashes full of pebbles stuck on a short piece of bamboo, which they shook in rhythm with the singing; they seemed perfectly insatiable of this form of amusement, and the dancing and howling lasted far into the night.

Below São Feliz we found another small launch towing a *batelão*, which in the course of a day took us to a camp called Tabajara. We had not gone more than a few miles the next morning when further progress was again barred by rapids. After a short walk we crossed a branch of the river in small dugouts and then started on a two-mile portage through the flooded forest. Another launch was waiting below the rapids, but within twenty minutes after weighing anchor we again heard the roar of troubled waters ahead of us; the river raced between high, rock-strewn banks. In the distance we could see flecks of foam dotting the surface, while a cloud of mist hung over the river; but from beyond the veil that obstructed our further view came the ominous roar of a great cataract, growing in intensity as we drew near. The landing was about a hundred yards above the brink of the first fall, but the current proved to be too strong for the launch's little engine, and we were in danger of being swept past; the moments that followed were exciting, but fortunately we managed to reach the bank. This same thing had occurred but a short time before, but the result had been disastrous; the boat was swept over the falls, and, of the thirty-one men aboard, twenty-seven were never seen again. The portage around these rapids, called São Vicente, was about a mile and a half in length and led over gently undulating country, all heavily forested. In many places the bed-rock had been uncovered by the torrential rains. This consisted of fine-grained, dark granite; usually there was a shallow layer of sand on the rock, with a thick covering of rich black mould. From the top of a high knoll we had a fairly good view of the falls and of the rapids below; after leaping over a twenty-foot ledge the river rushes through a narrow rock-filled gorge; enormous boulders tower out of the channel like so many black, unvanquished monarchs. Tongues of spray leap to a height of forty feet, and clouds of vapor rise in a constant stream. With the exception of the Salto Bello of the Rio Sacre and Utiarity Falls of the Papagaio, we had seen nothing to compare with São Vicente during our entire journey across Brazil.

That night we reached the first settlement, a small village named Doze de Novembro. We arrived tired and wet, for it had rained the greater part of the afternoon, but we congratulated ourselves upon having performed a remarkable day's work.

The place was overrun with ants, not the comparatively harmless *carregador* ants, which are content to carry away your clothing piecemeal while you sleep, but with endless armies of the fierce black carnivorous species that prey upon every living being. These ants are one of the scourges of the tropics; whether in the fever-stricken Chocó on the west coast of Colombia, at the base of Duida on the Orinoco, or in the wilds of Matto Grosso, the ravaging hordes seemed always the same. One moment they hurry along in solid formations, the next side-lines have been thrown out in all directions, covering many square yards of ground. Not one leaf or crevice escapes the alert scouting-parties, which ascend even to the top of the tallest tree. When a victim is discovered the news in some mysterious way is flashed to the main column, and battalions of reinforcements immediately rush to the encounter, charging the prey and clinging with vise-like mandibles to any part of its body that offers a hold. Usually the approach of the devastating host is preceded by a swarm of panic-stricken insects, crawling, hopping, and flying in their endeavors to escape destruction; large, hairy tarantulas crawl to the tops of bushes and leap from leaf to leaf, only to be discovered and routed, until in despair they spring to the ground, which by this time is one surging mass of ants, where they are despatched in short order. I have seen scorpions and centipedes eight inches long suffer a similar fate; no living thing seems to escape the avalanche of destruction. Flocks of ant-birds usually follow in the wake of the army, feeding upon the ants and upon the insects that have been driven from their hiding-places. One of the questions that naturally arises in this connection is how the callow young of birds escape from the ants, as caged birds are not immune from their attacks, and dead or wounded birds placed near the army's line of march are quickly discovered, torn to shreds, and carried away. While in British Guiana I had been watching the nest of an ant-wren containing two helpless young, placed in the crotch of a tree a few feet above the ground, for several days; one morning the whole region was swarming with ants and the nest was empty; however, not long after, and also on subsequent days, both parent birds were seen contentedly carrying food into a thicket fifty yards away. A casual search failed to reveal the new nest, but to my mind there was no doubt that the young birds had been removed upon the approach of danger; one of the adults was marked in a peculiar manner, so that there was no mistake in the identity of the pair. Doubtless this was an exceptional case, and in the vast majority of instances young birds perish in common with the other creatures which are overwhelmed by the ants.

On the day following our arrival at the little village we boarded a waiting launch sent from the Madeira to meet us—the *Jayme*, she was called—and started on the final stretch down the stream; within an hour we reached the boundary-line of Matto Grosso and entered the great state of Amazonas. The Gy-Paraná had assumed the proportions of a mighty river; it is doubtless one of the largest, if not the longest affluent of the Madeira, and frequently the distance between banks was not less than half a league. The water was yellow and there was little current; frequently we ran into drifts of floating trees, branches, and patches of grass that had been washed out of the flooded areas. There was no opening in the tall, tropical forest which stretched into the distance and disappeared in one long, unbroken vista of deepest green. Toward evening we reached the mouth of the Gy-Paraná and entered the vast, muddy expanse of the Madeira; we crossed to the other side and landed at a small port called Calama, the home of Senhor Asensi, owner of the rubber-camps we had passed on the last days of our journey down the river. Senhor Asensi very courteously placed his comfortable home at our disposal and suggested that we remain as his guests until we had in some measure recuperated from our rather trying experiences, and we were glad to accept his hospitality. Practically every member of the party had suffered from frequent and severe attacks of fever, although half a gram of quinine had been taken by each one daily, and some of the *camaradas* were so ill that they had to be carried ashore; the latter were sent to Manaos on the first available steamer for medical treatment. I was particularly eager to spend some time at Calama, as the locality appeared to offer unusual opportunities for zoological work. After a few days of thorough rest the Brazilian members of the party started up-river to Santo Antonio, for a tour of the Madeira-Mamoré railroad, while I remained to investigate the fauna of the region.

The country back of Calama is high and undulating, so that it remains untouched by the water that covers the lowlands during the wet season. A small space which had been cleared around the building was covered with a fine growth of grass and low bushes, and served as pasture for a few head of cattle; small birds, such as flycatchers, grass-finches, and tanagers teemed in the opening, and many thick-billed green parrots squawked in the tree-tops at the edge of the forest. A short distance below the landing there was an extensive swamp and many small brush-covered islands; masses of aquatic plants floated in the quiet, open pools, conspicuous among which was the beautiful *Victoria regia*, with leaves four feet in diameter. In the dense, tangled vegetation that grew out of the black depths of the murky swamp-water we found flocks of *hoatzins*, or lizard-birds, curious archaic creatures which retain some of the characteristics of their reptilian ancestors; they are about the size of a pheasant, of an olive color above and yellowish below; a high crest crowns the head, and they possess only a limited power of flight. It was the height of the nesting-season, and many of the fragile platforms of sticks contained two or three yellowish eggs, heavily spotted with reddish-brown; the wings of the young are provided with long, sharp claws which enable them to climb about over the branches like lizards; hence their name.

All travelling through the swamp had to be done in a canoe; and pushing the dugout through the almost solid mass of branches and creepers was a difficult task. Every twig seemed to swarm with small red ants, called fire-ants, on account of the intense burning sensation produced by their bites, and they were constantly dropping upon us in scores. Several times we blundered into *maribundi* nests, and in each instance the outraged wasps promptly retaliated. Large *iguanas* jumped out of the trees into the water with a loud splash as we passed underneath, and troops of woolly monkeys deserted the wild cashew-trees in which they fed and beat a hasty retreat. The swamp was full of life, but we rarely recovered anything we shot; the caymans and *piranhas* with which the water was infested usually snapped up our specimens before we could reach them. At night we set throw-lines and caught the great *pacu*, a fish of the *piranha* family; but unlike its bloodthirsty relative it prefers a vegetable diet. A *pirarucú*, six feet long and weighing two hundred and fifty pounds, was also taken; this is the largest fish that inhabits Amazonian waters; the scales are an inch and a half in diameter and of a silvery color, those of the latter half of the body being margined with deep scarlet. It is delicious, either fresh or salted.

The forest back of Calama contained about as much game as is ever found in one spot in South America. There were deer, agoutis, and peccaries, but it was impossible to penetrate far into the interior on account of the Parintintin Indians, who make this region their stronghold. These Indians have always maintained a hostile attitude toward the settlers. An attack was made on Calama one day at noon just as luncheon was being served; from out of the dead silence of midday there suddenly came a shower of arrows; this was promptly answered by rifle-shots from the house, and the Indians immediately fled. Thirty bamboo arrows were gathered up, many of them five feet tall, with barbs on each side of the head; some of the shafts were wrapped with hair and skin taken from the victims of previous raids.

The Parintintins are of medium stature and well built; they are frequently at war with their near neighbors, the Mundrucus; when hostilities are in progress, which is nearly always, the front of the head is shaven, leaving only a round spot of short hair no larger than a silver half-dollar in the centre; the hair on the back of the head remains long.

The Mundrucus have the curious custom of preserving the heads of the Parintintins slain in battle; one of these I subsequently saw, prepared somewhat in the same manner as those formerly so highly prized by the head-hunters of Ecuador. Apparently the head had been smoked, and the eyes had been replaced with balls of pitch; it was a weird trophy, suggestive of wild orgies and cannibalistic rites performed in the depths of the jungle by the light of flickering pitch-torches, and to the music of wailing reeds and deep-voiced tom-toms.

Captain Amilcar reached Calama about a week after our arrival. He had suffered a second accident, in which his canoe, all his personal effects, the instruments, and practically all of his scientific data had been lost. These incidents emphasize the uncertainty of travel and exploration on the great South American waterways, and the dangers to which everyone is constantly subjected who ventures beyond the beaten paths of steamships and tourists' routes.

On April 7 the *Fortaleza*, a good-sized steamer plying between Manaos and Santo Antonio, called at Calama on her down-stream journey, and we embarked for the last stage of our journey. We made excellent time, stopping only at long intervals for the purpose of taking aboard Brazil nuts. On the 9th of April we entered the Amazon, and the next morning found us steaming up the Rio Negro, with Manaos visible in the distance. It had been fifty-two days since the division of the expedition at the River of Doubt.

Upon reaching Manaos we found that as yet no word had been received from Colonel Roosevelt and his party, who were supposedly still on the *Rio da Duvida*. A steamer, provided with comforts which would indeed be welcome to the explorers after their long, arduous voyage in canoes, had been sent up the river; with each passing day the excitement grew more intense in Manaos, and many conjectures were made as to the probable date of the return of that expedition.

Before embarking on the unknown river Colonel Roosevelt had requested me to wait for him should I reach Manaos first, and in the event of his arriving in advance of our party he would await our return. I therefore spent a pleasant week in the city, and was treated with the utmost courtesy by the governor and the inhabitants.

I had become acquainted with a Senhor Ramos, who invited me to visit a ranch he was opening some distance up the Solimões, so I accompanied him, hoping to add new treasures to the large collections we had brought from the Gy-Paraná and the Madeira. After spending a profitable week at this *fazenda* we repaired to another locality on a different branch of the river.

The latter region proved fully as interesting as the first, but scarcely had we become well established in our new surroundings than we were awakened one morning about one o'clock by the sharp blasts of a siren from the river below. We reached the water's edge in a few moments, and there found a large steam-launch resting at anchor, the captain of which brought the good news that the long-absent expedition had arrived at Manaos. Half an hour later we were aboard, steaming at full speed down the river, arriving about seven o'clock in the city.

The story of Colonel Roosevelt's experiences on the unexplored river is well known. Owing to illness during the many weeks' struggle against all but insurmountable difficulties, he had wasted to a mere shadow of his former self; but his unbounded enthusiasm remained undiminished.

Shortly after noon on May 1 we boarded the S. S. *Dunstan*, on which we proceeded down the Amazon to Pará, and at that city transferred to the *Aiden* for the long, uneventful voyage home.

Chapter 17

Down the Coast of Peru—Lake Titicaca and La Paz—Through the Ancient Incan Empire to Cochabamba

The coast of Peru looked decidedly uninviting as day after day the S. S. *Palena* of the Chilean Line nosed her way southward through the placid water of the Pacific. The high, rocky shore stretched on interminably, it seemed; no graceful palm or speck of green of any kind gladdened the eye; there were only the barren cliffs, against which the swell dashed itself into snowy spray and, above them, slopes of hot brown sand.

This was in sharp contrast with the low Ecuadorian shore-line; that was bad enough, with its dense, dark jungles growing to the water's very edge, its overhanging masses of black clouds, and its breathless heat and silence that seemed to exude all the fatal maladies of a tropical clime. Nevertheless, there was a suggestion of life of some sort—inhospitable though it might be. It was not as if an outraged divinity had seared the land with withering breaths of hate, annihilating everything that possessed or gave promise of life, and leaving only the scorched desert as a fearsome reminder of celestial vengeance. But if the land appeared forsaken, the ocean teemed with life. Flocks of gulls always remained in the vicinity of the ship, and occasionally we saw petrels, shearwaters, and albatrosses; whales were not particularly plentiful, but porpoises appeared practically every day. Toward the end of the voyage seals also grew abundant.

There are numbers of ports along the Peruvian coast and the *Palena* stopped at many of them. The enormous swell coming from the south and scarcely felt at sea spends its violence along the shore, making landing very difficult, and often impossible. Steamships dare not approach close to the jutting rocks. All freight is unloaded into lighters; passengers are lowered in a chair operated by a steam-winch and dumped into the huge, flat-bottomed freight-carriers, together with their belongings. This always causes a good deal of excitement and not infrequently slight injuries are inflicted, as the boats are low one instant and come racing up the next on the top of a mountainous swell.

At noon on the eighth day out from Panama we reached Paita. The town lies on the beach and just below the edge of a high sandy plateau. This is the centre of Peru's oil-fields. Tanks were visible in the country near the town, and a thin film covered the water for several miles off-shore.

Salavery is a small town with flat, square board houses. In back of it rise high escarpments of rock and sand. It never rains, so water is brought from a little valley far distant in the foot-hills. A narrow-gauge railroad connects the valley with the port, and sugar is brought out for export.

It seems as if most of the coastal towns are merely ports or outlets for products from the interior. There are many fertile little spots between the ridges branching off from the main range; they are well watered by melting snow on the lofty summits, and a great variety of fruit, vegetables, cotton, and cane are grown.

After ten days the ship anchored off Callao; it is but a thirty minutes' train ride from this port to Lima. The route is flat and runs through corn, banana, and yucca fields and truck-gardens. We visited the creditable zoo and then accepted an invitation to inspect the medical college. The latter is surprisingly well equipped and had an attendance of over eight hundred students. The great cathedral next occupied our attention; the massive temple was in itself most interesting, but curiosity led me to spend the most of our limited time viewing the remains of Pizarro, which are exhibited in a glass-panelled marble casket. An inscription informs the viewers that the *conquistador* founded Lima in 1535; he died June 26, 1541, and was buried under the cathedral; in 1891 the bones were exhumed and placed in their present resting-place. If one may believe the statements of historians, a monument built of the skeletons of his helpless victims would be a far more suitable memorial to the bloodthirsty outlaw than the place of worship which his remains of necessity must defile.

We had heard a great deal about the difficulty of landing at Mollendo. At times the rollers from the south are so immense that ships do not attempt an anchorage, but continue the voyage down to Arica. We were relieved to find the sea perfectly smooth upon our arrival. The town differed from Paita and Salavery only in that it was somewhat larger. We found it possible to purchase through tickets to La Paz, and noon saw us on our way. The railroad started up the barren slope almost immediately; occasionally the incline was very gentle—so gentle, in fact, that the country lay like a great brown desert on each side of the track. These stretches were covered with crescent-shaped sand-dunes, some of them fifty feet high and several hundred feet from tip to tip. They creep slowly forward as the wind blows the sand up their rear slope to the crest, when it topples over into the centre of the half-moon.

At times the grade was very steep. The deep blue Pacific was visible several hours, sometimes on our right and then on our left, as the train wound up the mountainside, but always receding until it resembled a vast mist-enshrouded amethyst losing itself in the distance.

Alkali-dust entered the coaches in clouds and threatened to suffocate the passengers, but the impressiveness of the scenery more than compensated them for this annoyance.

Not far from Arequipa a deep gorge appeared with a stream threading its way through the bottom. Its banks were covered with trees and green vegetation—a veritable oasis amid the desert that hemmed it in on both sides. The Indians who now came to the car-windows when the train stopped to get up steam brought grapes, figs, oranges, guavas, and *empanadas*, or meat pies smelling strongly of onions. They were an unkempt, wild-looking lot and had apparently come from the green vale below. At seven o'clock we were up seven thousand feet, having ascended to that height from sea-level in six hours, and drew in at the station of Arequipa.

Figure 50. Country around Arequipa, showing Mount Misti.

There was no train for Puno the following day, so ample time was at our disposal in which to see the city and its immediate environs.

We found Arequipa to be a most delightful place. It was cool enough to permit the wearing of top-coats with comfort. The people were well dressed and healthy appearing. Electric trains provided adequate means of journeying from one part of the city to another, and if one preferred a carriage it also was obtainable. Beautiful plazas, ancient churches, and wooden buildings are distributed promiscuously among the rabble of low adobe or stone huts which predominate, and herds of llamas thread their way through the stone-paved streets. The atmosphere is so clear the year around that a spot near the city has been chosen for the site of the Harvard Observatory. One has a good view of the great snow-capped Mount Misti from every part of the city; the peak is conical in shape and nineteen thousand two hundred and fifty feet in height.

Continuing the trip from Arequipa, the first stage of the route passes over barren, gently rolling country. Small irrigated plots are not uncommon where some rivulet trickles down from the upper world of snow and ice; they support a limited population of Indians, which must lead a forlorn and miserable life among their desolate surroundings. Farther on, the slopes assumed a friendlier appearance; sparse vegetation in patches appeared and grew denser toward the snow-line, where there was naturally more moisture. Life followed closely in the wake of the grass and bush covered areas. Native hovels became more numerous, and flocks of llamas, sheep, and goats, with a sprinkling of horses and cattle, fed on the herbage.

Figure 51. The expedition en route via hand-car, Changollo to Arce.

The top of the divide is fourteen thousand five hundred feet above sea-level. As we approached it numbers of passengers became violently ill of *soroche*, or mountain-sickness. They acted very much like people aboard a steamship on a stormy voyage, although this illness seemed far worse than any seasickness I had ever seen. Several of the stricken ones rolled about on the floor and tried to tear off their clothing; a feeling of suffocation accompanies the nausea, and occasionally someone dies.

Beyond the ridge the country is level or gently rolling and there are numerous clear blue lakes—some of considerable size. Immense flocks of doves make this upland region their home, and ducks, gulls, and herons teemed about the water.

Just after dark we reached Puno, and a few minutes later embarked on the *Coya* for the trip across Lake Titicaca. The night was so cold and stormy that it was impossible to spend much time on the upper deck, and the cabins were so crowded that sleeping in comfort was impossible. The ship was small and overcrowded with people of many colors and nationalities; most of them spent the night in the dining-saloon drinking and gaming.

Dawn came at last, and shortly afterward the *Coya* slowly wended her way through the reed-grown marshes bordering the lake and tied up at the pier at Guaqui, on the Bolivian side. Indians in reed rafts with sails made of rushes dashed past and disappeared among the cattails, and water-fowl of several species—mostly ducks, coots, and grebes—paddled out into the ruffled water left in the wake of the boat.

It is unfortunate that this passage of the lake is made at night. Nearly every one visiting the vast body of water for the first time is eager to see as much of it as possible, both on account of its being the highest great lake in the world (twelve thousand five hundred feet above sea-level) and by reason of its associations with the nation of the Incas.

Guaqui is a garrison town. There were numerous soldiers in evidence on the streets, and a troop of lancers, under the command of a German officer, were giving a skilful display of their prowess on the lake front. Their mounts were not much to look at and the uniforms of the men were rather shabby, but both were well drilled.

The train for La Paz left at noon. It moved at a good rate of speed across the high, level upland. The scenery is impressive. We were always in sight of snow-covered peaks, although there was little snow on the plateau itself. Indian huts built of stone, some very ancient, are scattered about abundantly, but it requires some experience in locating them before they can be readily distinguished from their immediate surroundings. There were numerous fields of wheat and oats, and llamas without number nibbled the scant vegetation on the slope. In a few isolated spots small herds of cattle, horses, and pigs were visible. Indians came to the coach-windows to sell fruits when the train stopped; they were doubtless brought from the deep, sheltered fissures that have been cut into the range by snow-water from high peaks.

Within a short time we had reached the ruins of an enormous city called Tiahuanaco, which is said to date back many centuries before the Incan era. When discovered it was buried in the sand level with the surface of the plateau, but archæologists have excavated many of the larger buildings and brought to light ancient treasures of rare beauty. Later, in La Paz, we met a man named Poznaski who had done a great deal of work in this region. He had a remarkable collection of hundreds of skulls, pieces of pottery, gold ornaments, and well-preserved cloth. Among the ceramics was a "death's head" of exquisite workmanship, life size, and painted in gorgeous colors. He considered it the finest bit of pottery ever discovered in Bolivia and stated that a North American museum was negotiating for its purchase at a price that ran into five figures. This, however, did not seem probable. As we neared La Paz, the great mountains of Illimani, Murarata, and Huana Potosi loomed constantly more lofty and forbidding before us. They are the patriarchs of the Bolivian Andes, and are twenty-two thousand five hundred and eighty-one, twenty-one thousand, and twenty thousand two hundred and eighty-nine feet high respectively. The summits of all three have been reached by venturesome exploration-parties, but the task of climbing the steep, slippery slopes perpetually covered with deep snow and swept by frigid gales is a trying one that is not often attempted. Huana Potosi, the more distant of the group has a flat top, contrasting conspicuously with the sharp, pointed summits of the other two. The Indians tell a legend that explains this peculiar formation. In the days of long ago, when the world was young, vapors enveloped all the earth; suddenly the sun-god appeared and, beaming down from heaven, caused the mists to become dissipated and vanish. Illimani awoke to life and from his dizzy height beheld the queenly Huana Potosi smiling up at him. At the same time, however, Murarata emerged from the clouds and beholding the beautiful Huana Potosi fell violently in love with her. Illimani became insanely jealous and in a blind fury hurled forth fire, smoke, and stones of great size at his rival's head; the latter promptly replied in kind and fought valiantly. For days the earth quaked and trembled with the thunderous roar of the death-struggle, while heavy clouds covered the terrifying spectacle with a mantle of darkness. After a seemingly endless time the combat stopped; daylight returned, revealing an appalling state of affairs. Finding it impossible to vanquish the rival suitor, Illimani had beheaded his fair lady-love to prevent her from falling into the other's hands. The many streams of water rushing down the steep sides of Illimani are but the tears of grief and remorse over his hasty action; thus he is doomed to mourn and weep until the end of time. The legend has doubtless been handed down through many generations and obviously refers to one of the many volcanic disturbances that must have occurred when the Andes were young.

Shortly before sundown we came suddenly to the brink of a crater-like rent in the plateau and, on the bottom of the huge gash, thirteen hundred feet below, we could see the compactly built mass of edifices and green gardens of La Paz. The situation of the city is unique. One has no intimation of its nearness while speeding over the high, cold *plano alto* (which has an elevation of thirteen thousand three hundred feet) until the very edge of the fissure is reached. The sides are precipitous, but numerous footpaths make their way up or down the steep declivity. The far slopes of the Andes are checkered with cultivated fields; a roaring stream, the Choqueyapu, tears its way through the floor of the amphitheatre, and the series of snow-covered summits form a magnificent background for the unusual spectacle.

The steam-locomotive was taken off and an electric one substituted, and then the train slowly backed down along the face of the incline to the station below.

The impression of La Paz, gained from the first brief view above, is soon dispersed upon nearer and more intimate acquaintance. The streets are narrow, crooked, paved with small stones from the river-bed, and very steep. Walking any length of time entails a great amount of exertion on account of the high altitude; fortunately, carriages are not lacking, and a tramway also provides a ready means of locomotion, or I am afraid few travellers would ever see very much of the inner life of the city. With the exception of a few churches and government buildings that are worthy of note on account of their size and architecture, the buildings are low and of a primitive type, whitewashed and covered with tiles or thatched.

Ordinarily the streets are all but deserted, but on Sundays and fête-days a motley crowd throngs the winding thoroughfares. There are full-blooded Indians of the Aymará race, clothed in picturesque though not beautiful garments; half-breeds or Cholos are far more gayly clad in very full skirts and shawls of bright colors, round, flat-brimmed straw or felt hats, and imported shoes with high heels and tops that reach almost to the knees; the number of townspeople, creoles and foreigners, seems negligible compared to the throngs of Indians and Cholos; in fact, some authorities state that there are one hundred of the latter to one of the former. On market-days long lines of llamas, burros, and mules thread their way through the crowded streets, bearing fire-wood, charcoal, meat, and vegetables for the sustenance of the city.

About the most interesting place in La Paz to us, and at the same time the most repellent, was the Museo Nacional. It contained several dark, cavernous rooms crowded with a wealth of specimens, mostly in the form of ceramics, minerals, and mummies. They were piled promiscuously everywhere in the most slovenly and disgusting manner. Naturally, this treatment did not tend toward their preservation; rats had undermined the mounds of human remains, gnawed holes into the bodies, and made their nests in the interior; pottery had crashed from unstable shelves, and bird and mammal skins were badly moth-eaten. I trust that a more efficient management may rescue these treasures.

The plazas, of which there are four or five, are small and not particularly attractive. The cold climate prevents the growing of tropical decorative plants that are always so conspicuous in cities and towns of the lower country. The *gente decente*, or upper class, meet in the Plaza de Armas on Sundays for a chat with friends, a stroll to exhibit their finery, and to listen to the music.

The Aymarás who inhabit the entire highlands are of a treacherous disposition and have several times organized their forces preparatory to making war on the Bolivians. As their number is very great they are a menace that is very real and serious. When an uprising is threatened, the chiefs are arrested and punished, and then the rebellion dies down for want of leaders. These Indians still retain the *despeñadora*, or death-doctor, in the more remote and inaccessible regions. This person is a woman who possesses the knack of doing away with the aged and infirm of her district, and the office is handed down from mother to daughter. When anyone within her jurisdiction becomes too old to work, or is ill with a malady thought to be incurable, the *despeñadora* is called in; she straddles the poor unfortunate and ends his existence by deftly dislocating the vertebræ of the neck. Whenever government officials learn of the operation of one of these women, they are taken into custody and punished.

One of the favorite sports of the *Paceños* is to hunt wild cattle in the high valleys between the peaks. Numerous herds are still in existence and it is said that they are of a savage disposition and furnish good sport. I met an American who had been thrown from his horse and gored by a wild bull that charged him from a distance of several hundred yards.

The country between La Paz and Oruro is very similar to that we had crossed coming from Guaqui. There are the same vistas of barren plains, green fields, llamas, and asses on the slopes, and dazzling snow-fields in the background. The plateau is strewn with marine fossils, mostly *trilobites*, reminders of the days when Lake Titicaca was many times its present size. We covered the one hundred and twenty-seven miles to Oruro in six hours, and spent the night there. This city owes its existence to the many mines located nearby—some within the city's limits—and to the wealth they yield in tin, silver, and other metals. Next morning the journey was continued toward Cochabamba. Shortly before noon the level country was left behind and we started down the eastern slope of a ridge that leads into the lower country. This part of the road-bed is new; the greater part of it is laid on a narrow shelf of rock carved and blasted out of the mountainside. Perpendicular walls of granite tower above to a height of hundreds of feet on one side; in places the top of the huge masses seems to hang over the track in a tottering position and one expects the rumble of the train to set it in motion and bring an avalanche of destruction down upon one's head.

A small stream flows through the bottom of the gorge. During the greater part of the year it is a mere rivulet that trickles harmlessly over the shallow, pebbly bottom of its course; but when the torrential rains of winter fall it rises rapidly to the proportions of a mighty river and sweeps away sections of the railroad. Long rows of breakwater have been placed alongside the base of the road-bed to protect it from the ravages of the flood; they consist of loaf-shaped piles of stone bound together with wire netting; these would be effective against the water alone, but they cannot resist the demolishing force of the huge boulders that are rolled down from the mountains by the strong current.

A number of breaks in the line had been made by landslides just before our visit, so the train could not proceed beyond Changollo, a settlement of half a dozen Quechua Indian huts and a good-sized station, the elevation of which is ten thousand feet. We were met by a representative of the railway company and given quarters in the station buildings; the other passengers immediately engaged mules and llamas and started for Cochabamba. The reason for our delay was that we had just received a shipment of ammunition and supplies from New York, and some time would be required to repack them in parcels of equal weight suitable for transportation by pack-train.

Changollo was headquarters for the construction-gangs working on the line. About half a dozen Englishmen and Scotchmen were in charge of the work, and they showed us every possible courtesy and attention during our brief stay there. I regret constantly that it is not possible to give detailed credit to all the people, South Americans and foreigners alike, who treated us with such unfailing courtesy throughout our years of travel in the southern continent, and to whose assistance we are so heavily indebted for the success that attended our efforts; but to do so would fill the pages of a volume several times the size of this one without leaving space for my narrative.

All of the railroad men boarded with an Englishman named Cole and his wife. The Coles were a middle-aged couple who had spent the greater part of their lives together travelling around the world. Among other places, they had lived in India and in Africa. They had a score of parrots, cockatoos, and dogs that accompanied them in all their wanderings; caring for this miniature menagerie must have been a troublesome job while moving from place to place, but they took the place of children and were looked after just as tenderly. Cole claimed that he was the only man on earth who had been bitten by a black mamba—a species of giant cobra—and lived to tell the tale. He was following a path through the silent jungle one day at dusk when a black form lunged down upon him from some branches that overhung the trail; at the same time he felt a dull, throbbing pain in his left arm, and realized what had occurred. His first impulse was to flee in terror; however, better judgment prevailed and he opened and sucked the wound and applied a tourniquet above it. Then he hurried home and drank large quantities of ammonia and also applied some to his arm. He stated that he was very ill for several weeks but that persistent use of the ammonia overcame the effects of the poison and he gradually recovered.

Through the kindness of our new friends we secured hand-cars on which to resume the journey to the end of the line—about ten miles distant. The baggage was placed on some of them while we occupied another. The way lay down-hill and we dashed along at a great pace, taking curves without diminished speed. There were several short tunnels, the entrances of which loomed up like the black openings in a grotto; in a flash we were plunged into absolute darkness; a moment later we raced back into bright sunshine. Whenever an obstruction in the track ahead was sighted the brakes were applied and then everything was carried around and the trip continued. We met a good many Indians on the road-bed; they preferred its use to the rocky trail along the river, and even drove their burros and llamas on it. All employees of the company had orders to punish anyone found on the track, in order that they might learn to keep off it, as there would otherwise be a great loss of life when trains begin their runs over the line. The favorite form of chastisement consisted of pouncing on the Indians and taking their hats away from them. The head-gear was taken several miles down the track and thrown into the top of a cactus or thorny tree. If the offender resisted the seizure of his hat he was told that he might have it by calling on the foreman of the nearest construction-camp; when he arrived a good lecture was given him and in some instances a fine was imposed.

It took several hours to reach the end of the line, as landslides and the attendant portages around them had been numerous. We spent the remainder of the day and the night at Arce, an Indian village. Several hundred Quechuas had gathered, as it was market-day; they brought a good deal of cloth and beautiful blankets to sell, but their prices were several times those asked in more remote regions. At night the assembly played on reed flutes and native guitars, sang, danced, and drank *chicha*; the revelry lasted until the first gray streaks of dawn appeared over the mountain-top, and then the mob dispersed to their distant homes in the high valleys.

Our journey was continued the morning after reaching Arce. We had secured a train of good, strong mules and expert Indian *arrieros*. The trail lay along the river-bed, which was very wide and paved with small pebbles. At numerous points Quechua women had put up small shacks of stones and reeds; they squatted within the makeshift shelters all day long. A white rag floating above from a tall bamboo announced to the weary wayfarer that *chicha* was for sale within, and all the travellers we saw religiously stopped at each of these road-houses to slake their thirst. At one point a wall of rock rises from the stream to a height of three thousand feet; two condors were perched upon the very tip, their black forms clearly outlined against the sky, while two others circled swiftly above. We passed through the towns of Yberta and Sacamolla without stopping to rest, and after fifteen and a half hours' continuous riding reached the home of the manager of the railroad, a Mr. Taylor, with whom we spent the night. The place is called Parotani, and we subsequently spent some time there investigating its interesting fauna. At noon on the following day we reached Vinto, which marks the beginning of an electric tram-line to Cochabamba. We did not take advantage of this easy means of transportation, but continued the journey on mule-back, and two hours later found ourselves at our destination.

Chapter 18

CROSSING THE BOLIVIAN HIGHLANDS FROM COCHABAMBA TO THE CHAPARÉ

Cochabamba is one of the more important cities of Bolivia. In size it ranks next to La Paz, although it is not nearly so modern, and in point of activity it is far in advance of Sucre. The population is about thirty-five thousand.

The plain upon which the city is built was at one time the bottom of a lake, which fact is indicated by its ancient name of Oropeza, a Quechua word meaning "plain of the lake." On account of its high elevation, which is approximately eight thousand five hundred feet above sea-level, the region enjoys a mild climate; the average annual temperature is 66° F. Rain falls in abundance during the months from November to April; and during the dry months irrigation is resorted to for providing water to the fields of alfalfa and grain. The country is naturally of a decidedly semiarid character.

The city boasts a number of fairly modern buildings, although by far the greater number are of the low adobe type with thatched or tile roofs; delightful little plazas filled with tropical trees and shrubbery relieve the monotony of the rows of white edifices.

The shops are filled with provisions and dry-goods at remarkably low prices; the city market is supplied with a superabundance of produce, flowers, and articles of native manufacture; the people are courteous and obliging, and the great numbers of Indians and Cholos give a touch of gayety and color to the throngs which fill the streets.

Among the city's institutions deserving of special mention is the Cochabamba Institute, founded in 1911. The instructors are nearly all Americans of the type one meets all too rarely in South America, and who are really doing a great and noble work in furthering the educational and moral progress of the country. Several hundred students of both sexes, from many and remote parts of Bolivia, attend the literary and business classes of the college, live under the care and refining influence of its instructors, and, as I subsequently discovered, introduce into their homes the desirable and elevating qualities which they have acquired.

To the northwest towers the Cerro Tunari, a mountain over fifteen thousand feet high and of imposing appearance. It rises in majestic proportions above the uneven summits of the cordillera, in a manner befitting a snow-crowned monarch of the range.

We spent several days in the city, adding to our outfit and purchasing mules, and then started eastward *en route* to the Beni district; it was our intention to go slowly and stop at places which offered a suitable field for our operations.

Leaving Cochabamba on the afternoon of May 9, 1915, we rode the fifteen miles to the town of Sacaba, arriving there at dusk. The intervening country is thickly settled, and large areas are irrigated and planted in alfalfa, maize, wheat, grapes, and vegetables. Nearly all the inhabitants are Indians of the Quechua race.

Beyond Sacaba the trail adheres closely to the bed of a small stream, and ascends at a steep angle. Numerous little canals carry the water along the mountainside, and the country is dotted with small stone huts surrounded by carefully cultivated fields; this is made possible by the fact that the rivulet never dries, but, on the contrary, supplies a constant stream of water of sufficient volume to irrigate a large area. The canals have been dug with great precision; each family uses only as much as required, and at stated intervals, so there is enough for all.

The trail goes up steadily until an elevation of twelve thousand feet is reached. As we neared the top a strong wind sprang up, so that it was difficult to keep one's place in the saddle. The mountainside is covered with small, round rocks of uniform size, such as one would usually expect to find in a dry river-bed.

Beyond the high summit of the first ridge lies the high mountain valley in which is located the Quechua village of Cuchicancha (meaning "pig-pen"). There are several score of huts scattered about in little groups, and built of rocks, with thatched roofs. The Indians speak practically no Spanish, and live in much the same way as they did in the days of Atahualpa. In order to cultivate the land they have gathered the rocks which everywhere carpet the ground into huge piles, and also built fences of them; large quantities of potatoes, ocas, and avas are grown.

Each family owns a flock of sheep, which apparently replace the llamas of olden days, although flocks of the latter animals are still to be seen frequently; also a few pigs and burros. They have likewise taken to cultivating wheat, oats, and rye.

We decided to spend a week at Cuchicancha and succeeded in persuading an old Quechua man to rent us his hut for that length of time. He spoke not a word of Spanish, or at least pretended that he knew nothing whatever of that language, so all conversation had to be carried on through an interpreter. As our coming was a complete surprise to him, he asked if he could occupy the habitation with us for a few nights until he had time to find sleeping-quarters elsewhere; to this we, of course, consented. One night I was awakened by loud talking, and much to my astonishment found that the aged Indian, who had evidently taken too much *chicha* during the day, was restlessly tossing on his pile of sheepskins and blankets, and talking in his sleep—in excellent *Spanish*. After that we conversed with him without the aid of an interpreter, and he understood every word of it, too.

The weather at Cuchicancha was splendid; it was autumn, and while the nights were cold, the days were always comfortably warm. The Indians were friendly and brought us eggs, goats' milk, chickens, and bread. Each morning the children took the flocks to the narrow river-bed to feed on the sparse vegetation, and at night they brought them back to the stone corrals; they took a few boiled potatoes with them for lunch, and also their spinning for pastime. All spin except the men; and every one had an abundance of blankets and *ponchos*; even the bags for grain and potatoes are made of homespun wool.

The harvest had been gathered and everyone seemed contented. One day a party of Indians collected to thresh wheat; from a distance I could hear the boom of a drum and the shrill wail of reed flutes; as I approached, a strange sight met my eyes. Bundles of grain had been piled in a high mound, on the top of which sat the musicians; a dozen mounted Indians were driving a herd of mules and burros around the base. Around and around they went at a frantic pace, keeping perfect time with the music; as the animals circled the stack a man on top cast armfuls of wheat down in their path, so that in running over it repeatedly they naturally trampled out the grain. About a hundred men, each holding to a long rope, formed a circular fence around the racing mob and prevented any of the animals from escaping.

We were surprised at the abundance of life in this naturally barren region. There were practically no indigenous trees, but a long line of willows had been planted near one of the houses, and to these thousands of cowbirds, doves, and finches came each night to sleep. A short walk across the stubble-fields always revealed something new. There were tinamou which rose with a loud *whir*, reminding one of partridges; many species of brownish birds belonging to the wood-hewer family, one of them with a long, curved bill, but they ran about on the ground or perched on the stone fences; large flickers lived among the rocks, and condors soared above; and there were even flocks of gulls and plovers. The most unusual birds were two species of very small parrakeets which clambered about over the rocks and slept in holes in the high banks. Vast numbers of cavies lived in the rock-piles, from which they sallied at all hours of the day in quest of food, and many small rodents inhabited the grain-fields.

A good trail leads eastward from Cochicancha; the summit of the range rises about two leagues from the settlement. At the time of our visit the black, rocky peaks were covered with a mantle of snow and an icy wind swept through the cleft which serves as a pass. The elevation of the trail is thirteen thousand four hundred feet. At the base of the towering masses which rise several hundred feet above the passage, lies a placid little lake, and ducks, and gulls were swimming on its peaceful surface. Condors swept down from above to inspect us, and then soared back to their dizzy perches among the unscalable crags.

On the eastern side of the divide the trail leads downward abruptly, and the character of the country changes; at eleven thousand feet a sparse growth of bushes appears, growing denser with each passing mile. Suddenly we found ourselves on the rim of a gorge through which the Incachaca River rushes—a raging mountain torrent fed by snows melting in the high altitudes. The path is a mere shelf cut in the face of the cliff; to the left rise the smooth walls of frowning, black rock; to the right is a sheer drop to the river. We could peer over the edge of the precipice and see drifting clouds two thousand feet below, filling the chasm and shutting from view the bottom hundreds of feet lower down.

At seven thousand seven hundred feet the forest begins; a collection of half a dozen huts called Incachaca nestles in its inner border, and there we decided to remain for a few weeks. We secured space in a large house belonging to an organization which is engaged in digging a canal along the opposite side of the gorge; when this work is completed the water of the river will be turned into the artificial course and utilized for running dynamos to furnish electricity for the light and street-car service of Cochabamba. A power-house had been constructed at the bottom of the ravine, and the lines for transmitting the current had been strung across the mountains.

At Incachaca the river flows through an underground channel; while exploring the forest one day we came suddenly upon the narrow cleft in the mountainside, scarcely a dozen feet across, and with a great deal of effort were finally able to distinguish the roaring white torrent a hundred feet below. The edges of the cleft are so overgrown with ferns that one has no idea of its existence until the very brink is reached. A short distance below, the river emerges from the darkened cavern, and plunging over the face of a precipice, thunders into a pool in a sheer drop of fifty or sixty feet.

We found the upper limit of a subtropical fauna at Incachaca. Bird-flocks travelled hurriedly through the trees; they were composed of bright-colored tanagers, finches, and cotingas. Honey-creepers and hummers were plentiful in the flowering shrubs. Queer little ducks called merganettas disported in the pool below the falls, and dippers ran nimbly along the edge of the water. In one of the tall trees near the river we discovered the nest of an eagle. We found it impossible to climb the tree, but a German named Ricardo Marquardt, who was in charge of the workmen along the river, succeeded in reaching the huge mass of sticks seventy feet above the ground, and brought down a beautifully spotted egg. To my companion, Mr. Howarth S. Boyle, who accompanied me on the entire trip, belongs the credit of taking the rarest birds from this locality; they were a pair of white-eared thrushes (*Entomodestes*), which, so far as I can learn, exist in only two other museums. Among the lower growth lived many ant-thrushes (*Grallaria*), whose clear call could be heard at all hours of the day. This is one of the hardest of all birds to collect. The long-legged, tailless songsters never leave the thick growth of ferns and brush, and the only way to secure them is to enter the dense cover, sit quietly, and imitate the clear, ringing call in the hope of attracting the birds; sometimes this requires hours of patient work, and more often than not the effort is futile.

Coatimondis, or raccoons, roamed in the woods in small bands, sniffing in the damp mould and searching for insects; while feeding they uttered deep grunts, but when frightened they gave a succession of rapid bird-like chirps. These animals spend a good deal of their time in the trees, but are almost invariably found on the ground in the daytime; when pursued they are very pugnacious and it takes an exceptionally agile dog indeed to avoid being severely torn by the sharp teeth and claws. In captivity they become very tame, and make nice little pets, although their mischievous disposition often gets them into trouble.

From Incachaca to Locotál is a distance of only eight miles, but the scenery along a part of the route is as impressive as any to be found in the entire Andean chain; perhaps the gorge of the Urubamba, in Peru, alone equals it in grandeur and awe-inspiring magnificence. The bare, shattered, and split crags reach many hundred feet above the trail, and stand in a leaning position so that the tops actually hang over the narrow passageway as if threatening to topple over at any moment; below, the steep slope is covered with huge boulders which have fallen from the towering masses above.

At Locotál there are but half a dozen houses, occupied by Quechua families who subsist mainly on the profits derived from the sale of *chicha*. We stopped a few days in a hut owned by a kind-hearted old woman who gave us permission to use it; next day we found that we were occupying the schoolroom, and the teacher followed by his half-dozen ragged scholars came to take possession. He tried to show us how important it was to have the place at once, but we saw no reason why he could not conduct his class out under the trees just as well as under the shelter; this suggestion offended him very much, so greatly to the delight of the pupils he declared a vacation until the *gringos* should move on.

Chicha, the native drink of Quechuas and Bolivians alike, is a kind of corn-beer; it is made by grinding maize into a fine meal, after which the women and children thoroughly masticate a part of it; water is added to the mass and the thick liquid is boiled several hours, after which it is poured into jars to ferment; it is of a yellow color, has a tart, agreeable taste, and is intoxicating.

The forest at Locotál is somewhat taller than at Incachaca, but the birds are of a similar character. Very abundant and beautiful were the brilliant cocks-of-the-rock; the bright, orange-red creatures flashed through the deep green of the forest like fiery comets and, perching on the low branch of a tree, quietly surveyed their surrounding, or uttered hoarse, croaking calls. This bird is most conspicuous in its natural environment. Among the other large birds were green toucans (*Aulacorhynchus*); the natives hunted them on every possible occasion for the sake of obtaining the bill, which they use as *remedio*, the rasping sound made by rubbing the mandibles together being supposed to be an unfailing cure for epilepsy.

While pursuing our work at Locotál, a man named Quiroga chanced to pass, and begged that we pay him the honor of stopping at his house some distance below; it was a charming place, he said, in the very heart of the wonderful Yungas. We gladly accepted his invitation, and one morning loaded our outfit on mules and started down the trail. For a mile there is only a narrow ledge in the face of a rounded mountain of dark sandstone; a few stunted sprouts, and myriads of orchids covered with purple blooms, have secured a precarious foothold in crevices in the glazed surface; hundreds of feet below, but invisible, the river tears through a narrow gorge. At one point a strip of the shelf upon which we travelled had entirely disappeared; we could not see the bottom of the canyon—its depth was too great—but there were evidences we could not mistake, telling us the history of the gap in the trail. Vultures hovered over the spot and perched on the scant vegetation, and from below came an overpowering stench. What more was needed to reveal the fact that the missing section of trail, in its mad dash through space, had taken with it the pack-train of mules, and probably the men attending them, which chanced to be passing at the time.

Miguelito is only three miles below Locotál, and consists of three or four huts in the centre of a grassy clearing. The Quechuas who live there are friendly, and one may be sure of a welcome for a night's stop.

At five thousand five hundred feet the forest becomes taller and the trees attain a greater diameter. The vegetation of the subtropic zone reaches its highest development at this altitude. After crossing a ridge six thousand seven hundred feet high, the trail descends a long slope into the Yungas, properly known as the Yungas of Cochabamba. At the base of the ridge, and shortly before entering the cultivated area, we crossed the dry, narrow bed of a stream which was filled with rocks bearing the imprints of leaves, and also fossil shells.

Yungas is the name given to the fertile mountain slopes which have been cleared of forest and cultivated; it stretches along the sides of the Rio Yungas for a number of miles, and huts dot the roadside at frequent intervals. When we visited the region in June only the Indian caretakers lived in the habitations, the coca, which is the principal product, having been collected a short time before, and the *propietarios* having gone back to Cochabamba. The owners visit their plantations three times a year, supervise the picking and packing of the leaves and, after a month, return to Cochabamba to sell the drug and live on the proceeds until the next harvest.

After spending an hour in questioning the occupants of the various houses which we passed, we succeeded in locating the house to which we had been invited. It was a low, one-room board structure, open at both ends, and with wide entrances on each side, situated in the centre of a large banana-field. An Indian, so old that he could hardly walk, lived in the hovel and refused to admit us; however, we flourished our letter of introduction from the owner of the premises, took possession, and remained a week. When we left, the aged tenant implored us to remain, as we had daily provided him with all the game he could eat, and had provided him with some medicines that he greatly needed.

Figure 52. An Indian hut in the Yungas of Cochabamba.

The climate at this season, June, was most trying. Although the elevation is only three thousand five hundred feet, the whole region was covered with fog each night, and the cold and damp penetrated everything; during a part of the year the weather is good, and then life in the Yungas is more bearable. We had a trying time at Señor Quiroga's hut, and while the pleasure of investigating a new region is always intense, our joy at leaving was in this particular instance vastly greater.

There is no flat valley along the river, which is of considerable size, and all cultivation is done on the steep mountainsides. Coca is planted in terraces and occupies the greatest acreage; then there are red bananas, plantains, guavas, and sugar-cane.

The fauna of the country seems to represent a transition zone. There are birds typical of the higher country, and others which are common lower down; also, a number found at approximately this altitude only. Near the house, and on the edge of the banana-plantation, was a tall, isolated tree. Flocks of birds, in their flight from one side of the canyon to the other, would invariably alight in its branches for a few minutes' rest. There were many brilliantly colored little tanagers (*Tanagra*) which came to the tree in considerable numbers and chirped and quarrelled as they flitted about examining the leaves for insects, or reached out to pick the small fruits with which the tree was covered; one day not less than seven species of these birds visited this resort within a short time.

Giant orioles (*Ostinops*) were also very plentiful, and travelled in large, noisy flocks. One of the more interesting birds was a species of small, red-tailed parrakeet (*Pyrrhura*) which clung to and crawled up the sides of trees like squirrels; it was almost impossible to see them unless they moved, so well did their coloration, and more particularly their actions, conceal them.

We had travelled to the Yungas on mules owned by the expedition, and upon our arrival turned them loose to feed as usual. Next morning the animals were in a sorry plight; they had been visited by vampire-bats during the night, and bled so badly that we had to send them back to Locotál without delay. Severe as this attack seemed to be, it was mild compared to what we were to see later on. We discovered clumps of the small bats guilty of the execution spending the days under the roof of our hut, and despatched many of them, but this made no impression whatever upon their vast number. People, also, are bitten on any part of the body which is left exposed at night, and I have frequently seen Indians which had been attacked on nose, forehead, and arms.

After completing our work in the Yungas we returned to Cochabamba in order to await more favorable weather for the trip into the lowlands of eastern Bolivia, and to restock our outfit with articles which had been used, and others which it seemed necessary to acquire for the difficult undertaking ahead.

After spending several weeks in the vicinity of Cochabamba, we made arrangements with the mail-carrier which enabled us to travel jointly to the Chaparé. He usually made the trip at six weeks' intervals during the dry season, and, consequently, he knew the trail better than any one else. His *peons* were also accustomed to the country and knew how to adjust packs so they would meet the varying conditions of the road, which is an "art" that can be learned through long experience only.

On July 12 we left Cochabamba. Besides my companion and our personal attendant there were the mail-carrier and his three *peons*; twelve good, strong mules carried the luggage, and there were half a dozen riding and spare animals—quite a cavalcade for the kind of undertaking in hand.

Three days after starting we reached our old camping-spot in the Yungas, and, after stopping for a short chat with the old caretaker of Señor Quiroga's hut we proceeded into what was for us *terra incognita*.

Numerous huts of flimsy construction are scattered along the entire twenty miles or more of cultivated slopes; each has a fenced-in area paved with flat stones upon which coca leaves are dried. We stopped at a number of these dwellings in an attempt to buy fruit or vegetables, but unfortunately the men were all away working in the fields, and any one who has attempted to purchase anything from the average Quechua squaw knows how hopeless a task it is. Although they may have a superabundance of the article desired, they seem to take great delight in refusing to sell anything to a stranger; then the only method to follow is to take what is needed, offer a fair price for it and pass on, leaving them in the midst of their wild rantings; the men are easier to deal with.

The *peons*, and the *patrón* as well, stopped at each hut where the white flag announced that *chicha* was for sale, and attempted to drink enough to last them until their return; after their money gave out they left articles of clothing in payment for the drinks. It was therefore a great relief when the last abode of the intoxicating beverage had been left behind, and we plunged into the wilderness. Immediately after leaving the Yungas we ascended a precipitous slope, the top of which was seven hundred feet above the surrounding country, and then descended on the other side until the elevation was only two thousand feet; here the forest was more tropical in character, and some of the trees, especially the cottonwoods, reached a height of one hundred and fifty feet, and measured twenty-five feet through the buttressed roots at the base.

The day after leaving the Yungas we reached the most dangerous part of the whole trail. After crossing a number of steep, high ridges, we came to an abrupt slope, the side of which is seared by a huge gash where the treacherous white clay keeps sliding constantly into the river, many hundreds of feet below. Each caravan desiring to pass must first cut a ledge in the moving mass of soft, muddy earth, and then hurriedly lead the mules across, one at a time, before the newly made trail is obliterated. The spot is very appropriately named *Sal-si-Puedes* (pass if you can), for any one succeeding in crossing this slide is very apt to possess the ingenuity required to negotiate the remainder of the trail.

That night we made camp early on the banks of the Rio San Antonio, called *Chuspipascana* by the Indians, which means Mosquito River. The altitude of the site is only one thousand eight hundred feet above sea-level. The river was a clear, rapid stream one hundred feet wide, flowing through a rock-strewn bed a quarter of a mile across. Swarms of black flies, sand-flies, and other stinging or biting insects immediately came out to greet us. Birds were very abundant. In addition to the jays, ant-wrens and manakins, which remained in the forest, flocks of parrots and toucans flew across the open spaces. One of the most unusual occurrences was the great flocks of a new species of giant oriole; there were not less than one thousand of these birds in a single flock, and they roamed almost everywhere, coming close to camp to inspect the tents, and to discuss them in hoarse cries of curiosity or resentment. They were beautiful creatures, of a deep chestnut color with light olive-green head and neck; the face is devoid of plumage and of a flesh-color, while the tip of the bill is deep orange. The flesh is highly esteemed by the natives and we found it quite palatable.

As soon as the cargoes had been neatly placed in a pile and covered with a tarpaulin to keep them dry the *peons* cooked their supper; this consisted of a thick soup made of corn-meal and *charque* (dried beef). They had a meal in the morning and another at night; during the long walk throughout the day they chewed coca leaves. The mules were turned loose to shift for themselves, but as plenty of wild cane grows near the rivers, they had an abundance of food. One of the animals carried a bell tied to its neck, and the others would seldom stray out of hearing of the constant clanging. In the morning the men easily located the bell-mule and led it back to camp, the others following in single file. Should one be missing, which was a rare occurrence, it was only necessary to take the bell and shake it vigorously; this soon brought the stray member to the spot.

The remainder of the journey was through the heavily forested lowland; the last of the mountain ridges had been left behind.

Figure 53. The expedition in the Cuchicancha Pass.

During the dry season the caravans follow the courses of streams as much as possible. The water is low, and the wide, rocky margins serve as roads. This is far from being easy on the mules; the animals go stumbling and slipping along, but a good many miles are cut from the total length of the journey. Streams are encountered with frequency, and as one penetrates farther into the interior they become wider and deeper. We crossed not less than six fords in a single day, all between two hundred and three hundred feet wide, the water averaging from three to four feet deep. Although the current is strong, the mules are accustomed to this kind of work and usually manage to cross safely, often stopping in the deepest, swiftest spot to unconcernedly take a drink.

Occasionally, however, one of the animals slips on a moss-covered boulder and falls; then it is a difficult matter to assist the drowning creature to his feet, as the swift water may roll him over, and the weight of the pack keeps him down. In any event, the least result of such an accident is the thorough saturation of everything in the pack, and this means a day's loss of time while the soaked effects are spread out to dry. During the rainy season streams rise with startling rapidity, and parties have often been forced to camp on the river-bank many days until the water went down. To turn back is hopeless, as the last stream crossed is just as high as the one ahead; there is nothing to be done but wait.

Wild animals are particularly abundant in this section of the country. All day long we could hear the raucous scream of long-tailed, multicolored macaws (*Ara*) as they flew two by two overhead. Many hawks sat alertly on dead snags near the water, and black-and-white gulls flapped hurriedly up and down along the river. Occasionally we caught a glimpse of a small flock of muscovies, the largest of South American ducks, as the great, black birds flew heavily up-stream. There were also black guans, resembling small turkeys, which sat quietly in the tops of tall trees until we approached quite near to them; then, emitting a loud, mule-like bray, they set their wings and soared across the river or down into the underbrush. At night the forest was usually quiet, reminding one of "Pools of Silence." Occasionally, however, the still air was suddenly rent by the most unearthly noise that mortal man ever heard, and the woods rang with the wild, insane cackle of forest-rails (*Aramides*). Beginning with a shrill *oohoo-hee-cra*, the demoniacal chorus continued several minutes without interruption, swelling constantly and finally ending with a few low, explosive *cow-cow-cows*. A number of birds always sang together, and the first time one hears the performance it is enough to make the flesh creep and the hair to stand on end; but even after becoming somewhat accustomed to the noise, it falls short of conducing to peaceful slumber, suggesting as it does the agonized shrieks of some tortured spirit of the jungle.

Night-monkeys (*Douroucouli*) were apparently plentiful, but we never saw them in the daytime. After darkness had fallen they began to move about in the tree-tops; one night a troop selected the tree under which we camped for the scene of their frolic, and kept us awake the greater part of the night. They jumped about in the branches, and from the swishing noises which reached us it was easy to imagine them enjoying a good swing up and down upon some particularly springy limb. They dropped leaves and twigs down upon the tent-fly, probably through accident, but perhaps prompted by the desire to find out if anything would happen. At frequent intervals they drew together in a close group to chatter in low, grunting tones, and then, coming to the conclusion that the queer-looking objects below them must be capable of performing some interesting action, again began to tempt fate by showering down more twigs and leaves.

In many places the receding water of the river had left isolated pools; these were teeming with fish of many species; some of them were of large size. A number which we caught had practically the entire tail and fins eaten off; their cannibalistic brethren had no doubt taken advantage of the circumstances in which they were all placed, and begun to devour them piecemeal, at their leisure.

The trees were tall and straight, and there was dense undergrowth near the rivers only. Mosses and epiphytes, so typical of the subtropical zone, were almost lacking, but frequently the wind brought the delightful fragrance of ripening vanilla-beans and the perfume of flowers. Great clusters of scarlet trumpetflowers dangled from the tips of slender vines, and from the tops of many of the trees drooped long garlands of huge white-and-blue flowers that resembled sweet peas; some of these blooms were two inches in diameter. There were also clumps of terrestrial orchids on some of the rocks, with slender spikes of deep purple flowers waving daintily under the impulse of each passing breeze.

Seven days after leaving Cochabamba we came suddenly upon the little cluster of grass and bamboo houses known as Todos Santos; there were exactly seven of them, two of which were of large size, partially enclosing a wide plot of ground carpeted with soft green grass. Tall forest hemmed in the settlement on three sides, and the Rio Chaparé, flowing through deep banks, formed the boundary on the fourth side.

The largest building was occupied by the *corregidor*, or federal agent, who generously provided us with accommodations; in addition to the several living-rooms there was an immense wareroom stored with hides, salt, and other articles of commerce. The remainder of the houses were occupied by families of Bolivians who possessed land or concessions in the neighborhood, and owned numbers of Indians of the Yuracaré tribe; these latter lived in long sheds built in the rear of the dwellings of the people they served. There was also a small church, but no shops of any description. In spite of its inconsiderable size, Todos Santos is a place of importance because it serves as an outlet for commerce from Cochabamba and Bolivia in general, and is the port of entrance for hides from Trinidad, and merchandise entering by way of the Amazon and Madeira-Mamoré Railroad. A small steamer, the *Ana Catarina*, was tied up against the bank, waiting for the water to rise sufficiently for her to proceed down the river; this boat plies more or less regularly between Todos Santos and Trinidad, and requires three days for the downward trip, and five days coming up. From Cochabamba to Trinidad is a distance of approximately two hundred and sixty-five miles, one hundred and sixty-five overland and one hundred on the river.

During the dry season steam-navigation on the Chaparé is very irregular, but canoes of large size and native paddlers may always be had. During the rainy season there is a small steamer or launch each fortnight.

Several years before, the government had by law abolished the practice of keeping Indians in the condition of semislavery, and had ordered all owners to turn them over to the missions; this, however, had not been done, and each Bolivian family living at Todos Santos had a number of Yuracarés in its service. Not far from the settlement were a number of clearings, some of considerable size, where fruits and vegetables were cultivated for the benefit of the *amos*, as the owners of Indians are called; the Indians cleared the ground, cultivated it, and then brought in the results of their labor, receiving nothing in return. They seemed fairly contented, however, and did not appear to be suffering from ill treatment. They frequently spent days at a time in their shelters on the edges of the fields, or in hunting and fishing trips far from their homes.

Each Yuracaré woman kept a number of Amazon parrots which she looked after carefully and refused to sell, even at a good price. Upon asking the reason for this I was told that they reared them for the sake of the tail-feathers, which are in great demand by the Aymarás. Each parrot will grow three "crops" of feathers a year, each of which is worth fifty *centavos*. The Aymarás from the vicinity of La Paz send down agents at regular intervals to purchase these feathers, as they use them in making ornaments worn during their annual festivals.

In the branches of one of the tall trees near the village a neat little hut had been built of bamboo and leaves, reminding one a great deal of a Philippine tree-dwelling. Indians armed with bows and arrows would conceal themselves in this house, forty feet above the ground, and shoot many of the birds which came to feed on the fruit covering the tree; other Indians, hidden about the base of the tree, watched where the birds fell, gathered them up and skinned or plucked them. In this way quite a number could be shot without alarming a feeding flock.

The forest around Todos Santos abounds in wild life. Squirrel-monkeys (*Saimiri*) are very numerous and travel in troops of from twenty to fifty individuals; we saw them daily, playing about in the trees, and feeding on fruits, buds, and insects. They are delightful little pets, and one that we owned spent the greater part of the day catching the mosquitoes which infested our habitation. It searched every nook and crevice for insects, and one of its chief pastimes was to look through a pack of cards in the hopes of finding mosquitoes between them. Harpy eagles also are very plentiful, and feed on the squirrel-monkeys to a great extent, as they are easy to catch. However, monkeys are not the only animals which suffer; we one day found the remains of a sloth which had been dropped by an eagle, the entire fore part of which had been eaten away.

There were numerous trees covered with vivid-scarlet blossoms, scattered throughout the forest, and forming gaudy little islands of color, which stood out very conspicuously amid the green tree-tops. These trees are known as *madres de cacao*, because they are frequently planted in cacao-groves to shield the young plants from the sun. The flowers contain so much nectar that numbers of birds feed upon them, including parrots, macaws, and orioles; when the brilliant blooms fall into the river they are greedily snapped up by fish.

Of small birds there was such a variety that it would be impossible to mention all of them, but one in particular deserves attention. It is a species of manakin called the "child of the sun" by the Yuracarés, who look upon the tiny creature with reverence and would not harm it under any circumstances. The bird is not as large as a sparrow, but is of stocky build, with a bright orange-red head and neck, the remainder of the body being black. As it whirs from branch to branch it makes a loud sputtering, crackling noise which reminds one of a bunch of small, exploding firecrackers. The female of the species is of a dull-green color.

At Todos Santos, as elsewhere, local migrations of birds in the heart of the tropics were several times forcibly brought to our attention. We had been hunting in the forest a number of weeks and were pretty well acquainted with its inhabitants; suddenly a species entirely new to us appeared in great abundance in all parts of the region; each member of the expedition, including the native assistants, brought in specimens of it the same day. This can be explained only by the fact that flocks of these particular birds had arrived suddenly from some distant part, probably attracted by a fruit or insect which chanced to be plentiful at the time, and upon which they fed.

Several miles from port, and entirely concealed by the forest, stretches a lagoon of considerable size; it is connected with the Chaparé by a small, brush-clogged creek, but the water is stagnant and filled with decaying vegetation and detritus. Masses of bushes and swamp-grass grow all along the borders, and in some sections the surface of the water is covered with floating, aquatic plants. As may be supposed, many species of birds live both about the water and in the dense thickets that line the banks. Among the former was the rare little sun-grebe, but it was by no means abundant; the few solitary individuals we saw were always surprised out in the open water and, after giving a series of hoarse, loud cries, either flew or swam as rapidly as possible to the nearest clump of vegetation, which offered a secure retreat. Graceful jacanas stepped about daintily on the lily-pads; their toes are very long and give the feet a wide spread, thus enabling the birds to walk on the floating little islands of water-hyacinths and wild lettuce; for this reason the natives call them *pájaro de Jesu-cristo*, because they can "walk on the water."

Several species of flycatchers and large, noisy wrens (*Donacobia*) lived in the partly submerged bushes; we found several of the bulky, domed grass nests of the former, but it was almost invariably impossible to reach them as they always harbored colonies of biting ants, which rushed out in maddened frenzy when the nest was touched; however, the birds and ants seemed to live in perfect harmony.

In the tangles of tall bamboo growing on the bank and drooping out over the water lived flocks of hoatzins and numbers of several species of *dendrocolaptine* birds or wood-hewers; also an occasional water-turkey and cormorant. Many black-and-white ibises soared above in circles and at a great height; they acted not unlike vultures, but the long, outstretched neck and legs immediately gave a clue to their identity.

The forest was full of surprises. One morning my companion encountered a tamanduá ant-eater which was on the ground, and refused to realize that the close proximity of man meant danger; he was but lightly armed, and shot the tough, thick-skinned animal with the 32-bore auxiliary tube of his shotgun, and number 12 shot—an unheard-of feat.

It was, however, not always necessary to go into the forest to hunt; the open plot in which the settlement lay attracted many birds, such as scarlet tanagers, vermilion flycatchers, swallows, and others, which were never found in the forest, and small mammals in abundance lived in the houses. We frequently caught five species of rats in a single house in one night, and at least two species of bats lived in the palm-leaf thatch of the roof. Some of the rodents, particularly a large, spiny rat, lived under the floor, while others made the walls and ceiling their homes; each species seemed to adhere more or less to its own part of the dwelling, thus dividing the houses into well-defined "life-zones."

The natives are very fond of the flesh of the spiny rat and often begged for any which chanced to come to our traps.

Ocelots were not wanting in the neighborhood; they visited the hen-houses occasionally at night, but never entered by the doors, preferring to tear holes in the side of the structures; they killed a large number of fowls, on one occasion nearly twenty on a single visit, and prompted apparently by the mere lust for killing.

At night vampire-bats came out in hordes; they attacked everything from human beings on down; even the few miserable pigs kept by the Indians were severely bitten and kept up a continuous squealing as the bloodthirsty creatures settled on them, usually at the base of the ears, and began their painful operations. The worst sufferers by far, however, were our mules. As soon as the sun set our *peons* brought the animals to the corral and strapped canvas covers over them; this precaution was of little avail, for the bats attacked all exposed parts, causing the mules to kick and roll, with the result that their covers were soon torn off. We went out frequently to watch these obnoxious creatures at work; after circling above their prospective victim a few times, they dropped suddenly, usually upon the neck or flanks, and at once began to bite and suck, making a grating sound with the teeth all the while. They paid no attention to us, although we stood but a few feet away, but clung with folded wings to their prey, perfectly motionless and in an upright position; if we moved they uttered a few squeaks, but made no attempt to fly until we reached for them and came to within a few inches, when they reluctantly fluttered up, but almost immediately settled on the other side of the animal. Desiring specimens of them for our collection, we went one night to the corral armed with a butterfly-net and, approaching one of the mules on whose back were a dozen or more bats, made a hurried sweep with the net; as the large, white bag of netting scraped the back of the nervous animal he sank to his knees with a groan of despair, wondering, no doubt, what new monster had swooped down upon him to add fresh suffering to his already unbearable existence.

In the morning the mules were in a pathetic condition; blood continued to flow from the wounds made by the bats' sharp teeth, so that the ground was red and the animals were covered from head to foot. It was always necessary to take them to the river and wash them, then disinfect the numerous punctures; if this is not done flies attack the sore spots, infesting them with their larvæ, and the animals die of blood-poison. After three nights we were compelled to start the mules back to Cochabamba, as they were on the verge of exhaustion.

Figure 54. Vampire-bat from Todos Santos.

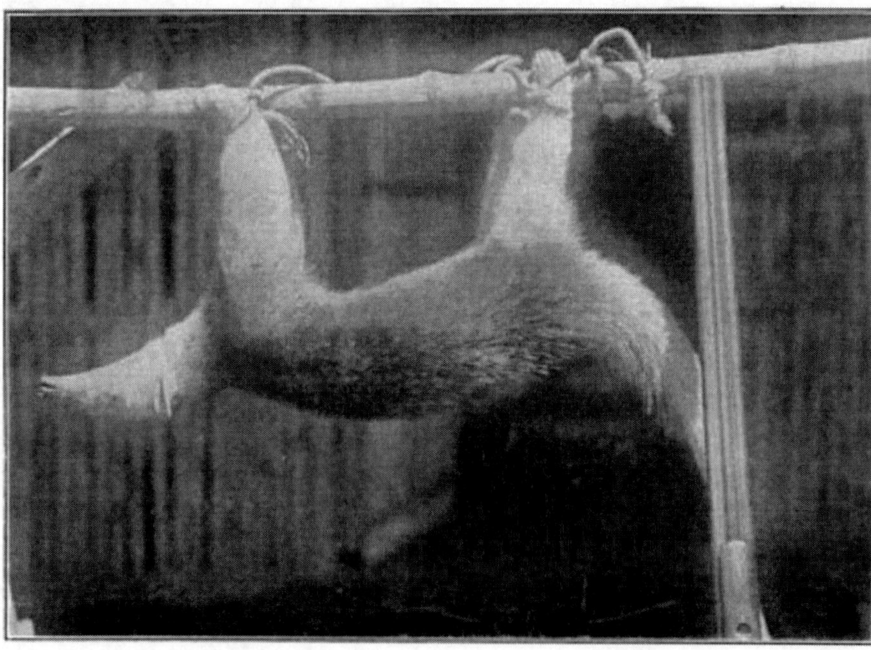

Figure 55. Tamanduá ant-eater.

While at Todos Santos we learned of a mission among the Yuracaré Indians about twelve miles distant, and near the Rio Chimoré. We expressed a desire to visit it, but the *intendente* told us that such a move was impossible. He said that the priest in charge of the mission was absolute monarch of the territory under his control; that he would permit no one to come near his retreat, and that this mandate had never been disobeyed. Such statements made the place seem of especial interest to us, and we were eager to go there at almost any cost; we devised many plans which we hoped would lead to an interview with the priest, but all of them failed miserably; finally, however, the opportunity came to us in an unlooked-for manner. A misfortune to one person frequently comes in the guise of a blessing to another, and so it happened in this instance. As we were pursuing our work one afternoon in the open corridor in front of our room, a long canoe drew up at the river-bank, and the priest, followed by a dozen Indians, stepped ashore and marched across the clearing to the *intendente's* quarters. We immediately recognized him as Padre Fulgencio, the missionary of whose despotic rule we had heard so much; but he did not even glance at us as he passed. While debating upon some diplomatic move which might serve as an excuse for an interview, for now or never was the time to obtain the coveted permission, he suddenly emerged from the house and came straight to us. A few curt remarks were exchanged, and then he began to relate his trouble. To make a long story short, he was suffering from a severe toothache; it had kept him awake many nights, and at last he was forced to come out of his retreat in search of a remedy. The *intendente* could do nothing for him; could we?

How I thanked my lucky star for a limited knowledge of medicine! After an examination, conducted with much formality, the trouble was pronounced curable. He submitted bravely to the injection of cocaine, and soon after was relieved of the aching member. Tears rolled down his cheeks as he expressed his gratitude, and then, taking note of the work upon which we were engaged, he suddenly asked: "Why don't you come to the mission; I have four hundred Indians who spend several days each week in hunting, and they can take you anywhere, and also bring you all kinds of animals."

We needed no urging, and within five minutes the day was set when porters in abundance should come to convey our equipment, and we should start on our journey to the mysterious stronghold of Padre Fulgencio, and the boundless jungles bordering the Rio Chimoré.

Chapter 19

Among the Yuracaré Indians of the Rio Chimoré

True to his promise, Padre Fulgencio sent the Indians to Todos Santos, and on the morning of August 2 we packed into canoes such of our equipment as was necessary for the trip and started across the brown water of the Chaparé.

On the other side of the river there was no clearing; the trees grew down to the water's edge, and the moment the canoes were left behind we plunged into the perpetual gloom of the forest.

An indistinct trail led into the heart of the jungle. The Indians adjusted our belongings on their backs, securing them with broad strips of bark placed across the forehead; then they set out at a good pace, a number of women and children carrying boiled yuccas and plantains, trudging at the rear of the procession.

There was not much undergrowth, but the ground, from which there is little evaporation on account of the dense canopy overhead, was very muddy. Every few rods we came to a deep streamlet which had to be crossed on the trunks of fallen trees; some of these slimy bridges were sixty feet long and almost impassable to us, but the Indians strode across as unconcernedly as geckos. Half-way to the mission the Indians stopped for lunch and a short rest, and by noon we reached the edge of the clearing, having covered a distance of twelve miles.

After a tramp of half a mile through weedy fields of maize and yuccas, we reached the mission-buildings—a few dozen low grass huts clustering around an open square. At one end rose two structures of large size which served as the church and general meeting-place. Near the centre of the clearing a stately cross had been erected, hewn from the heart of a giant ceiba.

The priest was delighted to see us and spared no effort to make us comfortable. We were soon installed in a room of one of the buildings which served as a boys' dormitory, and a short time later started out to inspect our surroundings.

At first the Indians were reticent and would peer at us from a distance. This was true particularly of the children, but as the days wore on we made friends with them, and from both the people themselves and the priest we learned a great deal about their history and habits.

The name Yuracaré, according to D'Orbigny, was given to them by the Quechuas, and means "white man"; this is most inappropriate, as they are of a decided brown color, although perhaps averaging lighter than the Quechuas. They were first discovered by Viedma in 1768.

At the present time, at least, the Yuracarés are a people of the hot, humid lowlands. Those who have not been captured and brought to the missions, or who escaped the unenviable fate of having been taken from their forest home by private "slaving expeditions," live along the smaller branches of the streams, which eventually find their way into the Mamoré; this includes particularly the Chaparé, Chimoré, the Ichilo, and the Isiboro.

There were about four hundred Indians residing at the mission. Although attempts have been made intermittently to civilize these people for more than a hundred years, there were long intervals when the work had to be abandoned, and the families naturally returned to their homes in the wilderness. Nearly all of the present aggregation had been brought together during the last few years. Newcomers are added to their number frequently. The priest, learning of other families far up some unmapped *quebrada* or streamlet, takes a few of the men who have learned to place confidence in him and whom he trusts, and starts forth on long canoe voyages in search of them. They approach the hidden dwelling suddenly, surround it, and persuade the occupants to accompany them immediately, giving them only an hour or two in which to collect their few belongings. Occasionally the Indians whom they seek learn of the approach of the emissaries and hide before their arrival; then the priest returns to the mission, his long trip having been made to no purpose. When, should the expedition prove to be successful, the families have departed to the waiting canoes, their huts are burned and the plantations destroyed. Knowing that neither home nor food have been left behind, they are not so apt to run away from their new quarters and go back to their old dwelling-places. I heard of no instance where they resisted this deportation.

The Yuracarés are a tall, well-built people of a rather docile disposition; however, the older generation never wholly becomes reconciled to the new mode of life, and remains at the mission only for reasons which I will explain later.

In their wild state they live in small family parties, obtaining their subsistence from the forest, which abounds in game, and from their fields of yuccas. Their native costume, a long, shirt-like garment called *tipoy*, is made from the fibrous bark of a tree; at the mission this has largely been replaced by cotton clothes. Each family has been provided with a separate hut of adequate size, where the parents and very small children live. The boys and girls over five or six years of age are under the constant supervision of the priest, and attend his classes; at night they sleep in separate locked dormitories, which prevents their returning to their homes, and also keeps the parents from running away, as they will not leave without their children.

Padre Fulgencio also explained that this kept them from observing and copying the customs of their elders. He recognizes the impossibility of reclaiming the forest-reared savage, and devotes practically all his efforts to the younger generation.

The Indians marry at an early age, the boys at sixteen and the girls at fourteen. In their wild state each family rears four or five children; at the mission never more than two, and frequently none at all. Should the first-born be a girl, she is permitted slowly to starve to death. The priest has inflicted severe punishment upon them in his efforts to break this custom, but so far all his work has been in vain.

As far as possible they are discouraged in the celebration of their native festivals, but it frequently occurs that the entire populace appear with faces gayly decorated with black and blue dots, and all join in weird songs and dances, the purpose of which remains a secret, as they cannot be induced to tell. They worship no divinity, being in this respect in a class almost by themselves.

Figure 56. Yuracarés chewing yucca-roots for making casire.

Food at the mission is abundant. The clearing comprises several hundred acres and is planted in maize, rice, yuccas, plantains, and sweet potatoes. Like most savages, they have an intoxicating drink, made of the boiled root of the yucca. The women dig great quantities of it, peel and thoroughly cook it, after which a certain per cent is chewed and expectorated into a huge earthenware jar; the remainder is mashed and thrown in also, and water added. The following day fermentation has started and the greenish yellow liquid is ready for use.

Figure 57. Yuracaré women and children.

At the mission the Indians have learned the use of salt, and this fact perhaps as much as any induces them to remain, for deprived of it they cannot long exist. A small amount is given to each individual at stated periods—only just enough to supply his wants until time for the next distribution. There are instances on record where families have escaped and gone back to their nomadic life for eighteen months, then returned voluntarily to promise future obedience, so great was their craving for salt.

The rites attending the death and burial of a man are among the curious and persisting ceremonies of the Yuracarés. When the husband dies the wife removes all her wearing apparel and casts herself upon his body, where she remains weeping and lamenting until the time of the funeral, which is a day or two later. All the women squat in a circle around the deceased, raise their voices in sorrowful wails, and recount the heroic deeds and good characteristics of the dead. The men drink *casire* and dig a deep hole in the ground; when the time for the burial arrives the body is carefully deposited therein, together with all his possessions, and the wife's clothing is placed on top, after which the earth is thrown in.

The weapons of this tribe consist entirely of bows made of *chonta*-palm wood, five or six feet high, and various kinds of arrows. The shaft of the latter is always composed of slender bamboo, but the points vary greatly; thus for large game there is a long double-edged blade of another variety of bamboo; slender, barbed points of *chonta* are used for birds, and a long, sharp spike of palm-wood for fish. They are wholly ignorant of the use of the deadly *curare* poison.

We were fortunate in timing our visit to the Chimoré for the dry season. Additions were being made to the already large areas under cultivation, and for this purpose the Indians were cutting down forest. They were required to work four days each week, the remaining three being devoted to fishing and hunting. All the men and boys participated in this work and seemed to enjoy it thoroughly. At first the undergrowth was removed; this naturally led to the discovery of many strange animals, all of which were promptly brought to us for examination. The number and variety of snakes was astonishing; even after having spent years in a similar type of country, I had never suspected that so many existed, which shows how inconspicuous they are until one actually goes over the ground with a comb, as it were. They captured green boas, several species of the fer-de-lance, and many others which we did not recognize. Some of them were poisonous, and others were innocuous. Among the former was one which in appearance closely resembled the green boa, but its attitude was defiant and even aggressive; examination showed that it possessed long fangs. One day several of the men came running into our room and shouted "*Pisisi*." We followed them to the clearing, and found that they had discovered a huge bushmaster coiled under a log. They tried to drive the reptile out with long poles, but it refused to move; finally the priest pulled the enormous creature bodily from its hiding-place with the aid of a hooked stick; it was very sluggish and made no attempt to strike. After shooting it we found that it measured nearly seven feet in length, with a diameter of five inches. The fangs, over an inch long, emitted about a tablespoonful of yellowish poison.

The bushmaster, called *surucucú* in Brazil, is truly a terrible creature. It grows to a length of ten feet or more, and attains a great thickness. A snake of that size has fangs an inch and a half long and injects nearly a tablespoonful of poison at a single thrust. The ground-color is reddish yellow crossed by black bands, sometimes forming a series of X's along the back. It does not take kindly to captivity and dies of starvation after a few months of confinement. It is one of the few snakes which are supposed to incubate their eggs. After selecting a hole in the ground or in a stump the reptile lays a dozen or more eggs; then it coils up on top of them and does not leave the vicinity until they hatch; at such times it is very irritable, and will strike with deadly results any creature which disturbs it. The poison acts rapidly, and I heard of a case where an Indian died in less than half an hour after having been bitten.

There were also small brown salamanders and lizards with spiny backs that resembled horned toads. Perhaps the rarest catch of all was a splendid example of the curious cane-rat (*Dactylomys*), an animal seldom encountered on account of its rarity and secretive habits. It resembles a large rat, being twenty-five inches long, and of a dark-gray color; the toes are divided into pairs in order to enable it to easily climb slender stalks, and instead of claws it has nails. The pupils of the eyes are elliptical, like a cat's; when annoyed it uttered a hoarse scream, a sound occasionally heard at night, but which we did not heretofore recognize.

After the brush had been removed for the distance of a hundred yards or more from the edge of the clearing, the Indians began to cut down the trees; some of these were of enormous size, especially the ceibas; one that I measured was twenty-five feet through the base, counting the supporting, bracket-like roots, and fifteen men hacked at it at the same time. When the tree fell they set up a wild cheering and took great delight in watching this monarch of the forest tumble to the ground.

Three days of each week were devoted to hunting and fishing. Usually the Indians went many miles away, in small parties, returning promptly at the expiration of their time. The children rarely accompanied them, and then only after having obtained special permission from the priest. Upon their return they brought baskets of fish and meat—enough to last them until their next journey into the wilds. Nearly all fish and game were taken with bow and arrow. To secure the former they selected a small creek up the shallow water of which huge shoals of fish went to feed, and then shot them. After a sufficient supply had been obtained they erected a framework of sticks, built a fire under it and slowly roasted and smoked them; later they were packed in baskets between layers of green leaves and taken home. They also brought numbers of freshly killed animals for our examination, for in keeping with his promise Padre Fulgencio had announced from the pulpit that all creatures taken by them were to be shown to us first, and we were permitted to select any that were of scientific value. In this manner a number of animals new to us were added to the collection.

The curl-breasted toucan (*Beauharnaisia*) is one of those birds of the Amazonian basin which is seldom seen by travellers, or even naturalists, who make every effort to learn something of its habits. Bates records having seen a number during his eleven years of exploration, and on one occasion he was attacked by a flock after he had wounded one of them. We therefore considered it an unusually good streak of fortune to find a large flock inhabiting a section of the forest several miles from the mission. They were wary, nervous creatures, and spent their time in the top of tall trees from which one of our men succeeded in shooting several with arrows before the remainder took alarm and flew away; they never returned to the locality. The bird is black above, with yellow underparts barred with black; the feathers on the top of the head are flattened and curled, resembling shining scales, and are drawn together to form a ridge. On the throat and breast the brilliant yellow feathers are tipped with glossy black dots, resembling beads of jet. Unfortunately they were not nesting, but the Indians reported having found the two white eggs in cavities in the taller trees. Another bird not frequently encountered is the giant frogmouth (*Nyctibius*), which, while not so rare, perhaps, is seldom seen, as it is nocturnal in habits and spends the days squatting horizontally upon some thick branch, where it resembles a gray lichen, or is altogether invisible. When the time for domestic cares arrives the bird lays a single white egg on the branch which has served as its perch, or at the junction of a limb and the tree-trunk, without making any sort of a nest. Doubtless many eggs roll off this precarious location and are broken. It feeds upon beetles and insects which are caught on the wing, and some observers say that it also catches small birds; this latter I am inclined to question. One individual that we collected was twenty-two inches long, with an expanse of wings of thirty inches. The mouth when opened measured five inches from tip to tip of the bill, and was three inches wide; but the œsophagus was less than half an inch in diameter, which would prevent it from swallowing anything larger than a humming-bird.

The nights at the mission were always pleasant. The priest usually conducted a short service in the chapel, and then we sat in front of his hut for an hour's chat, while the children romped and played before being sent to bed. Sometimes one of the boys brought out a queer drum; the ends were made of skin taken from the neck of a jabiru stork. He beat it in slow rhythm, swaying his head from side to side with each low thud. The girls placed their arms around one another's waists, forming lines of threes, and shuffled forward three steps and back, swinging their bodies all the while; suddenly they would whirl around once, take hold of one another's hands, and then the long line swept around at such a rapid pace that the individuals at the ends invariably went sprawling some distance away. After tiring of this or any other pastime upon which they were engaged, they lined up and said a "*Buenas noches, Padre*," in chorus. Then they ran away to the sleeping quarters.

After spending nearly two weeks at the mission we accepted the priest's invitation to accompany him on a short trip down the Chimoré. Twenty young men and boys were selected as paddlers; they started early one morning, taking all of our personal luggage with them; a large number of girls and women followed soon after, carrying baskets of plantains, yuccas, and other provisions. The missionary, Boyle, and I brought up the rear, and encouraged the few stragglers we met on the way, for the distance from the mission to the river is three miles, through the virgin forest.

The Chimoré is of about the same width as the Chaparé, although the water is in normal times somewhat clearer. It rises far to the south and is formed by the junction of the Blanco and Icona. Some distance below it unites with the waters of the Ichilo, a mighty river flowing from the south, through a solitary and unknown wilderness, and up which Padre Fulgencio had ascended a number of miles on a previous trip. In latitude 15° 30′ South, the Ichilo and Chaparé join, and form the Rio Mamorecillo, which lower in its course is known as the Mamoré.

The meaning of Mamoré, which is a Yuracaré word, is "mother of the human race." They have a legend to the effect that far away, at the source of the Sajta, which is the beginning of the furthermost tributary of the mighty river, there are three rocks of pyramidal shape that rise in terraces, one above the other, and in the heart of which the stream rises. In the very beginning of things this rock gave birth to the first people, for which reason it is called “Mamoré.” Later the name was also given to the river because its water, teeming with fish, supplied them with food and offered an easy highway for the dissemination of the race.

Arrived at the point of embarkation, the men began to load the five canoes which were waiting, and the women built a fire and cooked lunch. In a short time everything was ready and the canoes moved easily down-stream. The paddlers were adepts at their work, and as a good deal of rivalry existed between the different crews, they kept up an almost continuous race, with the natural consequence that we made good time. The scenery along the Chimoré is exactly like that on the upper courses of the many rivers of tropical South America; there is the same monotony of the yellow water highway, flanked by walls of deepest green. One thing that impresses the traveller as much as any other is the immensity of the silent, uninhabited areas; and also their comparative worthlessness. For days and even weeks one may enter deeper and deeper into the heart of the undefiled wilderness, and see always the same dark forest, the hurrying, mysterious streams, and the rafts of low, threatening clouds; hear the annoying buzz and feel the poisonous sting of the insect swarms, and swelter in the humid, enervating climate. The greater part of this country can never be cultivated to any extent, as the annual floods cover it to a depth of many feet; there are very few eminences safe from the inundations, and these are of inconsiderable size. The person who pictures the untrodden tropics as a paradise of fruits and flowers, teeming with gorgeous-colored creatures and inhabited by tribes of gracious Indians whose one desire in life is to serve the traveller or explorer, has yet to cut his eye-teeth in the field of exploration.

Our Indians were all well armed and frequently took long shots at some of the creatures that ventured to show themselves in the early mornings or just before dark. They were expert archers and even shot large birds on the wing as the flocks passed overhead. Occasionally an otter appeared, always a hundred yards or more away, swimming rapidly with only the head showing above the water. These animals were favorite targets, and from my seat in the middle of the canoe I had an unobstructed view of the arrows in flight as they left the bow of the man in front; he did not aim at his prey, but quite some distance above it. At the twang of the bow the arrow sped into the air, ascending slightly at first, and then dropping as it approached the mark; it described a curve exactly like a bullet fired from a rifle, and remained in a perfectly horizontal position during the entire flight.

When making camp on a sand-bank the Indians stuck their bows and arrows in the ground, near the shelters; this prevented their being stepped on and broken. In damp or rainy weather the arrows warped badly, but it was only a few minutes’ work to heat them near a fire and bend them back into alignment.

Tropical rivers are noted for their treachery. One can never be certain of their actions or character, even a few hours hence. We had a striking example of this on the Chimoré. Camp had been made on an extensive sand-bank one day at noon, as we planned to spend a few hours hunting and fishing in the neighborhood. The sun shone brightly and there was nothing to indicate a change of conditions in any manner whatever; but scarcely had the canoes been unloaded and a fire built over which we intended to do the cooking when we were startled by a dull roar that grew louder with each passing second; before we had time to hurriedly gather our belongings and throw them into the canoes a foam-capped, seething wall of water was upon us, sweeping down the river and carrying away everything in its path. As the tidal wave, several feet high, dashed over the sand-bank, the imprisoned air shot up from the great cracks and rents in the sun-baked earth, and set the raging mass of muddy water to hissing and boiling. In a few minutes only the higher mounds of sand projected above the roaring inferno, and against these hungry tongues of water lapped greedily until their bases were undermined. Then the whole mass crumbled and disappeared in the seething flood. Where our peaceful camp had stood but a few minutes before there was now a sea of agitated water. The explanation of this phenomenon is simple: A heavy rain had fallen in the mountains where the tributaries of the river rise, and the torrent of water dashing down the precipitous slopes had rushed into the lowlands. After this the water was so muddy that it was unfit for use without special preparation. In order to secure a supply for drinking and cooking we boiled a quantity of it; the sand was quickly precipitated to the bottom as the temperature rose, after which the clear water could be poured off the top. In some instances the amount of solid matter carried by the water was fully 50 per cent of the total volume.

Animals were not abundant on the river-banks, although we saw a deer or a small flock of curassows at infrequent intervals. If we went into the forest a short distance, however, we were sure to find game in abundance. On one occasion the Indians demonstrated their skill at calling up monkeys. A large troop of cebus and squirrel-monkeys were feeding in the tree-tops, but for some reason the men did not give chase as they usually do; they concealed themselves in the thick lower growth and whistled a few plaintive *kee-oows*. In a short time the animals began to evince a great deal of interest or curiosity, and many of them descended to the lower branches. Then the hunters shot a number with their arrows before the band realized what had occurred and took flight.

Large areas covered with an impenetrable cane-jungle are scattered all along the borders of the Chimoré. The tall stalks rise to a height of fifty feet or more, and are beautiful to look at, but impossible to penetrate until a trail has been cleared with hatchet or machete. The plant resembles the well-known sugar-cane of our Southern States, but grows much taller, and the stems are thin and hard. A large, white, feathery plume crowns each stalk. This plant is of inestimable value to the natives. The long poles are used almost exclusively in constructing their dwellings, and the leaves make an impervious thatch. Practically every stalk is infested with thick white grubs which live in the pith. These are extracted by the Yuracarés, who call them *chata*, and used for bait when fishing. Many runways perforate the matted growth; these have been made by capybaras, agoutis, and numerous other animals. Even tapirs seem to appreciate the protection afforded by the thick cover and resort to it in the daytime, while jaguars noiselessly steal along the paths in the course of their nightly prowls. One night we had an excellent illustration of how useful the cane-plant, or *chuchilla*, as the Indians call it, can be in an emergency. We had landed on a sand-bank rather early in the evening, spurred to this action by rapidly approaching black clouds, flashes of lightning, and the rumbling of distant thunder which bespoke the arrival of a tropical rain-storm. At first it looked as if we should be compelled to endure a thorough drenching, but Padre Fulgencio issued a few orders to the canoemen, and they hurried away to neighboring cane-brakes, with machete in hand; soon they returned, dragging an immense quantity of the plant; four of the strongest poles were firmly planted in the sand to form a square, about fifteen feet apart, and the tops bent over and tied together with strips of their leaves. These served as the corner posts of a shelter. Other stalks were laid across the top to form rafters, and firmly tied. The men then piled many more on top, binding each one to the rafter, until a complete hut had been built; although the height of the roof was fully eight feet, the ragged edges came down to the ground, entirely enclosing the sides and forming a snug retreat against which the elements raged without avail. After the first deluge had subsided other and smaller shelters were built. The Indians enjoyed the experience thoroughly; they threw aside all clothing, built fires over which fish and game were placed to roast, and squatted around the embers in a circle, doubtless indulging in pleasant reminiscences of the days before civilization with its restricting influences, and had come into their care-free existence.

Early next morning we were awakened by the reverberating howls of monkeys. The Indians rushed in a body from their shelters and, snatching up bows and arrows, ran in pursuit. A troop of red howlers had come to the *chuchilla* near our shelters; we could see none of the animals, but the tops of the canes waving as if agitated by a violent gust of wind told us of their whereabouts. Soon we heard shouts followed by the twang of bows and the snarl of arrows as they ripped through the flesh of the luckless victims. This continued until the creatures disappeared in the interior of the dense jungle, and then the hunters returned, dragging their quarry after them. We were eager to continue on our way, but in view of the efficient and willing service rendered by the men the night before it was decided to wait a few hours and permit them to have a feast. A huge fire was built, and the monkeys, after having been skinned and washed, were set on spits to roast. The Indians crowded around, sang and shouted, and tore off and ate chunks of the half-roasted flesh. In a short time our orderly Yuracarés had returned to the realms of savagery, and were indulging in a performance such as I had repeatedly seen among the wild Nhambiquaras of Matto Grosso.

Lower down we saw numerous islands, some of large size and of a peculiar formation. The river, which had risen so rapidly a few days before, had gone down to its normal level and left these obstructions in the channel exposed high above the surface. A matted mass of logs and branches of which a layer fifteen feet thick protruded above the water, formed the base of the islands; on this soil had gathered to a depth of five or six feet, and supported a luxuriant growth of vegetation. These islands are composed of deposits of driftwood which were left stranded on sand-banks during the season of high water, and while the edges are torn and jagged the force of succeeding floods seems to be of insufficient strength to wash them away. As we paddled along quietly near the banks the priest or the Indians pointed out many interesting and curious plants. One of these is the *palo santo*, or holy tree; it grows to be a great height, but the trunk is comparatively slender. The peculiar name is derived from the fact that it is as carefully guarded as any sacred object should be, but in this instance by myriads of fire-ants, which live in the hollow interior of the trunk. If the tree is struck sharply with a stick the ants pour out in endless files through minute openings. They are vicious insects, and the bite smarts and burns many hours after it is inflicted. The *tacuara*, a species of tall, feathery bamboo, is another interesting plant of this region. When the stalk is cut down the leaves shrivel and dry within a few minutes. Large numbers of cabbage-palms grew throughout the forest. The beautiful, plume-shaped leaves droop in a great umbrella-like mass from the top of a column sixty or seventy feet high; thick clumps of straight, tough roots branch out eight or ten feet above the ground and form a solid support to the stem. A delicious salad is made from the tender leaves, folded up in the bud; or if boiled the flavor is similar to that of asparagus. To secure the bud it is, of course, necessary to cut down a tree which has taken the greater part of a century to mature, but in a region where many millions are growing one is not inclined to be sentimental, and will only bemoan the fact that it requires an hour's hard work to chop through the steel-like trunk before the coveted morsel is brought down.

The country between the Guapay and Ichilo is probably as little known as any part of South America. This strip of land, covering approximately five thousand square miles, is heavily forested, and is the home of a tribe of savages known as the Sirionós. Judging from the accounts given to us by our canoemen and the priest, they must be a terrible and indomitable race. The Yuracarés fear them greatly, and as we neared the Ichilo they preferred to keep the canoes in the centre of the river and seemed reluctant to land; if they shot at an animal and the arrow missed its mark and dropped in the forest they did not go in search of it; a half-day of careful work is needed to make an arrow, and as a general rule Indians are very particular to hunt for any they may lose; but in this instance they preferred the loss of the arrows to risking their skins in the dense cover.

There were four Yuracarés at the mission, one, a girl of twelve years, who bore unsightly scars—the result of having been ambushed by parties of the Sirionós tribe; I was also told that occasionally some of them are killed.

The Sirionós seem to have no permanent homes and cultivate the ground to a very limited extent, if at all. They are a tribe of wanderers, and roam the forest in small parties, killing game for food. In appearance and stature they are not unlike the Guarayos, but in temperament they are totally different and have successfully resisted every attempt made to subdue them. Their weapons are bows and arrows, the former of great height and so powerful that they cannot be drawn with the arms alone. In order to shoot the Indian throws himself on his back, grasps the bow with the feet and draws the cord with both hands; the arrows, of which the priest had collected a number, are seven or eight feet long and made of wild cane or *chuchilla.* Apparently they are unacquainted with the use of poison.

Probably the Guarayos suffer more at the hands of the Sirionós than the Yuracarés, because the former two tribes come in contact more frequently. Padre Wolfgang, in charge of one of the missions of Santa Cruz de la Sierra, was on one occasion attacked and several of his men were captured. A few days later he found them nailed to trees with numerous long thorns. On another occasion he surprised a party of Sirionós and succeeded in capturing seven; these he took back to the mission, but they proved to be intractable. He found it necessary to tie them to posts in order to prevent them from escaping. They steadily refused food and water, and after a few days four died of starvation and sullenness. The priest took pity on the remaining three and released them.

After five pleasant days crowded with interesting and unusual experiences on the Chimoré we returned to the mission and spent a few days there packing the valuable zoological collections. We then went back to our base on the Chaparé; for this journey we decided to go by way of the Coni, a small stream emptying into the Chaparé, a few miles above Todos Santos. We followed a path through the forest for a distance of three miles, arriving at a large clearing which was planted in sugar-cane; but what surprised us greatly was the fact that the owner was a Quechua, who had deserted his home in one of the high valleys near Cochicancha, and had come to live in the hot tropics. He had constructed a crude wooden mill for expressing the juice from the cane-stalks, erected a still, and was making alcohol. We had gone to the mission with the intention of remaining a week, and filled with many misgivings as to the outcome of our visit; but the good missionary had proved to be one of the most kind-hearted and generous of men imaginable, and more than three weeks had flown before the many and imperative duties ahead forced us to return reluctantly to the port.

Padre Fulgencio walked to the Coni with us, and supplied men and canoes for the six hours' journey to Todos Santos. His regret at our departure was as genuine as our own, and I look forward with the utmost pleasure to another and longer visit to the mission and the boundless country of the upper Chimoré in the not far-distant future.

Chapter 20

THE CACTUS FORESTS OF CENTRAL BOLIVIA—COCHABAMBA TO SAMAIPATA

The journey from Cochabamba to Sucre presents difficulties, no matter which of the two available routes is selected. It is possible to take a pack-train to the beginning of the railroad at Cala Cala and proceed by train to Potosi, thence by cart or pack-train (or by motor-car in the dry season) to Sucre; but we preferred to go the whole way by pack-train, following the roundabout Santa Cruz trail, as this would enable us to see the country and also to stop at any time we chose to investigate the fauna of a promising region.

It is an easy matter to rent mules and *arrieros* in Cochabamba, either by the trip or month, and the latter way is the more satisfactory if one does not expect to spend too much time en route. We had been told, however, that it was better to secure the animals at Tarata, a small town southeast of Cochabamba, so we decided to make that the expedition's starting-point.

A narrow-gauge railroad connects Cochabamba with Arani, almost due east; Tarata is about half-way between the two. We took the train and sent our own mules overland, in charge of one of the men. It required but two hours to make the trip. The entire region is naturally of a barren, desolate nature; nevertheless it is densely populated with Quechua Indians. The low, earthen huts cover the desert-like plain and are so like it in color that it is at first impossible to distinguish them. During the short rainy season crops of wheat and other grain are sown, and their growth is later promoted by means of irrigation. They also cultivate grape-vines, and their small clumps of peach and apricot trees were in full bloom.

The train stopped at numerous little stations, and at each of them gayly dressed Quechua squaws sold fried eggs, boiled corn, and bread. Occasionally they also had stew or meat pies, but these were always to be regarded with suspicion. Our boy told us that cavies are ordinarily used in preparing the meat foods; but a woman tried substituting toads on one occasion, with the result that those who partook of the delicacy became violently ill.

Crowds of Indians boarded and left the train at each station. The accommodations at their disposal resemble cattle-cars from the outside, but have two long benches running through the centre. The fare is very low, and the Indians are fond of travelling, so the cars were invariably crowded to suffocation. In addition to the mass of humanity each person carried a huge parcel, pail, or basket, that filled the few interstices. The Quechuas and Cholos are a good-natured lot among themselves, and do not in the least mind being placed in such close proximity with one another.

We reached Tarata in two hours. It is a town of considerable size; the elevation is nine thousand eight hundred feet, and it is desolate beyond description. The inhabitants are largely Indians of an independent temperament, though living in abject poverty. We found it almost impossible to secure lodging, or to find help to carry our luggage up from the station, so appealed to the chief of police, who rounded up a number of men and placed them at our disposal. Perhaps our difficulty was due partly to the fact that the Indians were celebrating a religious holiday. They had taken an image of a saint from the church and were carrying it back and forth through the streets. A group of them preceded the procession and set off pinwheels and cannon crackers, while those following also employed explosives of various kinds with which to add to the din. The people are so fond of this sort of pastime that it is difficult to persuade them to desist long enough to perform any service, no matter how slight; and the guise of religious fervor gives them license to indulge in acts that would not be tolerated at other times.

Padre Fulgencio, with whom we had become acquainted at the mission on the Chimoré, had told me a great deal about the monastery of San José, located at Tarata, and had given us a letter of introduction to the abbot. We therefore called upon that personage at the first available moment.

The huge building stands on an eminence overlooking the town and surrounding country, and is said to be the largest of its kind in Bolivia. We were ushered through long, gloomy corridors, past rows of small, cell-like rooms, and finally into the quarters of the abbot. This good man received us in his cell, and cordially offered to assist us in any way possible. He also invited us to make the monastery our home during our stay in Tarata. A group of monks added their invitation to their superior's, but the edifice, with walls eight or ten feet thick, small, narrow windows, bare, gloomy rooms, and the chill damp as of a dungeon was not very inviting, and we preferred to return to the Quechua hut that seemed to belong more to the every-day world. One of the priests, however, secured an *arriero* and mules to take us the first stage of the journey.

Our man arrived about noon on Sunday, September 18. Much to our surprise we saw that he had but one arm, but this did not prevent him from being one of the best mule-men we ever employed. He had evolved a clever system of loading the packs that was admirably suited to his needs. Instead of the long ropes or thongs ordinarily used to tie on the cargoes he had strong nets that fitted over the packs, with loops that could be hooked over pegs in the pack-saddle. He lifted the trunks, each weighing one hundred and twenty-five pounds with his one arm, slipped them into place, and then tied them securely to prevent them from bouncing up and down as the animals trotted along.

Figure 58. The great *Puya*, a species of pine growing in the Bolivian Andes at an elevation of 13,000 feet.

The first afternoon's ride was short and ended at the *arriero's* house in a village called Uaiculi. There, as at Tarata, scores of yellow finches lived about the houses; they were fully as plentiful as English sparrows are in the United States, and acted not unlike them. The soil in this entire region is so arid and rocky that even cacti grow in limited numbers only. There are no streams, so water of a poor quality is obtained from deep wells. Nevertheless the whole country is thickly settled. The Indians are adepts at conserving the scanty water-supply, and at irrigating. They grow fruit and also cultivate small, isolated fields of grain, but the greater part of their subsistence is derived from the flocks of sheep and goats that seem to thrive in the desert-like country.

The climate is very cold and during the winter months there is a high wind. We could see funnel-shaped masses of dust moving across the plain all day long; occasionally a dozen or more, resembling small cyclones, were visible at the same time.

After leaving Uaiculi the way lay along the edge of the barren plain for some miles. A ridge of high peaks, some of them covered with snow, rises on each side. Then the trail ascended the slope to the east, rising gradually in a series of terraces, four to six hundred feet high. Sometimes low hills flanked the trail, and often we passed along the top of flat plateaus.

The slopes of the highest peaks were littered with fields of broken sandstone that resembled a quarry-dump for shattered rocks of large size; groves of gnarled trees, not over twenty-five feet high, grew in these rock-strewn areas, and we found them nowhere else. Where there were no rocks thick clumps of tall grass stood. When we reached the elevation of thirteen thousand four hundred and fifty feet we found a very peculiar plant belonging to the bromelias (*Puya*); the smooth, trunk-like stem was about eighteen inches through; this served as a pedestal for the dense clump of slender, bayonet-like leaves; a tall spike of small yellow flowers rose from the centre of the plant. Numbers of giant humming-birds (*Patagona gigas*) came to sip nectar from these flowers.

The mountains seem formed of solid sandstone. Here and there a ledge, worn and rounded by the elements, projects conspicuously and resembles an impregnable fort or castle of majestic though fantastic design. We reached a lone Indian hut late at night, and, while the *arriero* was loath to halt, the mules were too tired to go much farther. The neighborhood was in bad repute on account of a number of robberies and murders that had recently taken place there, and not long after our arrival we saw mysterious signal-fires spring up on the surrounding slopes. We therefore camped in a corral, the enclosing stone walls serving as a barricade, and alternately did sentry duty throughout the night. I believe one of the reasons for our being left severely alone was that each member of the expedition was advised to display casually his pistol to the inhabitants of the hut, and to acquaint them with its possibilities. This same ruse has prevented serious trouble in a number of instances. I have found that by far in the greater part of South America there is not the slightest necessity of carrying a weapon of any kind; but there are isolated regions where the moral effect on the natives produced by wearing a revolver of generous size in a conspicuous place is so great that one may tread with impunity what would otherwise be dangerous ground.

The next night was spent on the banks of a narrow creek called Usiamayo, the elevation being only seven thousand nine hundred feet. Many Indians live on both sides of the stream. They own numerous flocks of sheep and goats and cultivate extensive areas of maize and wheat. Their huts are low, primitive affairs, and of such small size that they resemble overcrowded rabbit-hutches. Freshly cut grain was piled in neat heaps that were surrounded by fences of thorny brush to keep the sheep and cattle away. Corn fodder was stored in the tops of low trees. From a distance these aerial storage-places looked as if they might be the nests of some giant bird. Invariably a little shelter or wigwam stood in the centre of each field, or in several instances it had been built in the branches of a stunted tree. These are the Quechuas' guard-houses; they are occupied by a watchman day and night so long as crops are in the field, and thieves have but slight chance, indeed, of escaping his vigilant eyes.

It was less than half a day's ride from the Usiamayo to Mizque, a town of small size, the capital of the Province of Mizque. The cactus-forest belt of central Bolivia has its beginning in this region, although it does not reach its maximum development until some distance farther south. A part of the surrounding country, however, is fertile and provides pasturage for horses and cattle, and areas of some extent are also cultivated; this is particularly true of the land near the small stream bearing the same name as the town and province. A great deal of the acreage is planted in peppers, for which there is a good demand throughout the country, and which fetch a high price. The seeds are sown in small, sunken squares or "pans," where the plants remain until several inches high. They are then transplanted to the fields. I saw numbers of Indians weeding in the plantations, and when they neglected their work or accidentally pulled up a few of the precious plants together with the grass, their employers did not hesitate to cuff or kick them. When the peppers are ripe they are dried and done up into bales of about fifty pounds each; the natives are very fond of them and eat quantities just as we eat an apple or other fruit.

The fauna of the Mizque region is typical of the arid highlands; but many species of birds belonging to a different zone were met with for the first time in Bolivia. I immediately recognized the white anis (*guiras*) that were so common near Asuncion, and there was also a species of puff-bird, or bucco, and a little finch of a deep-red color (*Coryphospingus*). The number of doves in the open fields was astonishing; they fed on weed-seeds, and when disturbed flew to the nearest bush or low tree which they covered much in the same manner that passenger pigeons are said to have done in this country not so many years ago. One could easily have secured thirty or forty with a single charge of number ten shot.

Near Mizque lies a narrow valley enclosed on both sides by ridges of low mountains. We repaired to this space and camped in a decaying structure that formerly served as a sawmill; for, strange to relate, this little valley was originally wooded. Most of the trees had been cut down and converted into lumber, and while a large part of the land was under cultivation, there were also extensive patches of brush and second growth. Tujma, as the place is called, deserved more time than we could give it. In addition to the birds found at Mizque were many species unknown to us; among them a blue-fronted and a red-fronted parrakeet, and a gorgeous Amazon parrot. There was also a kind of macaw (*Ara*) that we saw in that region only, and even there it was rare; the forehead and shoulders are of a blazing crimson, and the underparts of a pale-yellow color, the rest of the bird being green. Most members of the parrot family were feeding on cactus fruits that were ripening in great abundance.

Figure 59. The plaza at Mizque.

A hummer of rather modest attire, being of a grayish color, but larger than our own ruby-throat, had a dainty little nest, containing two white, bean-like eggs, suspended from the ceiling of our hut. At first the bird was greatly distressed at our appearance and darted out each time we came in; but finally it became more confident and returned, frequently hovering overhead to inspect us several minutes at a time, and then slipping quietly into the nest where it sat unconcernedly, its long tongue playing in and out of the bill, like a snake's.

Our next station was the large Indian town of Totora. We covered the entire distance of more than thirty miles in a day. The country is rough and the trail runs up and down over numerous mountain-tops, varying between seven thousand one hundred and ten thousand feet in height. There are a number of deep ravines filled with low, dry woods; they form the connecting-links with the lowland forest, and it is up these avenues that the new fauna we were constantly observing finds an easy means of invading the uplands. Before reaching Totora we had seen guans, and jays of a dark-blue color.

There were many Indians on the trail; most of them were driving burros laden with fire-wood, peppers, or sundry articles. When the tired animals stopped for a moment's breathing-spell, their owners beat them unmercifully with stones and clubs so that some of them dropped senseless in their tracks. The drivers also used sticks with sections of cactus stuck on the end as prods to urge on the worn-out creatures.

Totora is to me the most desolate and unattractive place in all Bolivia, and the inhabitants are quite in keeping with their town. It is frequently spoken of as the miniature La Paz because, like that city, it is built in a crevice in the mountains, and one does not see it until on the very brink of the precipice above. The inhabitants are practically all Quechuas, or Cholos of a low type who spend most of their time drinking, swearing, and fighting; then they unburden their souls of guilt by celebrating a religious fiesta. We witnessed one such performance the day after our arrival. Indians and Cholos formed the inevitable procession, headed by members of the clergy; they halted at each corner and sang a hymn to the tune of a few blaring brass horns. The *gente decente* stood on the upper balconies of their mud huts and showered home-made confetti and firecrackers on the heads of the sacred statue and the marchers.

The Indians of Totora make some of the loveliest blankets found in all Bolivia and—since the introduction of cheap German dyes—some of the most atrocious. They are woven of coarse yarn, are thick and heavy and of large size, being about seven or eight feet square. Usually there are wide stripes of two colors merging gradually into one another, and when some harmonious combination is used, such as dark green and yellow, the effect is very pleasing. The price of a blanket, requiring months to make and containing six or eight pounds of wool, was about three dollars.

Continuing our journey by way of Duraznillos and Lajma, we reached Chilón at the end of three days. A more tiresome trip is hard to imagine; the country is so uneven that one is constantly going either up hill or down, and the altitude varies from that of Totora, nine thousand eight hundred feet to ten thousand five hundred feet. The broken, arid landscape becomes monotonous, and the climate is trying owing to the heat at midday and the freezing temperature at night. The Indians scattered along the way are not of a particularly friendly nature, and are only indifferent at best.

At Chilón we entered the heart of the giant-cactus forest—and it can be properly known by no other name. The country, far as the eye can see, is covered with the thorny plants; some of the giant club-cacti rear their fluted columns to a height of sixty to seventy-five feet, and are of majestic appearance. There are also immense clumps of prickly-pear and several other varieties, while low, trailing kinds hug the rocky earth; the latter are rather unpleasant as one frequently strikes against them in walking, and the sharp spines penetrate shoe-leather and are extracted from the foot with difficulty; mules frequently get them into their noses while nibbling on leaves or the few blades of coarse grass, and are driven almost frantic with the pain. Many of the club-cactus plants bore an abundance of round fruit about two inches in diameter and covered with long, velvety down; when the outer covering was brushed off a smooth, red berry was revealed; it is very sweet and the flavor reminds one of strawberries.

Chilón is a settlement of twenty-five or thirty huts; its elevation above sea-level is five thousand six hundred feet, but the climate is very hot. We put up in one of the hovels where there was also a corral for the mules, and proceeded to work along the banks of the Rio Chilón, which is a small tributary of the Rio Mizque. The stream is rapid and shallow, and flows over a rock-strewn bed. Numbers of fish, including rays, were plainly visible through the clear water. The majority of the birds inhabiting the thorny jungle that grows on both sides of the watercourse, were still of the arid upland type; but there was a further encroachment of a foreign fauna, and the brown-shouldered orioles, coral-billed tinamou, and red-tailed parrakeets left no doubts in our minds of the origin of their distribution. They were the advance ranks of a stream of bird-life flowing up the valley of the Rio Grande and its tributaries, where conditions are at least somewhat similar to those obtaining in the chaco country to the east, which is their normal habitat.

Apparently the red-tailed parrakeets were mating. Large groups sat on the branches of some stunted tree, preening one another's plumage, and emitting queer ani-like wails. If one observed closely, however, it could be seen that the flocks were always broken up into pairs that were snuggled up as closely together as possible.

Comarapa, the next station, is very similar to Chilón, but somewhat larger. The town is built near the base of a high range that towers to the east. A stream of small size flows past the settlement; it is known as the Rio Comarapa, and is thought to be the headwater of the Ichilo and Mamoré. The Indians said that the river flows through a deep cleft in the mountains, impossible to follow or navigate; also that an exploration-party of Germans once crossed the range with the object of locating the Ichilo on the other side, but after spending several months in the wilderness they returned without having found the river. There was at one time a well-known trail across the mountains, over which war-parties of Yuracaré Indians crossed to attack the settlers, and later they came to work in the pepper-fields; but the location of this passageway doubtless leading to the Ichilo or some other navigable stream, has been forgotten.

A few of the older families of Comarapa possess wonderful collections of ancient silverware made by the Spaniards centuries ago. One finds it difficult to refrain from openly admiring the massive ladles, bowls, plates, and cups that are unostentatiously placed on the table before the guest, but such a procedure would be considered unpardonable, as any comment on such possessions is looked upon with suspicion. To attempt to purchase an article of this kind is regarded as a very grave breach of etiquette; but not infrequently the owners of these treasures experience the need of ready money and will offer them for a fraction of their value.

The elevation of Comarapa is six thousand six hundred feet. But a short distance away rises the first outlier of the Andean Range, eight thousand three hundred feet high; from its summit we could see two other ridges, both of greater height, that must be crossed before reaching the forested slopes on the eastern side; and there may be more. We descended one thousand seven hundred feet into a small valley called California, settled by a few Quechua families. These people were squalid beyond description. Their dilapidated huts swarmed with fleas, and vermin of many kinds was so numerous that during the three days and nights we spent in the valley, no member of the party found it possible to get an hour's sleep altogether. We left sooner than we had expected, as the insect plague drove us to the verge of exhaustion. Practically all the Indians we saw were suffering from consumption. Many of them had lost the sight of one eye, and I was told that in fighting among themselves they invariably try to gouge out one another's eyes with their thumbs.

From a short distance the valley and the slopes above California appear to be heavily forested, but a close inspection showed that there was but a dense growth of low, dry woods, the trees not exceeding forty feet in height; the interlocking branches were draped with long streamers of grayish moss. The ground was perfectly clean and one could see a long distance ahead in the greenish-gray light. The surroundings were almost weird; subconsciously one expected to find strange sacrificial altars, and bearded Druids officiating at some gruesome rite of a mythical religion. Beautiful little deer walked timidly among the column-like trunks of the garlanded sanctuary, sniffing the air, and nibbling daintily at a leaf or twig, and made the hunter feel like an intruder in a consecrated place.

Upon our return to Comarapa we met a gentleman representing a mercantile establishment in Cochabamba. He was making his semiannual tour of the region, taking orders for merchandise, and collecting for goods sold on the previous trip. Most of his customers paid with silver and nickel coins, so that he had several mule-loads of money in his possession. One night our Indian boy came to us in a state of great excitement. He had been drinking *chicha* in an Indian liquor-store together with the *peons* belonging to the merchant, and one of them, while under the influence of drink, boasted that he expected to murder and rob his *patrón*. A plan had been carefully formed to suddenly attack the man from behind, while riding along a lonely and precipitous part of the trail. The body was then to be thrown over the precipice into the river below, where no one would ever discover it, and the money taken by the highwayman and his accomplice. Naturally, we lost no time in imparting this information to the traveller, and he at once interviewed the would-be assassin. He first of all questioned the man carefully, and when he had succeeded in obtaining a partial confession, he mauled him back and forth across the room until he was tired out. Thereafter we all travelled together, and the plotter, as further punishment, was deprived of his horse and compelled to walk in advance of the party day after day. He had been in the merchant's employ six years, and the latter did not care to turn him over to the police, but was certain that the punishment inflicted was sufficient to inspire proper respect in the future.

A brisk canter of eighteen miles took us from Comarapa to Pulcina, also known as San Isidro. A tame condor was standing dolefully in the centre of the open square about which the houses were built; it was a friendly bird and liked to be petted and to romp, but was pretty rough at times, and picked off pieces of skin during the course of its rather too affectionate caresses.

As we unloaded the mules the bells in the tiny box-like church began to tinkle, and all the people rushed out of their houses, bearing lighted candles in their hands. They hurried to one of the huts where a youth lay dying, and crowded into the one dingy room, filling it to overflowing, and raising their voices in wails and lamentations; this continued for half an hour. No priest or physician was present; only the noisy mob of half-wild people, to whom death comes as a divertisement from the daily humdrum of half-lived lives, to speed the parting soul to the great beyond.

Pulcina was swarming with dogs. It seemed as if each family owned at least half a dozen. They were a hungry mongrel lot, that roamed at large, snarling at passers-by and rending the night with howls and fighting. It was impossible to keep them out of the houses, and no matter how often they were driven away they always returned to rummage among the luggage and attempt to tear open the provision-sacks. Toward morning, when the dogs had departed, pigs came to take their place. Each of them wore a long, forked stick over the neck, like an inverted Y; another stick was lashed across the bottom so that the pig's neck was enclosed in a complete wooden triangle. This arrangement would have kept the pigs from crawling through fences, had there been any. Some of the contrivances were so large that they had apparently been made in the hope that the animals would eventually grow to fit them; but as it was, they touched the ground and made the wearers think they were constantly about to step over something, so they walked along raising their front feet like well-trained circus horses.

A ride of thirty miles next day brought us to Pampa Grande. The town was anything but what the name led us to expect. Instead of a vast, grass-covered pampa, there was but a semiarid plain; nearby extended the wide, rocky bed of a river that contained not a drop of water. The inhabitants had dug deep down into the gravel and scooped up the small quantity of thin mud that had collected; it is a place about the size of Mizque but wretched-looking and forsaken. Formerly it had a population of sixty thousand and was noted for the brilliancy and gayety of its annual fairs, that drew crowds even from the Argentine. Epidemics of fever, it is said, killed off many of the people, and others fled from the threatening shadow of the pestilence, until to-day the once thriving city has all but ceased to exist.

At Pampa Grande we had a very good illustration of two extreme types of Bolivian character. When we entered the town, our travelling companion met an acquaintance who owned practically the only house of any size. The Bolivian greeted him in the friendliest and most polite manner possible, and insisted that all of us spend the night at his home; he directed us to the house and then excused himself, saying that he would return presently. We found the place without difficulty, but the wife refused to admit us and told us we might wait—in the street—until the return of her husband. The school-teacher, seeing our predicament, ventured to offer us the use of the classroom; he apologized because it was so small and the roof leaked; and the next day he refused to accept a single *centavo* for the accommodation. The first man had not returned home when we were leaving the following morning; from my experience with the same type of person, I am certain that had he returned and admitted us to his home, he would have made an exorbitant charge that courtesy demanded our paying.

There now remained but one day's ride to Samaipata, where the trail divides—one branch leading toward Sucre, and the other to Santa Cruz de la Sierra. The farther eastward one goes the greener the country becomes. Between the five-thousand-foot elevation of Pampa Grande and Samaipata, which is six thousand feet above sea-level, there are two peaks to be crossed, one seven thousand three hundred and twenty-five feet, and the other six thousand seven hundred feet high. The top of the former is known as the Alto de Mairana; it is a cold, dreary little plateau where half a dozen wretched Indians live. The town of Mairana is on the lower plain between the two peaks. Patches of low brush replace the cacti and thorny, arid-region type of vegetation; there is a sufficient water-supply; and the whole country seems to present a transition zone of reviving life between the alternately hot and frigid upland deserts and the green slopes stretching toward Santa Cruz.

Chapter 21

A Mule-Back Journey on the Santa Cruz Trail to Sucre

Samaipata is in no particular different from the towns through which we had passed during the previous two weeks. Perhaps provisions were somewhat more abundant, and a small number of mules and sheep grazed in the nearby pastures; but the general distress and dejection were very much the same, and never failed to give one the impression that the settlements were tottering on the brink of obliteration. Everywhere we heard tales of woe about the prevalence of malarial fever during a part of the year, and that this disease was the cause of the desolation and extermination of the people; but as none of the places was lower than five thousand feet above sea-level, and the country is of a semiarid type, I am unable to understand how malaria could work such havoc, and am inclined to attribute the dreadful inroads to some other little-known underlying cause.

Since leaving Cochabamba we had made very good time, although there had been several halts en route, the distance covered each day was comparatively great, ranging sometimes up to forty miles, considering that we always travelled with our pack-train. Such long rides were made possible by the fact that all *arrieros* were mounted; if they travelled on foot, as in Colombia, the distance traversed each day would be about half. Fast travelling, however, was hard on the mules. When we reached Samaipata our animals were in poor condition, so we left them in charge of an attendant and engaged a complete new outfit for a short side-trip toward Santa Cruz.

One of our main objects in undertaking this entire long, arduous journey was to attempt to determine the southern limit of the subtropical forest zone. This type of forest grows on the eastern slopes only of the Bolivian Andes; a section directly eastward would, therefore, take us through this zone and possibly enable us to find the solution to the problem. It was not intended to cover the entire one hundred and ten miles from Samaipata to Santa Cruz, but only to go far enough to secure the desired information.

The mountain range breaks down rapidly east of Samaipata, but the road to Santa Cruz is, nevertheless, neither an easy nor a level one. There are still four steep ridges to cross, called Cuevas, Negra, Herradura, and Guitara; between them lie small, well-watered valleys, planted in cane and fruits, and settled by Bolivians of Spanish extraction. There are no more Quechuas, nor is their language spoken; after many months we were once again in a Spanish-speaking world.

The trail, at least during the second day's travel, lies near the course of the Rio Piray, and the scenery flanking this watercourse is among the most picturesque found in the Bolivian Andes. There is a bewildering succession of dome-shaped peaks, unscalable cliffs and overhanging precipices, all of red sandstone. Many of the formations are spotlessly clean and smooth, as if scoured, or cut with a knife. The river laves the base of the rugged chain, and dark caverns worn into the frowning battlement open alluringly to tempt the adventurous spirit to explore their unknown depths.

Vermejo is the name given to a fertile region that may be called *yungas*, between the Negra and Herradura ridges. Several houses are scattered along the trail; the inhabitants grow maize, potatoes, and large quantities of cane that is used in making *chancaca* (brown sugar) and molasses. The people also make bread and a peculiar "food-drink" called *somo* to sell to passers-by. *Somo* is made of boiled maize that has been left standing until fermentation sets in, and is taken with molasses. To us, the taste was very disagreeable, but the natives were fond of it and purchased a bowlful at frequent intervals. *Chicha*, made from peanuts, was also to be had at some of the dwellings.

With the exception of the tracts cleared for cultivation, and the bare sandstone summits, the country is covered with light forest. There is practically no moss, but a dense undergrowth of climbing bamboo and a few palms and ferns. As a whole, the vegetation does not greatly resemble that of the true subtropic or cloud-forest zone, and as this was its upper limit and three thousand and five hundred feet above sea-level, it should have been of the subtropic type, if any exists in the region. We may, therefore, safely conclude that this marks the ending of the zone of cloud forest existing on the eastern slope of the Andes during practically their entire course north of this point.

Birds were not very common, and of comparatively few species; but the fauna is entirely different from that of the uplands. The brilliant little tanagers (*Calliste*), so typical of the mountain forest, are conspicuously absent. There were, however, several kinds of warblers, and wrens, parrots, and other birds properly belonging to such a region. A black-and-white guan (*Pipile*) was really plentiful, and while the distribution of the species is very great, I had always considered it a rare bird. It is about twenty-eight inches long, and of a bronzy-black color. The top of the head and a large blotch on the wings are white; the naked cheeks and a long throat-caruncle are of a delicate shade of grayish blue. The bird's rasping cry may be heard morning and evening, as it takes wing and alternately soars and flaps from one tree to another, or skims over the top of the forest. Adult birds weigh up to four pounds and are killed for food on every possible occasion, as the flesh is very good. The individuals I examined had been feeding on green leaves swallowed whole.

Jays in flocks followed us about in the forest and kept up a constant screaming and scolding. It was impossible to escape them without using drastic measures. They were a great nuisance, as their cries frightened other forms of wild life away; both the black-fronted blue and the green-and-yellow species mingled in the same flocks.

Figure 60. Vermejo on the Santa Cruz trail.

One day we rode to the top of the next ridge, the Herradura, which is six thousand feet high. The trail winds up along the face of the slope and is very poor in places; a row of wonderful crags and cathedral-shaped mountains stands like the ruins of a city on the opposite side of the ravine. On the face of one of the cliffs we saw what seems to be a gigantic serpent carved in the red sandstone directly above two massive stones that stand as if forming a gateway. The people say this is the entrance to a secret hiding-place used by the Indians many years ago; or perhaps it might have been a prehistoric shrine. The outline of the supposed snake can be discerned with ease, and the body is marked with transverse black bands. It seems that the natives have never taken the trouble to visit the spot, owing to the difficulty of crossing the wild gorge.

We continued to the crest of the ridge; from this point of vantage it was possible to secure a good view of the country to the east, but as it did not differ from that we had just left, there was no reason for going farther.

The vicinity of Vermejo had been headquarters of a band of brigands that preyed upon travellers and caravans going to and from Santa Cruz. They had had their rendezvous in one of the numerous caves, and for a long time conducted their nefarious occupation with impunity. Eventually, however, their depredations became so bold and wide-spread, that a body of soldiers was sent against them. The bandits, brought to bay among the hills, found it impossible to withstand the onslaught of their assailants, and surrendered. It was said that a great many horses and other property were recovered, and of the men captured a number were taken to Santa Cruz, and others to Cochabamba and executed. After that, thieving stopped for a while, but a new band was beginning operations at the time of our visit.

The amount of traffic along the trail was surprising. Most of the caravans were from Cochabamba; they took merchandise to Santa Cruz and brought back cigars and low-country products.

The language spoken by the *Cruzeños* is very peculiar; the diminutive *ito* is changed to *ingo*, so instead of saying *pocito*, *horita*, or *chiquito*, they say *pocingo*, *horinga*, and *chiquingo*, for instance. There are also other changes that sound either confusing or amusing at first. At any rate, they speak the language of the country, and do not copy that of the Indians. I have frequently wondered how any country, such as Paraguay or Bolivia, for example, could hope to advance when its inhabitants adopt the language and customs of its Indian population, instead of introducing their own mode of living and institutions which should, at least, be on a higher plane. The former procedure might be excusable to a limited degree in isolated cases when, for instance, a missionary goes among savages who have no reason for being interested in the white man, and who do not recognize his authority unless he can propound his doctrine in a way they can readily understand. It may be argued that a large proportion of the inhabitants of Paraguay or Bolivia are half-breeds and therefore naturally adhere to the ways of their Indian ancestry; but that only shows more conclusively than ever the weak, moral fibre of the Spanish half, that so readily succumbs to the Indian half. It is very safe to wager that if such a country were completely isolated from the remainder of the world for a few generations, savagery would again come into its own and obliterate the traces of to-day's civilization.

In the course of years of almost constant hunting one is compelled to have some very peculiar and unusual experiences. One of these occurred at Malena, Colombia, when the wounded macaw entered our room. Another took place at and near Vermejo. The evening before starting back to Samaipata, we noticed a flock of swifts soaring high above the hut. Boyle and I grabbed our shotguns and each took a quick shot before the birds disappeared; my companion scored a clean hit; apparently I had missed; but the next morning we were astonished to find a dead bird of the same species lying on a rock beside the trail, about two miles distant, and more than one thousand feet above the place we had left. I am convinced that it was the identical bird I had aimed at, and that it had continued flying until it died and fell in the spot where we chanced to find it. The natives do not shoot birds on the wing, because ammunition is too costly to take any chances with; under no circumstances would a charge be wasted on a small swift-winged bird;

and also, when I prepared the bird I found a number-ten shot in its head, which is what we used; such small shot is not to be had by natives, as none is used in the country. The coincidence of finding the bird is one that is not likely ever to be repeated.

From Samaipata we turned southward toward Vallegrande. It required two days' travel to reach that town, over the same monotonous, broken, barren country ranging in elevation from five thousand three hundred feet to eight thousand two hundred and fifty feet. There are a few trees near some of the small watercourses, but as a whole the country is unproductive. At Vallegrande, however, the ground is not so sterile. The town also is more attractive, and the more cheerful environment is reflected in the dispositions of the people. I was particularly glad to find that some of the inhabitants showed traits of character unmistakably alien to the average Bolivian, and it did not require a great effort to trace them directly to the wholesome influence exerted by the American College at Cochabamba. It was forcibly demonstrated that at least some of the students of the Cochabamba Institute introduce into their houses and home towns the admirable precepts of temperance, morality, and sincerity with which they have become imbued.

Travel in the highlands of Bolivia presents a succession of difficulties, chief of which is the scarcity of mules and also the lack of forage.

There is no natural pasturage, so the animals must subsist entirely on oats grown by the Indians in irrigated areas. The cost of keeping animals is prohibitive; instead of the one or two cents a day charged in the settled parts of Colombia, one is compelled to pay fifty cents or more. We should probably have been forced to remain in Vallegrande a long time, had it not been for one Señor Villazón who provided the pack-mules for the rest of the journey to Sucre.

The first day's ride took us to the village of Pucará. A part of the distance had been over a grass-covered plateau ten thousand feet high, cut in places by deep ravines filled with light woods. The second day we faced the unpleasant prospect of having to cross the Rio Grande. The few natives we met said that the river was probably very high and were inclined to be pessimistic concerning our ability to get across; they also advised us to return to Aiquile, near Mizque, as the stream is narrow and spanned by a bridge at that point; but as this meant retracing our steps the greater part of the way, we could not consider the suggestion.

One has the first view of the Rio Grande from the top of a rocky mountain nine thousand five hundred feet high, of which we reached the summit a few hours after leaving Pucará. Far below lay the dull, brown ribbon of water, looking like a painted streak across a grayish background. The descent to the watercourse is so abrupt that in many instances the trail consisted of a succession of steps hewn into the rock; toiling down the tortuous trail was life-sapping work for the pack-animals; we relieved the riding animals by walking. Downward, always downward, led the indistinct way, seemingly into a bottomless abyss. The mountainside is dry and cheerless; no dainty flower or blade of grass relieves the grim desolation of desert dust and shattered rock, and even the few grayish, stunted cacti seem to shrivel and die in the burning glare of a hostile sun. After hours that seemed more like days we arrived at the dry bed of a narrow stream and followed down its angular course. The aneroid showed that we were exactly one mile lower than our starting-point, but still the river seemed like a mirage, near, yet unattainable.

Although the declivity was now gentler, the lofty walls of gray sandstone flanking the dismal canyon through which we rode shut off any ventilating breeze that chanced to pass above, and made a stifling oven of the narrow fissure. For two hours we travelled over the rock-strewn stream-bed, and then suddenly entered a narrow belt of mimosas and cacti; the Rio Grande flows through the centre of the green little valley.

Although the river had appeared peaceful enough from the summit five thousand seven hundred feet above, we found it to be a wide, brown sheet of ruffled water racing over a boulder-encumbered bed. Our mule-drivers were filled with alarm and dared not venture into the treacherous flood. It was as we had feared; the spring rains had begun in the mountains, and the surplus water was rapidly swelling the lowland streams. While we were debating on the proper course to pursue, an Indian youth chanced along and consented to guide us to a ford about half a mile up-stream. Arrived at the spot, he stripped and waded cautiously into the river, which here spread over a wide bar. Fortunately the water was not over four feet deep; the youth returned to the bank and led the mules across one by one. When the river is too high to ford, the natives use tub-shaped boats made of ox-hide in which to cross; there is no way of controlling the craft, so the current may carry them a mile or so below the starting-point before it reaches the other side.

The water of the river was unfit for drinking. It contains about thirty per cent solid matter, although the reason for this was that it was rising rapidly and bringing down a great quantity of sand from the mountain.

Numbers of small ravines emerge from the barren slopes flanking the Rio Grande, and streams of inconsiderable size pour their water into the larger artery. All these openings are filled with brush and low trees; we followed up one of them and, within a few hours, reached a habitation called Bella Vista. The shambling structure stood on the edge of a clearing planted in sugar-cane. Dense jungles of wild cane and brush bounded the plantation. As I was already convinced that the Rio Grande is the avenue up which the chaco bird-life was penetrating into the higher regions, we determined to remain at Bella Vista sufficiently long to substantiate my views; it required only one day for this purpose. The species that had been found in limited numbers farther up, and that seemed to belong to a strange fauna, exist in abundance at Bella Vista; among them are brown-shouldered orioles, white anis (*Guira*), fork-tailed goatsuckers, white-throated toucans (*Ramphastos*) and many others.

Pigeons (*Leptoptila*) were so numerous that they suffered for lack of food. I am unable to say whether there had been an abnormal increase in the number of the species, or if the food-supply was unusually low; but one thing is certain—they were in a very emaciated condition and some of them had become so weakened that flight was impossible, and they fell an easy prey to the natives or predatory animals. I also noticed that all the pigeons were infested with parasites, but the weaker individuals were covered with them, including many winged, fly-like bird-ticks (*Hippoboscidæ*) that skipped among the feathers at bewildering speed, and finally flew away with a loud buzz; sometimes the repulsive insects settled on our hands or faces, when it was almost impossible to displace them, owing to their agile movements and to their clinging ability caused by the hooks on their feet.

This furnished a very good illustration of the survival of the fittest, and one that I believe is typical of what happens in many instances. Owing, perhaps, to unusual or long-continued favorable conditions, the species had become exceedingly numerous. So long as there was no shortage in the food-supply, the birds were able to hold their own and keep increasing; but, as the season of famine approached, as I believe it must occasionally do, though not necessarily at regular intervals, the weaker individuals were the first to feel the pinch of a reduced subsistence which automatically rendered them still less suited to obtain a livelihood. Their rapidly failing vigor also prevented them from coping with their natural enemies—whether parasitic or predaceous, so that they were soon eliminated and only those that entered the struggle in the strongest, healthiest condition stood a reasonable show of surviving.

While tramping through the cane-thickets, we found the nest of a pair of red-breasted thrushes. Both parent birds fluttered over our heads and with loud, angry cries expressed their resentment and anxiety. The nest was betrayed by the birds' very actions. It was cunningly concealed in a dense tangle of leaves and creepers, and was not unlike that made by our own robin; but the three eggs were heavily spotted with brown instead of being of a plain blue color.

When dusk overtook us on the first day out of Pescado, thirty-six miles southeast of Bella Vista, we were riding over a grass-covered plateau with a stream flowing along one side of it. It was therefore unnecessary to seek an Indian dwelling for the purpose of securing forage. We picketed the mules, and slept out in the open. The next morning a Quechua woman with a fowl under her arm passed along the trail; we asked her the price of the bird, as we suspected that she was taking it to some village to sell. "Four bolivianos," she replied promptly. The mule-driver remarked, very emphatically, that the price was exorbitant. "But," she protested, "this is a game-cock. It is a good fighter and can whip any rooster in the country." The arriero then informed her that we wanted the rooster to eat, and not to fight. "Oh," said the woman, "that is another matter; sixty *centavos*," and the sale was concluded without further argument.

Apparently the birds of the highlands were nesting. We saw numbers of newly constructed nests in the cacti and small-leaved vegetation; they belonged to mocking-birds, pigeons, and finches; but only a few of them contained eggs. The Indians had filled many of the little domiciles with stones before they were completed in an attempt to prevent an increase in the numbers of birds. Large flocks of several species gather in the grain-fields during the fall months and exact rather a heavy tribute, and it is for this reason that the Indians try to prevent their increase.

While riding along one morning we flushed a red-crested woodpecker (*Chrysoptilus cristatus*) from a hole in a stub near the road. The entrance to the cavity was about eight feet up, but the nest was down low in the hollow trunk. An investigation brought to view four pear-shaped, glossy, white eggs lying on a pad of chips.

This species is one of the commonest, and therefore one of the best-known woodpeckers. We found it very abundant throughout the uplands near, and south of Cochabamba, where there was a growth of cacti and low trees. Invariably there were two birds together, and not infrequently we saw flocks of four or five. It has a clear, powerful note, and a swift, undulating flight. I have frequently seen it on the ground in company with long-billed wood-hewers (*Drymornis*) and brown cachalotes (*Homorus*) searching for insects and larvæ among the débris always littering the ground beneath the giant club-cacti.

As we neared Sucre, a marked change was noticeable in the appearance of the Indians. Instead of the unattractive lot that we had encountered daily, they were a uniformly garbed, more primitive and more picturesque people. The greatest change was evident at Pulqué, which we reached a few weeks later.

Figure 61. Quechua habitation on the upland desert.

Tarabuco is the name given to a town of large size, located on a frigid *mesa* over ten thousand feet up. When we arrived there snow was falling and an icy wind blew at terrific velocity; but the natives seemed not at all discomfited by the blizzard, and were conducting the weekly market with the usual hilarity. Provisions of many kinds were to be had in abundance; mutton, bread, peaches, and eggs were particularly plentiful; but the lack of fruits and vegetables requiring a warm climate and rich soil was very noticeable. One could purchase all the necessaries of life in any of the numerous stores; most of them were imported from the United States and Europe.

We spent the night before reaching Sucre in a cluster of Indian dwellings called Cghilka. The pronunciation of the name is difficult to a foreigner, because two of the three "cliks" employed in the Quechua language are used in saying the word. Cghilka consisted of half a dozen low hovels, built of irregular stones and roofed over with grass. Flocks of sheep and a few burros nibbled the short grass, and goats clambered along the face of precipices unscalable to human beings; some of the latter also stood on the top of stone fences, or roofs, and several times we saw individuals that had climbed into the branches of a leaning mimosa and were unconcernedly browsing on the leaves.

The Indian women, it seemed to us, were everlastingly spinning in order to keep up the necessary supply of clothing. Those at Cghilka were no exceptions; but they also made unusually pretty blankets. In spite of the fact that many colors, such as red, blue, green, yellow, and white were used in the same blanket, the combination was so harmonious that the result was most pleasing. As a whole, the work somewhat resembles that of the Navajos, but the texture is not quite as fine. They also work attractive geometric designs into the pattern that immediately distinguishes the product of this region from that of any other. This is, perhaps, a retention of an ancient custom, for, it seems as if in olden times the inhabitants of each locality wore *ponchos* or blankets of a distinctive design; then, when the nation gathered in the holy city of Cuzco to celebrate some religious festival, it was possible to tell by these insignia from which part of the empire they came.

From Cghilka to Sucre is a distance of only eighteen miles, over a practically level plain, the elevation of which is in the neighborhood of ten thousand feet. There are few habitations until the immediate vicinity of the city is reached.

The approach to Sucre is quite attractive. We could see the assemblage of dazzling white edifices from a distance; and not long after we were galloping over the cobblestones between rows of neat, clean buildings on our way to the Hotel Español. In our journey from Cochabamba we had travelled nearly a thousand miles, and counting the several delays, had spent fifty-six days en route.

Chapter 22

SUCRE, THE RIO PILCOMAYO, AND THE UPLAND DESERT TO THE ARGENTINE FRONTIER

The inhabitants of Sucre insist that their city is still the capital of the country, and that the removal of the government to La Paz is temporary only, owing to the greater accessibility of the latter place. They are confident that with the completion of the railroad from Potosi the old régime will return, and with it the gayety and activities that such an event occasions. This, however, does not seem probable.

The city is built on a plateau over nine thousand feet up, on the site of an ancient Indian village known as Choquesaka. Its climate is that of perpetual spring. The streets are very wide, paved with cobblestones, and are kept exceedingly clean. The buildings are, for the greater part, low, although edifices of pretentious dimensions and imposing appearance are not lacking, and numbers of most attractive summer homes dot the surrounding country. The Medical Institute is well-known throughout the neighboring republics, and annually supplies them with thousands of tubes of vaccine. The markets are abundantly supplied with provisions of all kinds, at reasonable prices, including many fruits and vegetables of a temperate climate—brought from the eastern lowlands.

The inhabitants of the upper class are well educated, refined, and charming. There is a total population of about twenty-five thousand, but by far the greater part of it consists of Quechuas and Cholos. As a whole, Sucre is one of the most delightful spots in all Bolivia and, when the vast country to the east with its unlimited resources is made accessible, the city will unquestionably enjoy the growth and prosperity to which it is so well entitled.

However, South American cities, with few exceptions, possess little attraction for me. I touch upon them almost reluctantly, and am impatient to return to the wild, free life of the boundless jungle, desert, or plain.

Within a few days after reaching Sucre, our necessary business affairs had been looked after and we had decided upon the upper Rio Pilcomayo as our next field of operations. Pack-mules were not to be had; the few *patrones* who owned herds of these very necessary beasts were all en route to or from Cochabamba. A weekly motor-bus service is maintained between Sucre and Potosi, and the powerful cars passed within a stone's throw of the spot we decided to visit; but the list of waiting passengers was long, and even though a little monetary persuasion might have been helpful in securing an early passage for ourselves, the transportation of our luggage by that means was out of the question. We therefore secured the services of a *coche*. Six mules hitched to a lumbering vehicle that had seats inside for ourselves, with the luggage festooned about the exterior, took us thundering over the rocky, uneven road at a fast pace. The driver sat in front and diligently plied a long, thin whip that cracked with reports like those of a pistol, but inflicted little punishment on the mules, while a Quechua boy ran alongside and encouraged onward the panting animals with ear-splitting whistling and volleys of stones. I was never able to understand how these urchins could keep up the fast gait maintained by the mules, and at the same time have sufficient wind left with which to do the whistling.

Within an hour after leaving Sucre we had reached a point where the road ran along the rim of an attractive valley filled with trees, shrubbery, flowers, and pools; a number of queer structures combining Chinese, Arabian, Greek, and several other styles of architecture, were scattered about promiscuously and detracted greatly from the natural beauty of the spot. This place, known as *El Recreo* is the property of a Bolivian woman who calls herself a princess, and who for reasons unknown to me makes her home in far away Paris.

Soon after leaving El Recreo with its lovely vegetation and disfiguring minarets, stained glass, and other hall-marks of poor taste, the large town of Yotala was reached. Yotala is well-known throughout Bolivia for the excellent quality of the peaches and apricots that are grown and preserved there; and locally it enjoys the reputation of producing the best bread of the vicinity, although I could never agree with the latter assertion. The finest bread we had in all Bolivia was prepared by the hospitable señora living on the banks of the Pilcomayo, and in one of whose huts we resided the following eight days.

After an hour's halt at a house called Pulqué, where the mules were fed and watered, and where we refreshed ourselves with weak coffee at thirty *centavos* the cup, we resumed the journey, and 3 o'clock P. M., found us on the bank of the great river we had sought—having come a distance of nine leagues since 7.30 o'clock that morning.

The Pilcomayo at this point varies in width from a few hundred feet to half a mile, is crossed by a suspension bridge, and flows between high, barren, rocky hills. There was comparatively little water, but the current was strong. For me the Pilcomayo possesses an unusual fascination. While looking at the hurrying, muddy torrent underneath, I could not help picturing the awe-inspiring stretches of wilderness through which those same waters must flow before mingling with the less fearsome Paraguay hundreds of miles farther down: little-known savages indulging in wild orgies and cannibalistic dances on its banks, or paddling silently and mysteriously on its glassy bosom to some jungle rendezvous unknown to white men; jaguars eagerly lapping up a refreshing draft after a gory meal of deer or peccary; myriads of *pirañas* lashing its surface into spray in their mad struggles to tear the flesh off some struggling, despairing victim; lines of crocodiles sunning themselves on mud-banks or slowly patrolling the water's edge, like drifting logs, with only the ever-vigilant eyes showing the faintest animation; boundless wastes of pestilential swamps and lagoons, where mosquitoes and other obnoxious insects in clouds forestall the advent of man, but where millions of egrets, storks, cormorants, and other water-loving birds find a safe haven and lead their wild, joyous lives in blissful ignorance of despoiling plume-hunters; but, a shout of "*Ya está, señor*," from the mule-driver reminded me of the fact that day-dreams must soon end. The man had unloaded the luggage at a little hut surrounded by shade-trees and fields of alfalfa. He had been unable to find the owner, but thought we could arrange to stay there should that personage appear. Most important of all, he wanted his money—and then he was off with twenty-seven miles of up-hill road ahead of him, before reaching Sucre that night.

Figure 62. Rio Cachimayo at Peras Pampa, Sucre.

While taking stock of our outfit and arranging it conveniently in the little adobe hovel that was to serve as our home, an elderly Bolivian woman came from one of the alfalfa-fields nearby, and I rightly guessed that she was the owner of the property. To my request that we be permitted to remain, she promptly replied that she would consider it an honor to have us do so. I wondered if there are many places in our own country where courtesy to utter strangers is so universal as in Spanish America. Frequently, after long and trying journeys afoot or on mule-back (sometimes of hundreds of miles) our appearance was disreputable; but with one or two exceptions only during the entire course of my travels in South America, the kindness and politeness of the inhabitants was unfailing. When we left the Pilcomayo, the *señora* accepted not a cent of payment.

Figure 63. Bridge across the Rio Pilcomayo.

The country for many miles about was arid, excepting only the few irrigated flats near the river where fodder, grain, and vegetables grew luxuriantly. Cacti and thorny shrubbery dotted the slopes, but even these plants of the dry lands were not abundant. Numerous small streams empty into the river during the wet months; but now (November) their courses were dry and parched.

Birds were plentiful, but the species varied little from those typical of the uplands. However, they were nesting and this circumstance furnished a new and interesting field for study.

One of our first walks took us to an old mill, fallen into decay through neglect. There were hundreds of dollars' worth of machinery ruined through lack of care and the use of improper lubricants. I have frequently seen machinery of various kinds, ranging from typewriters and sewing-machines to Pelton wheels, seriously damaged because lard or tallow had been used instead of oil, and the wearing surfaces never cleaned. In one of the dust chutes a pair of chestnut flycatchers (*Hirundinea*) had built a flimsy nest of twigs and feathers. It contained two cream-colored eggs speckled with red. The birds remained in the vicinity all day long and paid no attention to the Indians working near by, but when a dog chanced to pass they darted at it furiously, making quick dashes at its head and snapping their bills with a loud, popping noise. Another pair of birds of the same species had a nest above the door of a near-by house.

Leaf-cutting finches also called tooth-billed finches (*Phytotoma*), were very abundant. The inhabitants destroyed them whenever possible, as the birds cut the blossoms off the fruit-trees and grape-vines. The bright, saffron-breasted male sat in the top of some thorny bush and uttered queer, unmusical wails that reminded us of the mewing of a forlorn alley cat, while his gray-and-black-striped mate incubated the eggs in a small but compact nest hidden farther down among the spine-armed branches. We examined numbers of the nests; each one contained three eggs of a deep-green color, marked with a few black lines about the large end.

Oven-birds built their dome-shaped mud nests on fence-posts or the larger branches of the few poplar-trees that had been planted about the huts for shade, and sang in unison from dawn to dusk as if their hearts were overflowing with happiness.

Parrakeets had excavated holes in the face of steep banks, and chattered and quarrelled noisily over their domestic affairs. I suspect that they also appropriated the cavities prepared by swallows, as there seemed to be frequent disputes between these neighbors.

Of humming-birds there were a number of species, including the giant hummer, which was truly monarch of all he surveyed, for when one appeared the smaller members of the group found it advantageous to depart to other regions. Doctor Frank M. Chapman, in Chile, saw an individual of this species pursue and catch in its claws a small humming-bird and fly away with it; for what purpose he did not know, unless from "sheer cussedness." It is a well-known fact that hummers possess a pugnacious disposition, are almost constantly fighting among themselves, and frequently pursue and strike at large birds such as flycatchers and even hawks, apparently for no reason other than the pleasure it affords them to torment their victims.

One afternoon we had the first indication of the coming rainy season in the form of a severe rain and thunder-storm. Before long the river was a seething, muddy torrent that continued to rise rapidly until well into the night. The next morning the water had subsided to its low level, leaving numbers of fish of several kinds stranded in depressions in the *playas*. A flock of caracaras appeared with daylight and, wading daintily into the shallow pools, extracted and devoured the stranded and helpless fish at their leisure.

Not long after we were fortunate in meeting an American by the name of Kolle, who was in the employ of a wealthy Bolivian owning estates in various parts of the country. To one of these we were subsequently invited, but before accepting the invitation of the affluent señor we decided to spend a few days at Pulqué where some variation in the avifauna from the upland type had been noticed. We had also seen numbers of Quechuas apparently living in much the same manner as their predecessors during the height of the Inca's glory.

As frequently occurs in semiarid country, and as I have stated before, birds were very abundant; but there was little else to indicate the close proximity of other forms of life unless one took into account the herds of goats clambering about on the steep ledges and seeming to delight in bombarding with showers of small stones everyone who passed below; or the caravans of burros and llamas passing on the main highway. A visit to the nebulous peaks of the adjacent mountains, however, revealed a different story. Patches of green dotted the isolated little depressions to which the name "valleys" can hardly be given, and thin pillars of smoke ascended from them straight into a cloudless sky. After long and patient looking a small, stone hut set among rocks would invariably be discovered, and sometimes we could even distinguish minute, moving forms which we knew were Indians. There, tucked away among the towering peaks they love so well, they were living a life of peace and plenty, apparently safe from the gaze of vulgar interlopers, and knowing or caring little about the outer world. It was as if one tore a page from the history of bygone centuries, or found himself suddenly transferred into the midst of a contented, pastoral community as must have existed in places unnumbered throughout the vast Incan Empire before its despoliation by the gold-crazed invaders. In this connection it might be well to go back briefly into the history of the events that brought about the present state of affairs.

The boundaries of the Incan Empire had been gradually extended until within five hundred years after the arrival of Mamo Capac and Mama Occlo, supposed Children of the Sun, it covered nearly one-third of the South American continent. Near the middle of the sixteenth century, when Pizarro and his insatiable band invaded the sacred precincts of Atahualpa's dominion, the star of the Inca seemed to have reached the apex of its ascendancy. Under the beneficent rule of their venerated sovereign the several tribes lived contentedly, if not always peaceably; agriculture thrived, arts and crafts were encouraged and, responsive to the efforts of many thousands of laborers, numerous mines poured a constant stream of precious metals into the kingdom, adding to its wealth and splendor.

We are all familiar with accounts of the advanced state of civilization, governmental organization, and fabulous riches of the ancient nation. Temples, palaces and forts—stately edifices of hewn stone—dotted the mountainsides and crowned the eminences; beautifully constructed highways connected many of the remote districts with the capital; countless herds of llamas fed on the slopes, and streams of water flowing through a system of aqueducts poured into the heretofore arid wastes and transferred them into fruitful fields capable of supporting a numerous population. The present-day republics of Ecuador, Peru, Bolivia, as well as a part of Colombia and Chile, were included within the limits of the vast kingdom.

Suddenly a dark cloud appeared on the horizon, and omens of evil import presaged the downfall of all this greatness and splendor. The fatal apparition quickly assumed the form of bearded strangers, some of whom were mounted on terrible beasts which filled the ranks of Indian warriors with panic, and who seemed to have succeeded in harnessing the thunder and lightning for the furtherance of their wicked designs. Suffice it to say, that before the avarice of the Spaniards had been abated, eight million subjects of the Inca perished and the organization of the nation was destroyed. With the single exception of the Aztecs of Mexico, who were practically exterminated by the same people, there has never been another example of such rapid and complete devastation in the history of the world.

The Quechua of to-day is a cowed, almost pathetic, individual; he has been kicked about by the descendants of the *conquistadores* until he has learned to become reconciled to his lot; but while it seems as if this recognition might, in many instances at least, give way to despair, such is not the case.

The partiality of the Quechuas for the high *puna* is well known—some of the ancient dwellings having been discovered at an elevation of more than seventeen thousand feet.

Those living near Pulqué seldom come down into the lower country; doubtless they are happier in their almost inaccessible fastness than if they lived nearer to their Bolivian neighbors. In appearance and dress these Indians differ greatly from the other members of the tribe living in the more populous sections of the country. Instead of the more or less conventional attire adopted by the latter, they still adhere to a form of dress at least a part of which may date back to the time of Atahualpa. The women wear a quantity of clothing—short, full skirts of dark blue, and shawls of varied colors. The men are garbed in loose, white knee-breeches, a gray or blue shirt, and belts which are neatly embroidered in gay colors and are very wide at the back so that they form a kind of sash; also, they wear the inevitable *poncho*. Strange as it may seem, the small children always wear very long clothing, and the little girls waddling along in their full, almost trailing skirts, resemble dwarfed aged women. All the apparel is made of woollen cloth of home manufacture. The men permit their hair to grow long and braid it in a queue which hangs down the back. Both sexes use peculiar little hats made of some kind of skin prepared by a process which renders it very hard; this head-gear reminded me of steel helmets. With the exception of huge spoon-shaped pins of copper, which the women used to fasten their shawls, we saw no metal ornaments or jewelry of any kind.

The home life of these Quechuas is tranquil and uneventful. Usually the little stone huts contain two or three rooms; potatoes, avas, and other produce are stored in one of them, and the rest are used for cooking and sleeping-quarters. In very cold weather a fire is kept burning day and night, and all the occupants of a house burrow deep into a pile of sheepskins and blankets close to the smouldering embers. We persuaded one of the women to bring goat's milk to camp each morning, but in this we had the greatest difficulty. Only by payment for a week's supply in advance could she be induced to perform this service. With past experiences with their fellow countrymen these Indians have learned to regard all strangers with apprehension. On several occasions we had ample opportunity to observe how the average *paisano* treats the Quechua. Should night overtake him on the trail, he stops at the nearest hut and demands food for himself and his horses. In the event that the owner has nothing to offer, he draws revolver or rifle and shoots any fowls that may be running about or, lacking these, a sheep or goat, and seizes whatever else he can find. Should he see an attractive blanket, it also is taken. In the morning a few *centavos* are thrown on the ground and he continues on his journey.

As a general rule, we found that if these Indians were treated in a frank, honest manner they were quite amiable. The little woman we had engaged to bring us milk trudged down from the mountain-top daily in faithful compliance with her obligations. She brought cheese also, and occasionally a few eggs. As it gradually dawned upon her that we really could be trusted, she became talkative and seemed to take an interest in our occupation. She spoke Quechua only in common with practically that entire tribe, which makes no attempt to learn Spanish; or, if they are able to understand it, will make no effort to speak the language.

Figure 64. Quechua Indians wearing the costume used during the reign of the Incas, five hundred years ago.

Upon seeing a woodpecker we had collected, she gave a sigh of satisfaction; for, according to the Indian's belief, they are birds of ill omen. If a pair of them make a nest near one of the huts, they are said to be excavating a tomb for a member of the family who will soon die. Oven-birds are looked upon with favor and are encouraged to remain in the vicinity of the dwellings. Should a pair of the cheery singers place their large, domed nest of mud nearby, good fortune will follow in their wake. Any one guilty of robbing a bird's nest will become violently ill; but as birds flock to the planted areas in such great bands that an appreciable amount of damage is done to the fruit and ripening grain, their increase in numbers is discouraged by filling many nests with small stones. After the seeds have been planted, a network of strings is stretched across the fields, and sometimes a dead hawk suspended from a post in the centre serves as a scarecrow to frighten away the marauding visitors. When the crops ripen, a small boy called the "*piscomanchachi*" is stationed in each sector. He is armed with a sling and keeps up an incessant fusillade of stones; fortunately his aim is poor, but he succeeds in killing a few birds each day.

These Quechuas lead a sedentary life. There are no more long, arduous journeys to far-away Lake Titicaca and Cuzco to participate in solemn festivals and gorgeous pageants. Their fields supply potatoes as of yore, and they still convert the tubers into their beloved *chuño* by simply allowing them to freeze and dry. From the wheat they have learned to cultivate, a splendid quality of bread is made. Their flocks provide flesh and milk, and the wool so essential to the preservation of human life and well-being in the high altitudes. *Tola* bushes and a peaty growth known as *yareta* furnish an adequate supply of fuel; but should these be lacking, dung is used. The demands of civilization, however, will alter this mode of existence until little remains to remind us of the contented nation which at one time willingly bowed to the rule of the Children of the Sun.

Birds were not quite so numerous as at the Pilcomayo, but we found several forms new to us. Among them was a large, white-fronted parrakeet (*Myiopsitta luchsi*) that we saw in no other place. It banded in flocks of ten to fifty and seemed to prefer the fruit-trees near the house. A species of humming-bird built nests in doorways and suspended under the thatched roofs of houses, often in the midst of a colony of swallows (*Atticora*). Tinamou were not uncommon in the dry ravines and provided a welcome change from the goat-flesh which is the staple meat of the people and the only kind we could purchase; the latter animals were killed when very young (about the size of a cat), and we could never become enthusiastic over this, locally considered, great delicacy.

In a region such as the country around Pulqué, there are few available nesting-sites, and nests are very conspicuous objects when placed in a cactus or thorny bush; however, the sharp thorns and spines with which they are surrounded protect them, alike from predatory animals and humans. The disused mud nests of oven-birds are collected as needed and made into a poultice that is supposed to cure stomachache. Judging by the quantity they gathered, this ailment must be of frequent occurrence. A bird of the wood-hewer family (*Upucerthia*) excavated burrows in banks and deposited two white eggs in a small, feather-lined nest placed in a roomy chamber at the end of the tunnel.

The *señora* at whose *rancho* we stopped complained that tiger-cats were killing her chickens; so one afternoon I set a steel trap at the base of a near-by stone wall, baiting it with a dead bird. I had not gone a dozen paces from the spot when the trap sprung with a loud twang, securely imprisoning the much-sought culprit. The cat's greed had overcome its discretion, at which we rejoiced, for it made a desirable addition to our collection. On another occasion one of these beautiful animals bounded out from under the roots of a huge tree and seized a bird as I was stooping to pick it up—and made a clean getaway to its hiding-place with the spoils.

Our hosts on the Cachimayo were awaiting us, in order that we might be present at the ushering in of the "month of baths," as December is called in this part of Bolivia. Whether or not they thought we were in need of the daily ablutions, I am unable to say; but this I do know, that we shocked the good people on numerous occasions by having a swim at every possible opportunity, even if the month was not in keeping with local traditions.

Peras Pampa is an immense estate on both sides of the Cachimayo, and but an hour's trip by motor from Sucre. We spent a delightful ten days amid pleasant surroundings, living in a comfortable bungalow, and passing the evenings at the *casa grande* where the elite of Sucre's society gathered for music, games, and dancing.

The grounds were a succession of orchards, fields, and vineyards. Scores of Indians lived and worked on the place, cultivating the ground, building stone fences, and taking care of the stock. At night they met and played very well on reed flutes of various sizes, each musician taking a separate part, so that the combined effort was somewhat like that produced by a well-organized band. Their favorite piece was "Red-Wing"—apparently learned from a phonograph record.

The evening parties were always enjoyable affairs. They began with a sumptuous dinner—including several kinds of wine; then a series of eight or ten well-chosen courses, followed by liqueurs and smoking. The women did not smoke.

After that there were charades, story-telling, music, singing, dancing, and perhaps a walk *en masse* in the moonlit grounds, through arbors of honeysuckle and other flowering vines and over paths bordered with hedges of roses. There were always more refreshments just before the party broke up at midnight. All the Bolivians we met at Peras Pampa were charming, and we heartily regretted that our time for combined work and play was not unlimited.

The majority of the people who formed the gay evening crowd lived in separate cottages on the estate—the guests of the owner. Each day they repaired faithfully to the river for a dip, although the water was usually very muddy, and there was about an even chance whether one would emerge without yesterday's coat of grime or with an additional one. December is chosen for this purpose because it is the warmest and most pleasant month of the year.

The time allotted us for work in Bolivia had nearly expired. We had thoroughly enjoyed our lengthy sojourn in the republic, and look forward to revisiting it in the future. Our schedule called for rather extensive work in the Argentine, so, after a great deal of difficulty, we succeeded in collecting a caravan of riding and pack mules for the ride of over three hundred miles to La Quiaca, on the Argentine frontier. Ordinarily the trip from Sucre to La Quiaca should not be undertaken on mule-back. One should go to Potosi in one day's time, utilizing the semiweekly motor-car service. A railroad connects the latter place with a small station a few miles this side of Tupiza, and from this point one may reach La Quiaca in two days by carriage. During the rainy season, however, both automobile and carriage service are suspended; and the difficulty of twice securing mules on which to cover the two stretches of road between railway terminals and the delays and other inconveniences are so great that we decided to travel the entire distance with a pack-train. This also gave us an opportunity to see the country.

The expedition left Sucre December 22. The caravan was appallingly large, for we were taking our entire outfit, and it required no less than six Quechuas to look after the mules and burros. All supplies, also, had to be taken with us, as very little is to be had from the Indians, who are virtually the sole inhabitants of the cheerless highlands. There are a number of large villages, it is true, but the person who relies on the natives for maintenance is as likely as not to have to live on coca and *chicha*, or suffer for his improvidence.

By noon we had reached the Cachimayo at a point where, ordinarily, it is fordable; but a heavy rain had caused the river to rise and we were confronted by a series of roaring cataracts covered with foam and débris washed down from the mountains. The mules were unloaded and driven into a corral. Soon other caravans arrived, until there were several hundreds of men and animals gathered on the river-bank. We spent the afternoon strolling through the adjacent apricot-orchards and vineyards. The former trees were laden with fruit, all ripening; it was small in size, but of delicious flavor. By seven o'clock the water had subsided many feet, and one of the *arrieros* having previously ridden across the river to test its depth, the caravan started across. The stream was three hundred feet wide and the current very strong, so that crossing it seemed an endless operation; the mules struggled onward gamely, but to the rider it seemed as if they stood stock-still while a maze of rushing water seethed and raged all around him in frantic efforts to sweep away everything in its path. Our own animals got across safely, although some of the packs were drenched; but a long train of burros laden with huge boxes of the popular *Sucrenses* cigarettes fared badly, and a number of the poor creatures were upset and whirled away down-stream. We continued onward in the darkness two leagues to Poste Escalera, a lone hut on a hillside, and spent a trying night at this flea-infested post. Next day we reached the Pilcomayo at a point where the river is divided into many narrow channels, although there is one main stream spanned by a swaying wooden bridge.

A detailed narration of each day's ride would mean the recounting of practically the same things. There were, however, a few things of unusual interest, and these will be mentioned later.

The country is dry, rolling, and unproductive. In some places there is a sparse growth of cacti and thorny shrubbery, but vast areas are rocky and barren of all vegetation. We crossed ridge after ridge, the elevation of the trail varying between eight thousand and twelve thousand feet. Travel in this type of country is most trying. Water is so scarce that long distances must be covered in order to find suitable camping-sites; in one instance we were compelled to ride thirty-six miles in the course of a day, between streams. The temperature varies 100° each twenty-four hours. At two in the afternoon the thermometer registered 132° F.; at night ice formed on the water in our pails.

Christmas day was spent at Puno, with every member of the party ill from the effects of the climatical changes. The inhabitants went about their occupations as usual, quite ignoring this all-important opportunity for a *fiesta*.

All the dwellings of the Indians were made of adobe. In the walls of some of them rows of disused earthenware pots had been used as building material. When the huts crumbled, a fine collection of pottery was covered up in the mound. This is probably an ancient custom and may account for much of the material found in old ruins to-day.

Two days later, the last of the long, weary miles across the cheerless upland had been left behind, and at noon we galloped briskly into Villazón, on the Bolivian side of the border.

Villazón contains about a score of scattered, low, adobe buildings. We arrived on a Sunday, when the customhouse was closed, but the officials in charge very courteously permitted us to proceed on our way. A brook three or four feet wide separates the two republics and, stepping across this, we found ourselves in La Quiaca and—in Argentina.

Chapter 23

Bird-Nesting in Northwestern Argentina

La Quiaca is similar in size and appearance to Villazón. There are a number of stores or trading-posts where miners from the surrounding mountains secure their outfits and provisions. It is also the terminus of the railroad from the south. One may go by rail directly to Buenos Aires. The settlement stands on a level, wind-swept plateau, and the weather was very cold. The neighboring peaks of the Andes are rich in mines, and multitudes of llamas and mules come down the steep trails each day, laden with copper, bismuth, silver ore, and gold ore. They discharge their burdens at the railroad-station, where it is loaded on cars to be hauled to the smelters in Buenos Aires.

Our object in coming to the Argentine was to continue the biological survey we had carried on in Bolivia; and also to secure specimens of a rare little bird (*Scytalopus*) which was thought to exist in the province of Salta. The acquisition of this bird was most important for the light it would throw on certain problems of distribution.

The little wren-like birds of this genus (*Scytalopus*), known commonly as "tapacolas," are perhaps among the most difficult to collect of any species in South America, and for this reason they are invariably only poorly represented in museum collections. Native collectors, hunting mainly with blow-guns, have gathered many thousands of birds, the greater number of which have eventually found their way to millinery establishments and scientific institutions in many parts of the world; but usually only those of brilliant plumage, and others which could be taken with little difficulty, have been collected. The small, slate-colored or blackish tapacolas, found only in the densest of subtropical forests or among the tangled vegetation bordering bleak, frigid *paramos*, have usually been overlooked. This is not surprising when we find how seldom even the trained field-naturalist of to-day finds it possible to lure the tiny, feathered creature from its secure retreat among the mosses, roots, and ferns to which its mouse-like habits confine it, and how rarely he succeeds in recovering the inconspicuously colored bird after it has been shot. Even after a long, patient search has revealed the specimen lodged somewhere in the deep stratum of matted plants, it is by no means sure of reaching the museum; I know of instances where birds, slipping from the hunter's hands and dropping at his feet, have been forever lost in the riot of vegetation which everywhere carpets the ground.

Our quest for this little creature was destined to extend over a period of months, and to take us into many an out-of-the-way place. We were eager to begin the search, so took the first available train which left La Quiaca two days after our arrival and started southward.

Leaving the desolate settlement, the railroad winds upward through a narrow, rocky gorge to the station Tres Cruces, the altitude of which is twelve thousand four hundred feet. There it descends at a steep grade—so steep in fact that a rack and pinion are used part of the way. The rocky knobs flanking the gorge are old and weathered and very picturesque. A small stream winds back and forth across a boulder-strewn course; the water is clear and cold. About mid-afternoon we encountered an abrupt change in the type of country. The bare crags and narrow, rocky floor of the gorge gave way to a wide expanse of brush-covered land and green pasture. This change was first noticeable at a small station called Leon (elevation five thousand feet); the vegetation grew thicker and the landscape more inviting as we continued the journey. At dusk we reached Jujuy, a city of some pretensions; the buildings are attractive, the streets are broad, and the people appeared clean and intelligent. Following Jujuy were numerous small towns and stations; also many truck-farms owned by Italians who were settling in Argentina in great numbers. There were also vast green meadows in which fine-looking cattle, horses, and sheep were grazing.

Our first stop was at Salta. The journey from La Quiaca had required fifteen hours.

Salta has about thirty thousand inhabitants, and is a modern city. It possesses wide, paved streets, buildings of imposing dimensions, electric trolleys, and lights, a zoological park, good hotels, and a college. The contrast between being in a city where comforts and luxuries abounded, and living on the bleak, Andean uplands amidst stolid Quechuas guarding their herds of llamas, was great, and we enjoyed the change to the fullest extent. After frozen potatoes and canned provisions, the inviting coffee-houses were irresistible; and the "movies" made us forget the miles of inhospitable desert. Fortunately there were enough of each of these attractions so that we could spend a whole day visiting them, alternating from one to the other, without repeating.

Our first headquarters in the Argentine were made at Rosario de Lerma, one hour by train from Salta. This is a most delightful spot and afforded rare opportunities for work and observation. The town contains about one hundred houses and is surrounded by fields, pastures, and patches of low, open woods. There is an abundance of water, and excellent meat, fruits, and vegetables may be had in abundance. The people are industrious and of good appearance, and treated us courteously.

We soon discovered that in Argentina we were not at liberty to carry on our work in any place or manner that suited our purpose; in other words, there were game-laws, closed and open seasons, and it was necessary to secure permits from the owners of all lands on which we proposed to hunt. Of all these restrictions we were ignorant, and spent a blissful three days doing as we pleased; then a sergeant of police called and notified us that we were under arrest, and to call at headquarters as soon as convenient. I lost no time in going to see the chief, explained the nature of our work to him, and then acting on his suggestion took the next train to Salta to get a permit which entitled us to hunt anywhere within that province. All this was accomplished within a few hours. The various officials with whom I came in contact were most courteous and obliging.

Our study of bird-nesting at Rosario de Lerma was confined largely to observing the parasitic habits of the black cowbird (*Molothrus b. bonariensis*), referred to by the Spanish-speaking people as the "tordo." The bird usually called tordo, however, is a species of oriole, highly esteemed as a cage bird on account of its not unmusical singing ability. This bird is of slender, graceful build, about the size of a red-winged blackbird, and of a uniform glossy, purplish-black color except on the wings and tail, which have a pronounced greenish sheen. The female is of a dark, ashy-brown color.

We saw flocks of them daily in the fields, on the backs of cattle grazing in the pastures, in the courtyards of houses, in corrals, and more particularly in the scattered trees, which were almost certain to contain at least one nest of the oven-bird (*Furnarius*) or of some species of brush-bird (*Phacellodomus*). Usually the flocks were composed of from ten to twelve individuals, the bright, glossy males outnumbering the dull, grayish females in the proportion of four to one. Azara gives the proportion of males to females as ten to one, but this disparity is too great for any part of the Argentine known to me.

The birds are noisy, keeping up a loud chatter, especially where a flock is on the wing, or when preparing for the night's sleep. The male bursts into a short, pretty song with frequency, dropping his wings and moving in a nervous manner while singing. Apparently the female does not sing.

It has been said that the females of this species lay eggs during a period of three or four months; to know how many are laid by a single bird would be interesting, as the number must be very great in order to make allowance for the incalculable numbers that are wasted, and still provide enough to keep the ranks of the multitudes at their normal level.

We did not find a single egg of *M. b. bonariensis* on the ground, although Hudson states that in the vicinity of Buenos Aires these birds "frequently waste their eggs by dropping them on the ground."

Dropping the eggs on the ground might entail a deliberate waste, as we know of no reason why the bird should suppose that they would be hatched and the young reared, if scattered broadcast over the country. On the other hand, this might merely indicate that the birds had found no suitable place in which to deposit their eggs. The form of waste caused by the birds laying in old, disused nests, or by laying such a large number of eggs in a single nest that it is impossible for the rightful owner to incubate them and rear the young, can hardly be said to be deliberate, as it is doubtless caused by a lack of intelligence; if the bird designedly scatters its eggs broadcast on the ground, it is wantonly wasteful; if it merely lays in disused nests, or overcrowds nests actually occupied, the bird may simply be stupid.

It would be impossible to say what per cent of eggs laid by this species of cowbird is wasted. Hudson estimates that each female lays from sixty to one hundred eggs in a single season, and it does not seem to me that this statement is an exaggeration. One female which I dissected had laid three eggs within the few preceding days, and a fourth was almost ready to be deposited.

The bird which suffers most from the parasitic habits of the cowbird in the vicinity of Rosario de Lerma, is the oven-bird (*Furnarius rufus*); however, of the great number of eggs laid in the nests of the above-named species, our observations tend to show that the greater part are lost. Among the scores of oven-bird nests which we examined, only two were still occupied by the owners, the desertion being apparently due to the invasion of the cowbirds. So persecuted were the oven-birds that it is difficult to understand how any of them survived in this immediate locality. The nests were common enough, it being not unusual to find several of them in a single tree, but the birds themselves were not abundant. It is possible that some of the pairs may have built several nests each in their vain attempts to escape the attentions of the cowbirds.

In no instance had the walls or top of the oven-birds' nests been broken or perforated in any manner, in order that light could penetrate to the interior; they were not tampered with in any way, and the cowbirds seemed content to use them just as the oven-birds had constructed them.

I believe that the greater number of *M. b. bonariensis* that reach maturity are reared by the smaller birds, such as finches, warblers, and vireos, in whose nests only a few eggs are laid, which increases the favorable chances of their incubation. Also, the larger and heavier eggs of the cowbird frequently crush at least a part of the smaller eggs which naturally have a more fragile shell, thus forestalling to a marked degree the competition that might arise between the young birds in the nest.

We collected about two hundred eggs of this species, nearly all of them at Rosario de Lerma, and a great variation in marking exists; there is also some difference in color. As a general rule the eggs are greenish or bluish, rather heavily spotted with reddish-brown; in a very few specimens the background is of a pale flesh-color, and in a small number of others it approaches white, having, however, a dull grayish tinge; of the entire lot, four only are so lightly marked as to appear unspotted. Not a single egg is pure white or has a pure white background (my standard of comparison is an egg of the oven-bird) "like the eggs of birds that breed in dark holes"; the majority of these eggs were taken from the darkened interiors of oven-birds' nests.

A type of egg not uncommon is heavily and evenly marked all over with fine dots and larger spots of reddish-brown. Judging from the material at hand I should say that there is a characteristic type of marking running through the eggs of the species if we except the two extremes, *viz.*, those almost unspotted, and those so entirely covered with heavy blotches that they appear to be of a uniform chocolate color.

However, the eggs of each individual seem to vary in some respect from those of any other, as it is impossible to find two exactly alike in comparing series from different places. Frequently, two or more eggs found in the same nest resemble each other so closely in size, shape, and coloration, that I think it reasonably safe to say that they were laid by the same bird.

The nests of the smaller birds contained from one to four eggs of these parasites, in addition to those of the rightful owners. On January 12, I opened an oven-bird nest and was surprised to find fifteen cowbirds' eggs in the dark interior. This I considered a record, but Boyle brought one in on the same day containing twenty-six of the speckled eggs. In the days that followed, we discovered numerous "sets" of from ten to twenty. The nest that contained the final record number was found January 16, it contained thirty-eight eggs—one of the oven-birds and thirty-seven of the cowbirds.

Later, we again met these old acquaintances wintering in the rice-fields and rush-grown marshes of Tucuman.

The white ani (*Guira guira*) or Guiraca, first seen near Asuncion, and later in Bolivia, was plentiful at Rosario de Lerma. The bird was usually found in small flocks and fed on the ground.

We found several of their nests near Rosario de Lerma. They were large, loosely built of sticks and placed in the crotch of a cactus or other thorny plant, at no great height from the ground. However, the nest is not conspicuous in spite of its size.

Pablo Girard, an Argentine naturalist, informed me that these birds frequently nest in communities and that a number of females lay their eggs in the same nest, although this is not always the case. The natives verified this statement. This seems probable as I at no time saw the groups split up into pairs; on the contrary, there were always numbers of birds in the vicinity of each domicile. Our record set contained twelve eggs.

After ten days at Rosario de Lerma, we returned to Salta and then took the train to Perico, a ride of three and a half hours northward. At this station a branch railroad runs northeastward into Argentina's vast Chaco region. The track was being extended as rapidly as labor and material can be obtained for the work, and we desired to go to the end of the line where is located a station called Embarcacion. Before starting on this journey, however, we spent some time at points noted on the downward journey from La Quiaca. Perico is a busy little town, owing its activity to the traffic occasioned by the railroad junction. The buildings are low and dilapidated, and most of them consist of a shop, or *venta*, in front, with living-rooms in the rear.

The shops are always worthy of exploration. In some, huge piles of watermelons were displayed for sale; others offered fruits and vegetables, and still others groceries and dry-goods. Drinking-places were abundant.

We were particularly interested to find numbers of rhea eggs on sale in the outdoor market. They brought forty *centavos* each and were delicious; the contents of each was equal to about a dozen hen's eggs. I was told that they were gathered from the nests of wild birds in the Chaco. Each nest contains from ten to twenty or even thirty eggs, which are more than one man can carry. When fresh, the shell is of a deep cream-color; after incubation has started or if the egg is addled, the color is pale, ashy gray. The birds are killed and eaten—the flesh resembling that of a goat's in flavor.

One day a number of Indians arrived from San Pedro. They brought huge baskets and crates of young amazon parrots. These birds are taken when very young from nests placed in the cavities of trees, and are reared by hand until they are able to eat unaided. Usually two are found in a nest—occasionally three. They also brought a tame coypu rat and several three-banded armadillos.

Perico is surrounded by miles of cattle lands, light woods, and limited areas covered with vegetation of a semiarid type. In the latter places small deer or brockets are not uncommon; they hide in the low, thorny growth of Spanish bayonet until one is within a few yards of them, then dash away at great speed; the inhabitants hunt them with dogs trained for the purpose, and rarely fail to bag their quarry, though usually after a long chase.

We found the coral-billed tinamou not uncommon in the wooded districts. They are essentially birds of the tree-covered regions and are difficult to secure on account of their terrestrial habits, and also owing to the fact that they adhere closely to the densest cover. I have on a number of occasions seen captive specimens, but they seem to not take kindly to the restricted life of a cage or aviary, and spent most of the time dashing wildly about, injuring themselves so seriously that they did not long survive.

A number of the birds of this locality are not included in the avifauna of Rosario de Lerma, but belong to the Chaco type, and I recognized some species which were common near Asuncion, Paraguay; among them a large blue jay and a brown-shouldered oriole.

Our next station was at Volcan. About the only attractive thing about this place was a great lake almost entirely surrounded by high hills, and teeming with water-fowl. The Quechua boy we had brought from Bolivia was the first to find the lake. He rushed back to us excitedly with the information that there was a large body of water nearby with a huge, white duck on it; he had shot at the queer bird, that had a black neck, a number of times but failed to hit it. Fortunate for all of us that his marksmanship was poor! The "duck" was of course a black-necked swan belonging to the owner of the terrain, and its untimely demise would have cost us dearly. There were, however, hundreds of ducks; teals, ruddies, shovellers, and pintails; also, many coots, grebes, and rails.

The body of water had an area of over a square mile, and in its edges a tall fringe of cattails grew. Marsh-wrens and military flycatchers haunted these swaying green thickets, and grebes stole silently in and out of their ragged borders. There were many disused nests of coots and ducks; but while making our way through the high, tangled growth we came suddenly upon the nest of a giant grebe (*Fulica gigantica*); it consisted of a huge mass of reed stems, slightly concave on top, and extending about a foot above the water; in it were four pointed, brown eggs, heavily dotted with deep brown and black. This was apparently a second clutch, the first, perhaps, having been destroyed. There were scores of other nests, but all were empty and falling into decay.

Figure 65. Ploughing at Rosario de Lerma.

We spent a busy day tramping about the borders of the hidden lake, watching the flocks of coming and departing ducks and bagging such as we needed—whenever a duck or cormorant plumped into the water Boyle swam out and got it; this was risky work that I did not encourage, as the water was ice cold and many fathoms deep, and the ensnaring under-water growths of reeds and cattail stems formed dense, slimy masses capable of holding a man who might become entangled in them until he became exhausted and drowned.

While at Volcan we stayed at the house of an Italian trader. He asked if we had any recent reports of the war, and then expressed the hope that it would last years longer, as he owned part interest in a copper-mine, and was receiving war prices for the much-needed metal. We decided not to accept his hospitality any longer and took the train to Tilcara. I have often met foreigners in South America (including some from the United

States) who were representative of anything but the better class of citizens of their respective countries; it is unfortunate that many Latin Americans base their estimate of a people upon the appearance and doings of these few misguided and objectionable characters.

Figure 66. Tilcara, showing the stream and valley and the snow-capped Andes in the distance.

At Tilcara we lived with another Italian family, but of an entirely different type. The village, the elevation of which is eight thousand feet, stands about half a mile from the railway-station. We were engaging *peons* to carry our luggage there when the man stepped up and offered us the use of part of his humble home, which stood within a hundred feet of the spot. We accepted the invitation, and during our entire stay were treated with great courtesy.

There is a narrow valley between high, rugged, barren peaks, some of which are snow-capped. Parts of the depression are dry and semiarid; others, marshy and covered with high, rank grass. Small Indian huts built of stones or adobe are strewn about, and there are numerous fields from which the rocks have been gathered through years of effort so that the land may be cultivated.

There were many birds. They represented a fauna intermediate between that of the high, cold plateau and that found lower down at Rosario de Lerma. Large red-breasted meadow-larks (*Troupialis*) were common and always found in pairs. Of hummers there were numerous kinds, attracted by clumps of flowering shrubs that grew alongside the fences; the giant humming-bird and the gorgeous coppery-tailed comet were particularly plentiful. The former are very stupid. They came fluttering along like awkward swallows and often settled comfortably on a branch near to us, from which they would inspect us at their leisure, while they chirped and darted out the tongue like a snake. One of the comets that we collected had eaten quantities of gnats and small ants.

The walls of a deserted Quechua hut had been appropriated by a flock of bay-winged cowbirds (*Molothrus badius*) for their nesting-sites. Dozens of small, round holes penetrated the thick, earthen walls, and some of them extended entirely through; the latter were not occupied. Apparently whatever birds had drilled the cavities, frequently surprised themselves by emerging suddenly into the daylight they were trying to get away from, at the far end of the burrow. However, not to be discouraged, repeated other attempts were made, some of which were successful as the walls varied in thickness. A small, flat nest of sticks lined with a few feathers comprised the bay-wing's domicile. Some of them contained young birds, and one had five eggs in it. The adult birds always remained in a flock in the vicinity and kept up a shrill screaming while we were near.

Large, blackish rails inhabited the reedy marshes; they came in flocks to feed in the velvety green islands interspersed among the weed and water covered areas. Watching from a concealed position, we could see them strut unconcernedly about, flicking their tails over their backs and jerking their necks and picking up the tiny mollusks and insects that were so abundant. When alarmed they craned their necks, looked about inquisitively, then gave a few hoarse cackles and ran into the weeds; within a few moments they returned, one at a time, and at first slowly and cautiously; but soon, forgetting that danger might lurk nearby, they rushed for the spots where food was most abundant. Rails are peculiar and interesting birds. The body is narrow and compressed like a flea's; this enables them to slip through the dense reeds and water-plants in which they live. The comparatively long bills make it possible for them to pick up food in shallow water. Their long, slender toes, giving the feet a wide spread, make walking on floating vegetation and soft mud easy; nevertheless, at least some species are good swimmers.

Flocks of night-herons spent the days in a small clump of willows fringing the marsh. At dusk they grew very active and we could hear them croaking from afar. They are splendid eating.

As at Pulqué and the Pilcomayo, birds were hard pressed for nesting-sites. Giant club-cacti apparently were at a premium. The old, disused nests of brush-birds (*Synallaxis*), or *leñateros*, were inhabited by mocking-birds which built a nest of their own within the huge structure of twigs; and, when the mocking-birds were away, cowbirds slipped in and deposited a few eggs. One mocking-bird had been so unwise as to place its nest in a thorny bush covered with dense foliage so that it could not be watched from a distance and defended from cowbirds; before the owners were ready to use their new home, it had received many visits from the black parasites (*M. b. bonariensis*) who left their cards in the shape of fourteen speckled eggs. We collected this "set" but have the idea that this only encouraged the cowbirds to increased efforts.

The abundance of ducks in South America in places where one least expects to find them, is a source of never-ending surprise. A small stream flows through the valley at Tilcara. It is nowhere more than twenty feet across, and two or three feet deep, but flocks of green-winged teals visited it regularly at dawn and dusk. They swam in the rapid water, and then lined up on the rocky bank for a quiet nap.

The inhabitants of Tilcara shot many, but others came to the same place daily.

Of mammals there were but few. Cavies, as usual, lived among the rock piles and in the stone fences, and a few other small rodents inhabited the grain-fields. One day we secured a fine specimen of the rare, elusive yellow cat called *gato pampero*, or pampas-cat. It was stealing cautiously along the river-bank; but I am unable to say whether it had come in quest of fish or merely for the purpose of quenching its thirst. Our work in this region being completed, we returned to Perico, and prepared for the journey to Embarcacion.

Chapter 24

THE CHACO, SUGAR-PLANTATIONS AND RICE MARSHES—A SEARCH FOR A RARE BIRD

The train for the Chaco left Perico at 9 P. M. It was composed largely of second-class coaches crowded with immigrants, mostly Italians bound for various parts of the great land that is being rapidly thrown open to colonization. There was, however, also a compartment-car in which we had taken the precaution of making a reservation some time in advance. The darkness prevented our seeing the landscape through which we passed, but on our return we noted that there was little change from that around Perico as far as San Pedro. There were, however, numerous fields of sugar-cane, some of very great size. Beyond San Pedro the country is all of the Chaco type; that is, vast stretches of pampas liberally sprinkled with islands of forest. The Vermejo, a river about the size of the Wabash, was crossed on a steel bridge three miles before reaching our destination, which was at six o'clock the following morning.

A group of newly erected shacks, low and so lightly built of packing-cases and corrugated sheet iron that many of them resembled mere skeletons of houses; narrow, crooked streets; shops loaded with fruits and conducted by Italians, and others festooned with bandanna handkerchiefs, gaudy wearing apparel and cheap jewelry, and owned by Turks or Syrians; gambling and liquor houses; a motley crowd of slovenly, not overdressed people, and a tropical sun blazing down mercilessly on the whole assemblage. That is Embarcacion, the "farthest east" to date in the north-eastern part of Argentina's vast Chaco. I was told that as the railroad is extended farther and farther into the interior, many of the residents take down their abodes and ship them to the new station where they are re-assembled; and so a great portion of the town moves bodily at different intervals.

On all sides lies the seemingly limitless Chaco. There is practically no cultivation and but few herds of cattle had been introduced to date. In addition to the great possibilities for cattle-raising, the country also possesses enormous wealth in quebracho-wood; at present quantities of it are cut for use as fuel in the locomotives of several of the railroads. Within a short time, no doubt, these assets will be utilized in a manner that will be advantageous to both the exploiters and the country at large.

One of the interesting discoveries in the pampas was a wintering-place for bobolinks. The extent of this bird's migration had been shrouded in mystery, and but a single specimen in winter plumage had ever been recorded. We found them in flocks of thousands, perched in the top of the tall grass or picking up seeds from the ground. Their cheery song was conspicuously absent. They were in spotted plumage. Small red-breasted meadow-larks (*Leistes*) mingled freely with the bobolinks.

Another place that never failed to attract us was a small lagoon flanked by forest on two sides, and by prairie on the others. This region was the resort of many birds. Flocks of Brazilian cardinals (*Paroaria*) numbering up to thirty individuals congregated in the bushes, their flaming red heads reminding one of clusters of brilliant flowers. We discovered a nest of the species, a shallow affair of grass stems, placed in the end of a branch twenty feet above the ground; in it were two eggs resembling those of the English sparrow. Small black-and-white flycatchers (*Fluvicola*) found the lagoon a most attractive spot. Their pear-shaped bag nests of interwoven grasses and feathers were scattered about in the overhanging bushes and also fastened to the stems of aquatic plants, sometimes but a foot above the water. There were also numbers of grebes, coots, and gallinules, and occasionally a pair of beautiful Brazilian teals visited the quiet, secluded body of water. Night-herons kept well to the tops of the taller trees; and everglade kites flew gracefully and swiftly overhead, usually singly, and rarely in pairs. We heard the weird call of chachalacas almost daily, but these birds had been persecuted by native hunters until they had acquired enough wisdom to avoid hunters and human beings in general. In one wet strip of woods we found limpkins in limited numbers. They did not seem to ever come out into the open country. There was not time to study the frogs, fish, and small snakes that we saw occasionally; nor to more than admire the myriads of flowers and curious plants growing on all sides. There was, however, another naturalist (José Steinbach) working in the locality at the time of our visit, and fortunately he devoted practically all his energies to the study of the very things we had to omit, so between both expeditions the fauna and flora were pretty well covered.

Many of the available trees were burdened with the huge stick nests of the leñateros (*Synallaxis*). Some of the structures measured six feet long and two feet through. They were built of thorny twigs, at the ends of branches. A heap of material is first placed at the very tip of a limb, and as the weight causes it to sag downward, more sticks are added until the huge mass hangs suspended in a vertical position. The thorns cause the whole affair to hold together so well that opossums and other predatory animals find it impossible to burrow their way through the walls to the interior cavity where the four or five white eggs, or the young birds, are cleverly concealed in a downy cup. There is usually a second chamber near the top of the nest; this is the male bird's night quarters while his mate is incubating or brooding in the lower story.

Figure 67. The lagoon in the Chaco, Embarcacion.

Figure 68. Paramo above Tafí.

Blue-headed tanagers (*Thraupis*) preferred to nest in the trees and bushes near to some human habitation, while blue grosbeaks selected more secluded sites in some little woods or thickets. The latter birds breed before the male changes his brown nestling plumage to the deep indigo-blue coat of the adult.

The most beautiful of all South American birds' eggs are laid by the tinamou. They are placed in a depression in the ground, usually under a tuft of grass or near a log or stone. Their color varies in the different species, running through turquoise and deep blue, lavender, brown, green, and gold. The shape is rounded or broadly ovate and the shell is very smooth so that it glows like a varnished or highly polished sphere. In spite of the glossy texture of the surface, minute scrutiny will reveal the fact that it is pitted like that of the eggs of the rhea to which the tinamou are closely related.

We saw a fox occasionally, slinking across a trail and always well out of gun-range. Each morning there were tracks of cats and large cavies in the dusty paths, but mammals were scarce and few came to our traps.

There were no mosquitoes during the day, and only enough at night to make the use of a net desirable. Sand-flies, however, often appeared in considerable numbers and were troublesome. The climate was intolerably hot during the greater part of our stay. Each day the thermometer rose a few degrees higher until we found even the lightest and scantiest amount of clothing uncomfortable; all through the long afternoons we sat shirtless with streams of perspiration pouring down our backs, preparing the specimens that spoiled within a few hours unless properly preserved. About every fourth or fifth day the weather broke and a deluge of rain falling throughout the afternoon and night brought with it a lowered temperature and welcome respite from the oppressive heat.

Our greatest problem was dealing with the hosts of small red ants that persisted in getting at our specimens. We kept the latter on a table the legs of which stood in tin cans half full of kerosene; but a trailing thread, a piece of paper blown by the wind, or any one of a dozen other trivial things that happened daily furnished bridges over which the insatiable hordes promptly swarmed to destroy our hard-gotten trophies.

We next headed toward Tucuman and upon our arrival there were pleased to find a city of sixty thousand inhabitants, delightful from practically every point of view. The people were particularly interesting. We saw few of them on the streets during the daytime, but in the late afternoon after the shops and offices had closed and bolted their doors, the men appeared in crowds, all well and neatly dressed. They congregated in the saloons and cafés fringing the plaza, and drank beer and small cups of strong, black coffee until about seven o'clock. In many instances the tables were arranged on both sides of the pavement so that one walked through a lane between rows of sleek-combed youths twirling gaudily banded straw hats or canes, and noisily discussing—what-not, and grave-faced men with gray hair and beards everlastingly talking politics. After going home to supper they reappeared with the womenfolk, the wealthier ones circling about the plaza in carriages or motor-cars, the less opulent afoot. The band played every other night.

The great Province of Tucuman, of which the city bearing the same name is the capital, is one of the most fertile in all Argentina. Its principal products are sugar, rice, and cattle. Land values are high—too high in some cases, but it cannot be denied that there is good reason for the rapidly rising scale of prices.

In Tucuman we found the chief of police a hard proposition to handle when it came to securing the hunting-license. To begin with, we had great difficulty in entering his sanctuary. The door was guarded by a mammoth negro who rushed into the inner chamber each time the *intendente* rang for him. First he always jerked a huge club out from under his coat—ready perhaps to take the first whack at the official if some one started anything, instead of defending him. Finally we succeeded in entering the holy of holies, and found a small, rather elderly man sitting behind a large, flat desk, sipping tea while several *secretarios* hopped wildly about him and yelled into an ear-trumpet held in position by one of his hands. He failed utterly to understand our request, and curtly refused to have anything to do with any *millinery establishment*. We argued in vain, then retired to think of some new move, for the permit was necessary if we wished to keep out of jail, and I must admit that such was our ardent desire.

There being no United States Consul in Tucuman, I appealed to the British Consul for assistance. He very kindly spent many hours calling on various officials, from the governor down, explaining our mission and asking that the small matter be arranged for us. Our quest seemed hopeless until one day a copy of one of the large daily newspapers arrived from Buenos Aires, and in this I found an account of how representatives of Latin-American countries who were attending the scientific congress in Washington had been received and entertained at the American Museum of Natural History during their visit to New York. Armed with this clipping I again invaded the *palacio*. Ordinarily I should not have done such a thing, as there are many reasons why it is not commendable, but the situation was desperate and called for aggressive tactics. Suffice it to say that this visit was the last. A mild comparison of how their people were treated in our country, and the difficulties we had in theirs was sufficient, and when I left the building the permit was in my pocket.

The Sierra de Tucuman, a range of comparatively low mountains, rises directly west of the city. This we found to be covered with a growth of tall, dense forest, so we lost no time in moving there. We left the city by rail and proceeded southwestward to a small station called San Pablo, a short distance away. This is in the heart of the sugar region and vast fields of cane stretch on either side of the railway. Here and there the tall brick chimneys of a refinery rise above the waving green fields, and wide, deep canals divide the cultivated areas into sections and supply water for irrigation.

A good cart-road leads from San Pablo up the side of the mountain to the very summit, four thousand six hundred feet high, where the little town of Villa Nougués is situated. This settlement is a favorite resort of the wealthier class of people who come up from Tucuman to spend the summer months in pleasant châteaux, thereby avoiding the heat of the lower country.

The view from the top of the range is superb; the country to the east is perfectly level, and is laid out in symmetrical fields of cane as far as the eye can see. A small, muddy river, threading its way through the ocean of green divides it into two sections and vanishes into the horizon in a haze of purple mist. To the west stands the stern Andean chain, barren and precipitous, its summit hidden in banks of cold, gray clouds.

We made a first camp in the forest below Villa Nougués, at an altitude of four thousand feet. From the very first day we had heard the shrill little call of a bird which we attributed to the much-coveted tapacola (*Scytalopus*) we were looking for; but the elusive creature always remained in concealment among the ferns and mosses and not once did we get a glimpse of it. Then we secured ox-carts and moved to the other side of the mountain, where, we were told, hunting was not so difficult.

Birds were not abundant, the fall migrations having left the forest almost deserted. The few species which remained, however, such as wood-hewers, thrushes, tanagers, and jays, were plentiful, and several kinds of humming-birds added life and color to the sombre green of the vegetation. After many days we succeeded in tracing the mysterious chirp to its source, and found, not the bird we were seeking, but a dainty little wood-wren of the shyest possible nature. The minute, secretive creature seemed to spend its entire time among the buttresses, roots, and moss-draped undergrowth, where no ray of sunlight ever penetrated to dispel the chill and semidarkness, or give a touch of warmth to the soggy mould. Its glimpses of daylight must be brief indeed, and at infrequent intervals. We had come to the mountains in a state of enthusiasm and expectancy, for here it seemed we should succeed in ending our long quest for the tapacola. As the days passed, thrilling excitement gave way to exasperation, and finally disappointment alone remained to fill the void created by the flight of the other emotions.

We returned to Tucuman for a brief time, and then struck for the forest farther south. This time we left the railroad at a station called Acherál two and one-half hours from Tucuman, and camped in the forest at the foot of the ridge. Again we were doomed to disappointment. Birds were more abundant than at Villa Nougués, but the tapacola was not forthcoming. There were, however, numerous other interesting species. Pigmy woodpeckers (*Picumnus*) selected the patches of high brush and second-growth woods just without the edge of the forest proper. They are little larger than a good-sized humming-bird, dark or black above, white underneath, and have a red cap. Their industrious hammering always advertised the presence of a pair as they hopped quickly along the trunks and branches, tapping for worms or excavating a nesting-site.

The woods were undermined with tunnels made by the queer *tuco-tuco*, or *oculto* (*Ctenomys*), a species of which we had come in contact with in Brazil. We set a steel trap in one of the subterranean runways, carefully covering with a log the opening we had made; soon a series of low grunts emanated from the spot, and we found a fine, large specimen of the strange rodent safely held by the steel jaws.

Bottle flies were so numerous as to prove a most disagreeable pest. Blankets, clothing, food, and specimens alike were covered with "blow" if left exposed for but a few minutes. We were lucky in possessing enough netting with which to rig up covers for everything, but even then numbers gained entrance, and we had to clean the infested articles frequently by passing them over a fire or by scraping and brushing.

After a few days we concluded that a visit to the top of the range, which at this point attains an altitude of over ten thousand feet, was necessary. We secured a pack-train of mules from Acherál, and one morning at one o'clock started up the steep slope. A full moon showered a flood of light upon the earth, but the overhanging branches formed a thick canopy over the trail, impermeable to the silvery radiance save when an occasional breeze stirred the leafy arch, thus permitting fitful shafts of light to pierce the darkness of the tunnel, and to fall in quavering, dancing blotches on the ground. We could almost feel the impenetrable blackness which closed in from all sides like water in a deep, dark pool. The light touch of a streamer dangling from the moss-festooned branches overhead, or the velvety swish of fern leaves protruding beyond the protecting walls of tree-trunks, made it seem as if the forest were peopled with hovering, invisible forms. No sound disturbed the brooding silence of the night except the dull hoof-beats of the mules as, guided by some mysterious instinct, they cautiously picked their way through the muddy and rock-strewn lane.

Hour after hour we followed blindly in the wake of the bell-mule, winding back and forth along the mountainside, but mounting ever upward. The latter part of the way seemed to lie near the course of a small mountain torrent, for we were almost constantly within hearing distance of rushing water. Finally, we emerged from the forest, and, just as day was breaking, reached a brush-covered strip of country, the elevation of which is five thousand feet. This continued to the top of the ridge, two thousand feet above. Then there was a depression of considerable extent, filled with rank, low vegetation and infested with swarms of bloodthirsty flies which render it uninhabitable.

After ascending another ridge, the trail led gently downward into a level valley a dozen miles long and from one to two miles wide. Herds of cattle were grazing on the abundant grass; a few small areas had been enclosed within stone walls and planted in maize; and at the far end, half concealed by willows and fruit-trees, lay a village of whitewashed houses. At half past four in the afternoon we reached the settlement, called Tafí del Valle, and soon after were comfortably ensconced in a hut hospitably provided by one of the inhabitants. After the fifteen and a half hours' uninterrupted ride over a difficult trail we were ready for a journey into a still more remote region, and the sun was shining brightly the following morning when we again returned to the stern realities of this world.

Tafí del Valle is a most delightful place. Even though the altitude is seven thousand feet, the surrounding peaks shut in the valley and protect it from the icy winds. There is no natural forest in this region, but groves of willows have been planted near the houses; to these, large numbers of birds came to spend the night. Hawks were especially abundant and of many kinds—we collected no fewer than seventeen species during our ten days' stay; then there were also burrowing owls, larks, flycatchers, thrushes, and many other birds. Some species which ordinarily live in brush-covered country had become adapted to their barren surroundings and were nesting in holes excavated in banks of earth. When the birds had reared their broods, rats, mice, and pigmy opossums occupied the old nesting-sites.

A clear, cold stream, which flows through one side of the valley, spreads out at the lower end over a large area, forming lagoons and marshes. Geese, ducks, coots, night-herons, and sandpipers made these places a favorite resort. Pectoral sandpipers were not uncommon, and were so fat that they were unable to fly and could be taken with the hands. There were also flocks of stilts; they are beautiful creatures, either when flying in compact formation, with measured wing-beats and outstretched necks and legs, or when standing motionless in the shallow water, their snowy underparts reflected in quivering outlines. Lapwings screamed and cackled in resentment of our visit and frequently frightened away flocks of water-fowl which we were stalking.

Apparently our *Scytalopus* was not a bird of the open highlands. We even began to wonder if it existed at all, because, so far, the most thorough search had failed to reveal any trace of it. There remained, however, the high *paramo* above, and to this we next turned our attention.

Our sudden arrival at Tafí had caused much comment among the inhabitants. They found it impossible to believe that we had come to that remote region in search of a small, dull-colored bird, and after a few days it became an open secret that we were regarded as spies—though just what nature of information we sought, could not be determined. They even went so far as to refer to the matter occasionally in a good-natured manner; and when we were away on hunting excursions, it was their custom to put our cook, a Bolivian, through a sort of "third degree" in an effort to compel him to confess the real object of our visit. Therefore, when we planned to move to the high peaks bordering the little valley, the natives considered their evidence complete; we were going, they said, to prepare a diagram of the country from our new point of vantage. The only person who really understood the purpose of our mission was a man from Tucuman who had been sent up to vaccinate the Indians. He started out each morning accompanied by two or three soldiers, rounded up all the Indians of a given locality, and vaccinated them. The natives did not at the time realize the significance of this act; but when, a few weeks later, the inoculations had had time to become effective, they grew frantic, and grim-faced little parties began to scour the country in search of the person who had "poisoned" them. Fortunately, none of the scouting-parties came our way, for to them all strangers look very much alike, and there was the possibility that one of us might have been mistaken for the doctor.

The *paramo* above Tafí is a bleak region, almost perpetually enveloped in mist. Work in this type of country possesses its disadvantages, for in addition to the intense cold and the lack of fuel, there is always the possibility that one may be trapped far from camp by banks of clouds which roll in unexpectedly! The cold, penetrating mist is so dense that it is impossible to distinguish objects but a few yards away, and the most familiar landmarks assume strange and fantastic outlines. In the event that one is overtaken by this phenomenon, there is nothing to do but wait until the mist lifts, which may be in a few hours, or perhaps, not until the next day. Strange to say, the inhospitable *paramo* supports a varied fauna. Herds of wary guanacos feed on the tall, wiry grass growing in the more sheltered places; when alarmed, they flee to the inaccessible rocky slopes. The *paja*, or grass, harbors also a species of large tinamou, but the bird is loath to leave its safe cover, for no sooner does it take wing than hawks, which are always hovering about, swoop down and carry it away.

Numbers of deep ravines have been worn in the mountainsides by water coming from the melting snows on the higher peaks. These are filled with a rank growth of shrubbery. The sides are so abrupt that we could find no spot where a descent was possible without the aid of a thousand feet or more of rope. After a number of days, however, a narrow fissure was discovered leading to one of the ravines from which came faint bird-calls that we at once recognized with a fair degree of certainty. On account of the high altitude and tangled plant-life it was slow, tiring work to follow along the bottom of the jagged gash; there was also the unpleasant possibility of breaking through the matted brush and falling into deep crevices among the rocks.

As we struggled along slowly, high-pitched, whistling calls rang clear and loud from numerous places nearby, but still it seemed as if our efforts might be of no avail; for among that chaos of vegetation it was impossible to move without causing great disturbance and frightening the birds away. Then there recurred to us the old saying about Mahomet and the mountain and we resorted to quiet concealment.

Presently there was a crisp little chirp and a rustle among the mosses a few yards away; one, two, five minutes passed; then a minute, shadowy form darted out of the darkness, perched on a moss-covered boulder, and turned a pair of bright, inquisitive eyes upon the strange monsters which had invaded its snug retreat. The white throat gleamed conspicuously among the deep-green surroundings as the bird paused a moment to complete its inspection; then up went the short, barred tail, straight into the air, and a succession of low, scolding notes emanated from the feathered mite as it hopped about in angry excitement.

We found that the bird existed in numbers; once we had discovered a way of entering its stronghold, it was possible to make the desired studies. Thus our difficult search, covering so many hundreds of miles, came to a pleasant and successful close.

Our work in the Argentine, however, was by no means completed. After a short return trip to our base, we went some distance farther south to Aguilares, a village similar to San Pablo and Acherál. Persimmons and tangerines were in season, and at each station women came to the car-windows offering great bunches of the fruit for sale. The former were most attractive while on the trees; they were as large as a hen's egg, of a deep-red color, and were evenly distributed among the dense, green foliage. The flavor of both was excellent.

Within an hour after reaching Aguilares we had been invited to visit the estate of a wealthy rice-grower named Da Costa, and soon after we were on our way, his son taking us there in a carriage while the luggage went in a cart. At the ranch we found a large, rather dilapidated house occupied by the family of the caretaker. On one side were great rice-fields; on the other, totora marshes, pastures, and woods. The place was most attractive, and the people altogether delightful, so that we spent over two weeks busily engrossed in the abundant work at hand.

The marshes covered many acres and were filled with cattails except for a few narrow lanes of open water. Coypu rats had their runways crisscrossing in every direction—sometimes neat, rounded tunnels with the bottom just under water, and again, wide trails where the vegetation had been trampled down. They look like very large musk-rats and their skins, known commercially as *nutria*, are exported by hundreds of thousands each year to be manufactured into felt hats of the better quality. We caught several that gnawed down all the stalks within reach and piled them into neat islands on which to sit. They feigned death until touched with a stick when they attempted to bite and fought viciously. Jumping mice and large, light-brown, woolly rats used the same paths as their bigger relatives. One afternoon a fine individual of the great red wolf we had secured at Corumbá appeared at the edge of the rushes for a moment only to vanish into the dark marsh at our first movement; a few minutes later he was seen loping into the brush several hundreds of yards away.

Ducks came to the region daily, mostly teals and rosy-bills, but in small numbers only. They were hard to get, as wading in the waist-deep, ice-cold water and mud was slow work and they invariably took warning and left while still out of range. At night flocks of painted snipe (*Rostratula*) ventured to the open borders to feed. While we were quietly waiting, a dusky form appeared and began to probe the mud frantically, to be joined by others in a short time. They skipped about on the flats adjoining the reed-beds in a most erratic manner, reminding one of the actions of water-beetles, and upon the first sign of danger promptly disappeared in the labyrinth of stems and grasses. They seldom took wing, and then it was but to flutter up above the tallest reeds and immediately drop out of sight in the thick cover.

It is to this region of dense totara marshes that the cowbirds revert to spend the winter season, arriving from all directions in comparatively small flocks, but increasing in numbers until there are tens of thousands.

As the rice was ripening about this time, the birds did an enormous amount of damage. All day long, men on horseback rode back and forth through the fields, armed with slings and a bag full of pebbles; they hurled stones and shouted themselves hoarse in a vain endeavor to frighten away the marauding hosts.

The birds, in bands of a few individuals to several hundred, arrived each morning at daybreak, flying low and swiftly, and making a "swishing" sound as they cut through the air. When immediately over the rice-fields, the band would suddenly swerve as if to circle, but drop almost instantly and eat greedily without a moment's delay. Upon seeing a flock approach, the men threw stones and shouted, often succeeding in making it pass straight over or leave the vicinity after circling once or twice. Should the birds alight, the hail of stones soon put them to rout, but not until a few grains of the much-coveted rice had been secured by each individual.

As the day advanced the birds spread out over the surrounding country where they were not persecuted, and spent most of the time on the ground near the cattle and horses, often perched on the backs of the grazing animals. At nightfall they returned to the cattails, and in passing over the rice-fields again took toll from the planters. The flocks in the marshes assumed tremendous proportions, and the babble of voices resembled a rushing wind; the roar of wings, if the masses were suddenly startled by the report of a gun, was not unlike the roll of distant thunder. Before finally settling down for the night they spent some time hopping about on the mud-flats and eating minute animal and vegetable matter.

Carlos S. Reed, F. Z. S., Director of the Natural History Museum, Mendoza, Argentine Republic, gives the results of his investigations as to the food of *Molothrus bonariensis* in a paper in the *Revista Chilena de Historia Natural*, año XVII, No. 3, 1913. The following is a translation, as literal as possible, of a part of the original paper, which is written in Spanish:

> "In the summer of 1910 there occurred in various departments of the Province of Mendoza, a great invasion of *Isocas* (larvæ of a lepidopteran) and in various inspections which I realized in the infected countryside I was able to confirm that a number of birds occupied themselves in eating the larvæ and adults of these *Isocas* (*Colias lesbia* Fabr.) and among them *Molothrus bonariensis* predominated.
>
> "It is also a voracious destroyer of the white worm (larva of *Ligyrus bidentulus* Fairm.) when these are exposed in ploughing furrows in the vineyards. The '*bicho de cesto*' (*Æceticus platensis* Berg) is also very much persecuted by the bird with which we are occupied.
>
> "The corn-fields suffer damages by reason of *Molothrus bonariensis*, but only during the period between the beginning of the ripening of the ear and its collection; certainly, one ought not to take this damage into consideration when, during eleven months, *Molothrus bonariensis* has fed in the cultivated country on other products, not on maize,

and among these has predominated the larva of *Chloridea armigera*, the most formidable enemy of the maize-fields.

"I have examined the stomach contents of more than sixty specimens of *Molothrus bonariensis*, freshly shot, in the various seasons and have encountered about 90 per cent of substance of animal origin and the rest grains, principally maize, but the maize they have generally obtained from the offal of horses and mules, as in Mendoza a good deal of maize is given to working animals, and, as the grain is fed entire, a goodly percentage of it is eliminated without having been digested. It is for this reason that one frequently finds this bird scratching among and turning over the offal. This custom is why it has been given the name of *virabosta* in Brazil. Therefore, *Molothrus bonariensis* may be looked upon as a bird helpful rather than destructive to agriculture."

Rice is planted in "boxes" about twenty-five feet square. Water is supplied through a system of canals some of which are many miles long, and its level is regulated by sets of locks and gates. A few of the fields had already been cut over and the sheaves piled in stacks to dry. Small rodents—rats and mice—were so numerous that they worked great havoc. We ran over our traps thrice daily and always found all of them filled. At dusk short-eared owls came to the vicinity and perched on the mounds from which they could swoop down and capture the mice that teemed in the stubble below. I fired several heavy charges of shot at these birds one evening, and the weather being clear and quiet, the sound carried to the village about a mile and a half away. Early next morning a police sergeant rode up and informed us that we were under arrest. We thanked him for the information, and he left while we went on with our work. At noon another orderly came to repeat the message of the first, and to add that we were expected to report at the police-station immediately. The next day we went to see what all the trouble was about. The "jefe" was waiting for us at the entrance to the jail, surrounded by a curious audience of townspeople. He looked sad, grave, and offended as he began: "*Señores*, I heard five shots night before last." "Yes, *señor*," I interrupted, "I fired at least eight or ten." "*Pues*, that is absolutely prohibited here; one may not shoot under any circumstances whatsoever, so I am compelled to place you in confinement." At this part of the proceedings I merely flashed our permit and asked him why the governor of the province should give out such a document, and charge two *pesos* for it, if one could not hunt under any circumstances. He was taken completely by surprise and did not know what to say, so we wished him good morning and went home, much to the amusement of the crowd which had a good laugh at the *jefe's* expense.

The Argentinians are inveterate drinkers of *maté*. It is taken from a *bombilla*, as in Paraguay, and all classes of people indulge in the habit. I heard that a law had recently been passed requiring each person to use an individual tube as the old system of everybody's using the same one indiscriminately had caused the spread of various diseases, among them cancer of the mouth, at an alarming rate. Our good friends at Los Sarmientos were very fond of their daily brew, and usually took nothing else for breakfast. They at first very generously passed the steaming bowl to us, but soon grew accustomed to our refusals and refrained from extending further invitations to drink.

The weather grew rapidly colder and rain or snow fell almost daily. A mantle of white completely covered the Andes stretching in an unbroken range to the west of us; the picture presented in the early mornings was one of great beauty, as the sun lit up the snowy summits with a rosy light, while a thin bank of purplish vapor enveloped the foot of the range in a soft mantle of regal splendor.

Hunting in the marshes grew most difficult on account of the cold, and the thin ice through which we had to crunch to reach the better collecting-grounds. We therefore decided to seek a friendlier clime, and returned to Tucuman to prepare for a visit to the desert regions of Santiago del Estero.

Chapter 25

VIZCACHA-HUNTING IN AN ARGENTINE DESERT—GIANT SNAKES

Our stay in Tucuman lasted but a few days. During this time our Quechua boy, who had been with us constantly since our first arrival in Cochabamba, spent most of his time at the zoological park. The lions, the tigers, even the camels did not interest him greatly; but the elephant! It was impossible that there could be any such animal. He spent hour after hour seated on the ground silently contemplating the great creature. I wondered what his people would say to him when he returned to them and attempted to describe what he had seen.

As our next efforts were to be directed toward a new province, it was again necessary to secure the very essential permits. This time there was no trouble. At Santiago del Estero, a backward city of small size and not particularly attractive appearance, we were required merely to be photographed and have our finger-prints taken, after which we received certificates stating that we had no police record in that state and were assumed to be respectable and trustworthy; the licenses to hunt were attached. We wasted no time in the city and took the first available train to Suncho Corral, about a five hours' ride.

Suncho Corral is a collection of perhaps fifty adobe shacks, and its inhabitants seemed to be mostly Turks and Syrians. We paid our respects to the local *jefe* without delay and he secured for us permission to camp on the landholding of a friend of his; the place was about a mile distant. We pitched the tents in a delightful grove on the bank of the Rio Salido. All the country is covered with a dense growth of cacti, shrubbery, and tall, thorny trees; it was unlike any we had seen before. There were a few small areas cleared of the indigenous growth and planted in corn, which thrived; water was supplied by irrigation. However, the people, who lived in widely separated huts, seemed to subsist mainly on their flocks of sheep, goats, and the limited number of cattle. There were so many dogs in the neighborhood that they were a plague. Each night numbers prowled about camp, barking, fighting, and trying to tear open bags of provisions. We did not know how to get rid of them without killing them, and this we did not wish to do; but our boy found a way. One night we heard series after series of yelps followed by frantic rushes to distant parts. Next morning we discovered that Antonio had set a dozen large, powerful "rat-killers" around the tent, baited with tempting morsels of meat. When a dog attempted to take the food it received a terrific blow across the nose—hence the yelps. We of course stopped the practice, but the dogs did not return in sufficient numbers to be troublesome.

The water of the Rio Salido is brackish and unfit for drinking. There were few fish—catfish and a species of *pacu*. We had no time for angling, but occasionally saw a string caught by some villager.

About the first bird to attract our attention was a species of wood-hewer with a curved bill three or four inches long. They were always seen in pairs or small flocks, often in company with the very common woodpecker (*Chrysoptilus*). Occasionally there were half a dozen of the former and twenty or more of the latter in a single party, on the ground, feeding on insects and larvæ that lived in the litter of bark and leaves under the giant cactus plants. They formed a noisy group, especially if alarmed, when they took to the trees or cacti and kept up a continuous chirping. They tried to keep on the far side of the trunks and branches, but curiosity prompted them to peep around the edges frequently to see what was going on. The giant wood-hewer (*Xiphocolaptes major*), as large as a mourning-dove and with a long, powerful beak, was far less common. Another bird frequently found in company with any or all of the former was a species of brush-bird the size of a blue jay, but of a brown color; it built stick-nests three feet across that must have weighed up to fifty pounds. We also saw for the first time a bird whose habits reminded us greatly of the road-runner. It ran along the ground with crest erected and tail held high, and was so wary that one could not approach it within shooting distance. But the moment it reached a thicket and hopped up into the branches it lost practically all caution and we could get to within a few yards of it. Perhaps the bird's chief enemies are terrestrial—hence its extremely suspicious nature while on the ground, and the apparent feeling of safety when in a bush or tree.

Next, we again headed for the Chaco, having as our goal a station called Avia Terai, about half-way to Resistencia on the Paraguay River. The train was packed with Italian home-seekers; they were a noisy, quarrelsome lot. Many of them were drunk or ill, and so many unsavory things were occurring constantly in the coaches (there were no compartment-cars), that we remained in the buffet-car. An aged bishop, accompanied by two priests, were fellow passengers. The prelate got off at each stop to bless the crowds that had collected to see him, and then as many as possible knelt to kiss his ring before the train pulled out. After the trio returned to their table, the two priests promptly fell asleep while their venerable superior read from a small prayer-book. I wondered why he tolerated such sleepy, uninteresting companions. At midnight we reached Añatuya and changed to another train. This place was one of wild confusion. There were mountains of luggage piled on the platform, and mobs of excited people rushing wildly about in vain attempts to locate their belongings. I was alarmed over the safety of our own possessions, so stationed the faithful Antonio near the door of the baggage-car with instructions to let me know when unloading began; we then secured *peons* to immediately carry the trunks and bags to our train, thus avoiding their being dumped on the huge piles, and perhaps lost.

In the early morning we reached Quimilí, at which place a siding branches off to Tintina; most of the immigrants went in this direction. The country was all flat and covered with grass. Later on clumps of forest appeared which grew larger and denser as we went farther east. There were numerous stops but no towns of any importance. At 2.30 P. M., the train halted at Avia Terai, and we were soon encamped in the rear of one of the two huts comprising that station.

About all we could see from our abode was an immense area covered with tall weeds, surrounded by dense forest. Sand-flies, called *polvoriños*, filled the air like flecks of dust so that we had to keep a smudge going most of the time. The people said there was a great deal of malaria in the neighborhood, and one look into their faces was ample to substantiate the statement. Usually it was very hot; it rained most of the time, but occasionally the nights were very cold—an altogether disagreeable combination of weather.

One of our trunks, containing all the instruments, had mysteriously disappeared from the baggage-car, so we had only a pocket-knife with which to work; but, by putting in longer hours we managed to keep up to our average daily number in preparing specimens. We gave the conductor of a passing train a tip of several *pesos*, and on his next run he brought us the missing trunk, saying that he had found it at a station a few miles below.

It was impossible to explore the country as thoroughly as we should have liked on account of the almost incessant rain. When the downpour did stop, which was at dusk, flocks of large, white-bellied night-hawks appeared and circled above the grass, catching insects. They were beautiful creatures, and always came back to the same restricted areas to feed on small black beetles that flew up in great numbers from the grass. As darkness settled over the Chaco the flocks suddenly dispersed and they disappeared singly in all directions. We found them spending the days in open places—out in the hot sun or rain. The railroad-track, or small plots where there was not even grass, were the favorite sleeping-sites chosen, and sometimes two or three were found together.

After a week we returned over the route we had come to a station called General Pinedo. This was a new settlement and several dozen board huts were being constructed on both sides of the track. Here there were seemingly limitless stretches of fine pampas with occasional small clumps of red quebracho-woods. Numbers of cattle grazed in the rich grass, and this place was much more attractive than the one we had just left. As might be supposed, the fauna was typical of the open country and included an abundance of short-eared and burrowing owls. The latter sat on fence-posts or on the mounds near their burrows all day long; at night they became very active and flew back and forth over the fields grabbing up beetles and small rodents with their feet. Their long, tremulous screeches pierced the darkness all night long.

On Sunday all the men congregated at the two rum-shops and tested their capacities for strong drink. Often the day ended in a series of brawls when knives and machetes were plied freely—once with fatal result to one of the *compadres*. I asked one of the guards what would be done with the murderer, who had promptly been arrested. He said that if he could give two hundred *pesos* to the *commisario* and ten to each of the guards, the matter would be dropped. Later I was told that the matter had been "fixed up" satisfactorily, but of course could not verify this.

June 14 found us in the village of Lavalle, in the heart of Argentina's desert regions. When the train from Tucuman pulled out, leaving ourselves and our belongings on the station platform, we at once began to regret that we had come at all. The place looked decidedly uninviting. There was only the small cluster of adobe hovels, while all around stretched the cheerless waste of sandy desert. That there could be any considerable amount of wild life in the region seemed impossible; but, as we soon discovered to our unbounded delight, it was only one of the instances where first impressions are deceptive.

Our first care was to find a place where we could put up as we had come prepared to remain a week; so we inquired of the station agent if there was a *posada* in town. He promptly said that there was none. Then we called on the judge, to whom we had a letter of introduction. He took us to the home of a kind-hearted old woman who immediately agreed to give us a room and board; and here let me insert that in no place in all South America were we treated with more courtesy and consideration than in the home of this venerable old woman, during the entire month we finally remained. Learning of our mission, her three daughters became very enthusiastic and plied us with information about the country, and the vast numbers of animals to be found within a short distance of their very doors. They told us that the country was teeming with vizcachas—large rodents that weigh up to twenty-five pounds and come out of their burrows only at night. We wanted to go out and hunt them at once but, unfortunately, there was no moonlight during the first part of our stay, so it was impossible to go in quest of them. We therefore devoted our time looking for other things.

Investigation showed that the country was not quite so barren as it had at first appeared. A short walk took us into a region where there was a dense growth of cacti and thorny shrubbery—so thick in fact that it was almost impossible to get through; many of the former plants were in bloom, the spiny columns being covered with large white, waxy flowers. Here and there a native hut adorned the top of a small rise in the landscape, and nearby we were sure to see the inevitable flock of goats nibbling on the leaves of acacia and mimosa, and guarded by bad-tempered dogs. A little distance away from each hovel was a pond of considerable size; these fill up during the short rainy season and their contents are used to water the stock and to irrigate the small patches of corn and potatoes.

Figure 69. The great crested tinamou.

Everywhere we came across evidences of the animals about which we had heard so much. The country was dotted with huge mounds out of which large tunnels opened. From the mouths of the burrows lead deeply worn paths and in these the ground had been trampled into dust six inches deep. The mounds are built up by the vizcachas, of the earth thrown out of the tunnels, and they take advantage of the hillocks thus created by using them as observation-posts before going far away from their homes. The tops are often strewn with skulls and bones of the large rodents that have died in the burrows and which have been thrown out by the survivors. Burrowing owls sat on the mounds, and swallows flitted in and out of the openings below. There were also the telltale little foot-prints of numerous small animals which appropriated the vizcacha's dwelling for their own use and apparently lived on peaceful terms with it. We wondered how far the tunnels ran underground, and how many species of animals occupied them, but there was nothing to give us a clue to the answer of either conjecture. As the time flew by, however, we learned many things, and one at least was of a startling character.

Figure 70. A burrowing owl.

The days were cold and the sun shone at infrequent intervals. Desirous of taking some photographs, we selected one of the brightest days, and, armed with guns and cameras, we sallied forth. After a time we found a vizcacha mound which was conveniently situated, and walked around it a few times in order to find the best spot from which to take the picture. We noticed nothing unusual about it, and finally set up the camera and began to focus. While thus engaged, with my head under the black cloth, I was suddenly startled by a wild yell from my companion and looked just in time to see him make a long jump to one side. The reason was apparent. There, not three feet away, lay a huge boa emitting a hiss that resembled a jet of escaping steam. Why we had not seen it before is hard to understand, as it lay fully exposed on the bare ground; but probably it was because the great reptile had lain motionless. Now it was slowly crawling, and the broad, mottled back glistened beautifully in the sunlight, with a purple iridescent sheen. We poked, and finally touched it, but as it did not resent these advances we took its picture; then it seemed to grow weary of our attentions and made for the nearest hole, whereupon we shot it. Upon taking the snake to the village the natives told us that they were very abundant and lived down in the burrows with the vizcachas. During the cold season they crawl out at noon for a sun-bath, but are very sluggish. Subsequently, we saw many more, and even kept a number of them alive; they grew tame and friendly almost at once and never attempted to bite.

There are two distinct species, namely: the boa-constrictor, or land-snake; and the anaconda, which spends the greater part of its life in and near water. This latter attains the greater length. A fully grown boa-constrictor does not exceed twelve feet in length; ten or eleven feet is the usual size attained. There is a great difference in the tempers of the two species. A boa soon becomes very tame, and in many places the natives keep them running at large in the huts to catch rats. The anaconda is of a restless disposition and easily irritated. Both will bite if annoyed, and while they are not poisonous, they hold very tight with the strong, curved teeth so that if one tried to pull away from them the flesh would probably be torn to shreds.

Of course it is a well-known fact that snakes are descendants of the lizards; they have lost their legs, but in the boa two good-sized claws are still found on the under-side, near the tail, extending out a little distance from between the plates.

We collected a number of the giant reptiles for their skins. Skinning a boa-constrictor is not an easy undertaking. We always made an incision all along the under-side, from the neck to the end of the tail, and then loosened the skin from the tail end with a knife. This would leave enough of the body exposed for a good hand-hold; after this, one took hold of the body, and the other of the skin; then a real tug-of-war ensued as the skin very slowly peeled off. Sometimes it was necessary to throw a hitch around a tree in order to get a better grip on the body. After the skin was removed, it was scraped and tacked out on the wall and left for a few days to dry; it could then be rolled up and packed for shipment.

The skins tan beautifully, and make very desirable decorations for the mantel for den or library.

Other days we spent hunting tiger-cats, deer, jack-rabbits, rheas (South American ostriches), and others of the animals which were so abundant.

Early morning was the best time for cats. They could then be found in the open paths stalking cavies, with which the country swarmed, or tinamou. They are prettily spotted, and somewhat larger than a house-cat. Upon being seen they pause for a moment to gaze at the intruder, and then vanish into the bushes in a single bound. Small deer with spike-horns are not rare but are hard to get. They hide in the thick cover and can usually hear a person coming far enough away to disappear from the neighborhood without being seen. Rheas travel about in bands but are wary; it is almost impossible to approach them on foot, and they soon learn to regard a man on horseback with suspicion. The natives kill large numbers with rifle and bolas; they eat the flesh and sell the feathers. Three years ago I saw sixty tons of rhea feathers in a single warehouse in Buenos Aires, all of which had been taken from killed birds and were destined to be used in making feather dusters. However, the bird is still abundant. Many large flocks are kept on cattle-ranches. The eggs, the contents of which are equal to a dozen hen's eggs, are sold in the markets during the laying season.

Figure 71. Skinning a boa.

At last the long-awaited time arrived when the full moon lighted up the landscape, so we made preparations to go in pursuit of the wily vizcacha. The judge sent word for us to be ready early one afternoon as he was going to accompany us, and we could spend a few hours beforehand to advantage looking for other things. Two o'clock found us clamoring at his door, and a few minutes later we had started on our excursion.

The judge carried a double-barrelled shotgun of European make; his ten-year-old son, whom he always called the *secretario*, had a "nigger-killer," a large bag full of pebbles and a machete; he was a fine little fellow, always friendly, always polite, and nothing suited him better than to tramp at his father's heels on the long excursions into the country. I had my Parker which had served me so splendidly in many places.

Figure 72. Boa sunning itself at the entrance to a vizcacha burrow.

For an hour or two we tramped broad reaches of cactus desert; but it was silent as the very sphinx, and we saw nothing. However, as the sun began to drop slowly out of sight, things began to stir. At first we heard a shrill turkey-like gobble some distance away, and holding up his hand to command silence, the judge whispered: "*Chuña*; they are right over there. You and the *secretario* go down this little path, and I'll go on this side; *quien sabe?* we might head them off." His fine Spanish face beamed with excitement as he turned away.

We sneaked along for a distance of a hundred yards, and presently I saw a pair of gray forms moving swiftly away underneath the thorny growth. They looked like fleeting shadows, and there was time for a hurried shot only. The *secretario* rushed forward and triumphantly brought back a large, crested, crane-like bird of a uniform gray color, the common name of which is seriema. In some ways the bird resembles a hawk. It lives on the ground and eats grasshoppers, cavies, mice, and almost anything it can catch and swallow; at night it roosts in the trees. Its flesh is excellent. Perhaps no bird is more wary or harder to hunt in this entire region, so we were highly elated with our first shot.

Many birds began to appear now; there were the long-billed brown wood-hewers we had seen at Suncho Corral; Argentine "road-runners" which perked their tails and sped away into the thickets; large brownish *leñadores*, singing on the edges of their huge nests; there were also woodpeckers, hawks, cardinals, and doves.

The judge suggested that we visit one of the reservoirs as we might find ducks there, and calmly floating on the very first one we came to was a small flock of shovellers; they saw us just too late, and one was added to the bag as they rose from the water.

We now cut across a little field from which the corn had been gathered, and here we were kept busy for some time picking off the swift-winged tinamou as they rushed away at our approach. I know of no bird which furnishes better shooting or better eating, and the pity of it is that it does not exist in our own country. After we had shot a number, the judge suggested that we might try for a fox as they would soon be prowling about, so he tied a string to the foot of one of the freshly killed birds and the *secretario* dragged it on the ground after him as he walked along. Some time later we sat down to rest, and much to my surprise a fox appeared on the trail of the bird; as he stopped short, at sight of us, the judge bagged him, and he proved to be the largest and the finest of the dozen or more we succeeded in getting during our entire stay. These foxes, which are of a rich gray color, silver-tipped, spend a great part of their time in dens in the vizcacha burrows, but seem to feed principally on tinamou and other birds.

Cavies without number ran about under the low bushes, and uttered queer little squeaks as they became frightened and dashed into the holes which honeycombed the ground; but of the giant cavy we had not a glimpse until we entered a dry, little gully; there we were just in time to see a fleeing, rabbit-like form, which was added to our assortment.

Figure 73. Oculto, or *Tucotuco*, a rare rodent with mole-like habits.

As the sun set, large flocks of blue-crowned parrakeets flew screaming overhead to seek their sleeping-quarters in the tops of the gnarled, stunted trees; and gray-throated parrakeets hurried to their bulky stick nests to chatter and quarrel before settling for the night. The latter species is an abundant bird in the Chaco of Brazil and Paraguay as well as in the Argentine. In the Argentine its range extends eastward into the province of Tucuman, while it is most plentiful in Santiago del Estero. They are extremely noisy and live and travel in flocks of a dozen to several thousand individuals. Should one approach a tree in which a band is feeding or resting, all chatter is hushed. But the birds crane their necks and noiselessly clamber to points of vantage from which they suspiciously eye the intruder. Then there is a sudden burst of wild screams as the whole colony takes wing and swiftly departs at great speed. They feed largely upon the thistle and on cactus fruits as well as on grain when it is to be had.

Figure 74. Gray fox, abundant in the semiarid regions.

The nests vary in size from those containing not more than an armful of twigs, and occupied by a single pair of birds, to huge structures weighing several hundred pounds and harboring a dozen families. Frequently three or four nests are placed in the same tree, and usually a number of trees in a given area are occupied. The ground beneath the domiciles is strewn with a thick litter of old nesting material that has fallen from the disused bulky masses above.

The nesting cavities are in the under-side of the "apartments," and entrance to them is gained through tubular openings underneath, which prevents opossums from entering them. It is not unusual to find a family of the marsupials living in a cavity in the upper part of the structure, but so strongly are the twigs interlaced that they are unable to tear their way through the thorny mass to the toothsome morsels that are so tantalizingly near. The birds occupy the nests throughout the year and it is rare to find them entirely deserted at any hour of the day. The eggs are white, slightly glossy, and of an oval shape; five to eight comprise a set. With the approach of darkness small birds seemed to disappear among the cacti. The *secretario* kept up a constant fusillade with his sling, but he was a poor shot and did no damage.

Finally the judge suggested supper, so we sat down on a fallen cactus trunk from which the spines had decayed, and enjoyed the bread, sausage, and tangerines which the boy fished out of the bag containing his pebbles and sundry articles; then, in answer to our call, a plump Quechua squaw brought a gourd of water from her near-by hovel; we gave her a cigarette in return, which pleased her so much that she showed us a wonderful vizcacha village not far distant, which, she said, harbored the largest and fattest of the rodents to be found in the district; she also agreed to take charge of our game so that we would not be hampered with it the rest of the evening.

The great silvery moon now began to peep above the cloudless horizon, and in a few minutes the whole country was flooded with light. Not a plant grew on the broad acres the Indian woman had pointed out; there was only the dead stump of a cactus here and there, but these loomed tall and ghostlike in the mellow light. Soon we heard deep, guttural grunts, followed by shrill squeaks, and in a low tone the judge said "vizcachas." Then he dug down into his pockets and produced some beeswax and cotton, so each of us fixed a small fluff on the sight of our guns, and were then ready for business. We had not gone a hundred yards, after this, when the judge pointed to a mound ahead, and there, looming high above the yellow earth, sat some great, restless creature, squeaking and grunting. My companion had explained to me the business of stalking, a score of times, so I set out as directed, making a wide détour in order to get behind a cactus stump; but I am afraid the excitement was too great and I went too fast, for the first thing I knew there was a glimpse of a fleeting, shadowy form, a sharp, shrill squeal, and the mound was bare.

A few minutes later we could make out another animal some little distance away, so the judge went after it; he crept up cautiously, pausing at frequent intervals; then there was a bright flash, followed by a loud report, and we all rushed forward to pick up the first vizcacha. His disappointment was great when he found that he had "potted" a nice little cactus stump.

It was not long before we saw another of the animals. It being my turn, I began to stalk, profiting by past experience. The creature was outlined clearly, and frequently it sat up to look about; then the white breast showed distinctly. When the vizcacha sat up, I stopped; when it dropped down on all fours, I crept on. At forty yards I took the shot, and this time luck was with us. When we reached the spot the animal was tumbling about, and the judge yelled not to touch it, as they can inflict serious wounds with their sharp teeth and claws. At this stage of the game the *secretario* came in for his share of the work; he followed the dancing form in its erratic course, and finally dealt it a blow on the neck with the blunt side of the machete, killing it. It was a splendid specimen, weighing a trifle over eighteen pounds, as we later discovered. It differed from the species found in the high mountains in having a shorter tail and coarse fur, besides being much larger; the appearance of the former had always brought to my mind a combination of a squirrel and a rabbit; this creature was, well, simply a vizcacha; there is nothing else like it. The color is slaty-blue on the back and white underneath.

After that the animals began to appear on all sides as the village was very large and there were numerous mounds; it was therefore a comparatively easy matter to secure the half-dozen we wanted, although I am compelled to admit that each of us shot at least one more stump before the evening was over. If not killed by the first shot, the creatures frequently tumble into their burrows and are lost. The males leave the hiding-places first, and after spending a short time looking about from the top of the mounds, spread out over the surrounding country to feed; the females follow a short time later, and both return at the break of day. On account of their great numbers in some parts of the country, they destroy vast areas of pasturage and are therefore looked upon as vermin. We heard the reports of guns frequently, not far away, indicating that other hunting-parties were out. The flesh of the animal is greatly esteemed by the natives.

Figure 75. Long-tailed vizcacha of the high Andes.

Another of the *secretario's* duties was to carry the game; but this was soon too heavy for him, so we helped. Then he made the discovery that the animals were covered with fleas, ticks, and other parasites, and that this host of unwelcome guests preferred him to the dead creatures he was carrying; we made the same discovery, so hired an Indian to lug the trophies home for us.

While homeward bound we crossed a small open place where not a plant grew, and the sand shimmered with a dull glow. Coming directly for us was a white, plume-like, waving object which could hardly be distinguished from its surroundings, but when both the judge and the *secretario* shouted *Zorino* I knew enough to shoot, and shoot to kill. We waited a moment to see whether the animal was dead, then approaching carefully, I picked up a fine skunk. Just then his mate put in an appearance on the edge of the opening, and there was no choice but to add her to the collection. When it came to carrying home these additions to the bag, even the Indian balked, so I tied them to the end of my gun-barrel and carried them in this manner. Early the next morning the entire town came to see the *Zorinos*; the scent had penetrated into the furthermost hut, and they had unerringly traced it to its source.

Figure 76. Short-tailed vizcacha of the Argentine lowlands.

Few things could be more delightful than the tramp home across the desert; the clear moonlight, the crisp air, and the tremulous wail of an owl, all added enchantment to the night's outing; and, above all, we had had a capital good time, and cemented a friendship, as only a trip of this kind can, with our kindly Argentine host. He is a splendid fellow, a peerless companion; and one of my fondest hopes is that I may some day again tramp the moonlit Argentine deserts in his company.

Chapter 26

The Lake Region of Western Argentina—The Heart of the Wine Country

Inhabitants of the vine-growing districts of Argentina claim that their country produces more wine than California; and, judging by appearances as we entered the Province of San Juan, there seemed to be abundant evidence to support the belief that the yield of grapes is enormous. The soil is sandy and the seepage of snow-water from the mountains is ample to make up for the lack of rainfall.

Many of the vineyards are of great extent. Grapes of numerous varieties are grown, and for size and flavor they are unequalled by any I have ever seen anywhere. Wines of many kinds, and grades, are made, and they are of uniformly excellent quality. Even the champagnes are good. The price at which they sell is low in that part of the country—so low in fact that even the laboring class drink them with their meals. In Buenos Aires they cost as much as the imported article, owing to the fact that freight between San Juan or Mendoza and Buenos Aires equals or exceeds shipping charges from Spain, Portugal, and France.

The city of San Juan reminded us of Salta; perhaps it is not quite so large or up-to-date, but it is nevertheless not unattractive; we spent little time there as we had been invited to a *finca*, where there is a lake of considerable size, to shoot ducks.

One of my ambitions had always been to find a place where ducks and geese were really plentiful—in fact abundant enough to furnish an interesting pastime, observing them under conditions that were not too trying, and where they would also furnish good sport. We had heard of the wonderful shooting on Lake Titicaca, but upon our arrival the season was closed and there was little besides coots and grebes; however, at certain times of the year there is an abundance of water-fowl and sportsmen from La Paz get enviable shooting opportunities.

The marshes along the Cauca River, Colombia, had given better results. Teals, tree-ducks, ruddies, and an occasional pair of big muscovies could always be found; but the ducks were loath to take wing, and going after them in the dense grass and thorny shrubbery growing in the marshes was very trying work.

Then we had reached the rice-growing district around Los Sarmientos.

"Ducks?" they said. "Why, *hombre*, they are bringing them into Tucuman by the thousands. The government is paying a bounty of five cents a head on them as they are destroying all the rice. They are swooping down by the tens of thousands; all the lakes in the south have dried up, so they are coming here. When the flocks rise from the fields, the earth trembles."

That was certainly good news; but when we arrived, the birds had departed for regions unknown.

Leaving San Juan at 6 P. M., we reached a station called Media Agua (half water) two hours later. Our new friend had sent a *peon* to meet us, bringing a wagon; so as soon as we could extricate our luggage from the pile on the station platform, we loaded the vehicle and started on the long drive across the cold, barren country. It was very dark and there was not much of a road anyway, so the wagon jolted along over the rocks or dragged heavily through deep sand. The cold was intense; we wrapped up in heavy Indian blankets, which, however, did not give complete protection from the stinging blasts.

At midnight the driver refused to go any farther and drew up at a lonely hut, where we spent the rest of the night. Early next morning we were off again. We now passed through large irrigated fields where wheat was grown, and also a good deal of maize. Then the desert began again, and from appearances there was not a drop of water within many miles.

We questioned the driver about the lake, and whether there were any *patos* (ducks); but he only shrugged his shoulders and said: "*Quien sabe?*"

Suddenly we saw the shimmer of placid water ahead, and soon drew up at a board shack some little distance from the lake. Our man had told us to take nothing but our guns and ammunition, as his caretaker, who lived in the hut we had just reached, would provide everything else. We took a tent and a few provisions anyway, just to be safe, and it was lucky that we did. Not only had nothing been provided, but the tenant had not even been advised of our coming. He had only one dirty little room, but this he very generously placed at our disposal; however, we preferred to camp outside, although it was bitter cold. His wife consented to do the cooking.

The tent was hastily put up; then we hurried to the lake, leaving the family busily engaged in slaughtering a goat for lunch.

All the surrounding country is a wind-swept desert, there being no trees and but a few thorny bushes. In spots the sand and alkali dust is several feet deep. It seemed impossible that there could be a lake in such a parched-looking locality; but there lay the glistening sheet of water, stretching away into the distance as far as the eye could see. Along the edges were vast, shallow marshes, covering hundreds of acres; in these, sedges grew abundantly, forming shelter and providing a limitless feeding-ground for water-fowl. Half a mile from the bank stood great clumps of *totoras*, or cattails, rearing their tough, slender stems to a height of seven or eight feet above the water. What was infinitely more interesting to us, the whole surface of the lake, from its marshy edge to the rows of *totoras* fading away in the distance, was teeming with water-birds.

There were no boats to be had in the neighborhood, as the natives use reed rafts. They cut quantities of cattails, bind them into long, thick bundles and, lashing several of them together, form a craft that will support a man, although his feet are always under the icy water. Shooting from such a contrivance, unless it is larger than any I have seen, would be impossible. Therefore we started to walk along the muddy banks in the hope that something would fly over.

After having gone a short distance, a commotion among the sedges attracted our attention, and a moment later a large gray fox appeared and trotted away. A charge of No. 4 shot stopped him; he was in splendid fur and made a desirable addition to our lot of trophies. Later, we saw them frequently; they haunt the edges of the marsh and feed upon coots and wounded ducks. Carrion-hawks, also, were always about in considerable numbers and reaped a rich harvest.

Coots of several species were running around everywhere. They wandered far away from the water, apparently to pick up toads or lizards, and as we approached, scurried back to the marsh or hid in the dense, low bushes, where they remained motionless until the cause of their fright had passed. Ducks were all well out of range and refused to fly over. I hesitate to estimate their numbers, there were such countless thousands, but in many places the water was covered with them, and there were large white geese and black-necked swans. Black rails of good size darted about or waded boldly out in the open, jerking their tails and clucking.

It did not take us long to discover that we were too late in the day for ducks, so we started back to camp, cutting across the country. Several tinamous got up singly, with a loud whir of wings; they flew straight and fast, a great contrast to the slow, wavering flight of the forest-inhabiting species.

When we reached camp, some of the goat-meat had been roasted and we had a feast! The rest of the day was spent in straightening up camp.

Our eight by twelve foot "balloon silk" tent had been put up under a shed adjoining the house; this protected it from the wind on at least one side. To be of any use in the tropics, the tent must of course be provided with a ground-cloth and bobbinet curtains; it should also contain a window, screened with netting, in the roof. We did not need the curtains, so tied them back. A brazier was kindled, and after it was filled with glowing embers, it was taken into the tent; it warmed the tent thoroughly within a few moments and kept a fire all night. The window, which was always kept open, served its mission splendidly as a means of ventilation.

The owner of the hut had gone away to look for a boat, and that night returned with one of ample size; but next morning a furious wind was blowing, so hunting was out of the question. The air was so filled with dust that one could not see anything more than a few yards away, and huge waves rolled in from the lake and tore hungrily at the sandy banks. These storms are very common during the winter months and blow up several times a week.

The third day of our visit was beautiful. We pushed the boat out of the tangle of sedges and made straight for the cattails. The birds were stirring, and flock after flock passed overhead. When we paddled quietly into the midst of the green islets, we seemed to enter a new world, filled with surprises and wonderful beyond description. The tall, graceful stems of the totoras swayed gently with the swell made by the passing boat, and cast long shadows in the narrow lanes of glassy water they enclosed. Coots and grebes, like shadows, paddled silently away and lost themselves among the reeds; ruddy ducks popped up here, there, and everywhere, stared a moment, and then dived again with a splash; they seemed to spend a good deal of their time under water, and the fishermen frequently caught them in gill-nets set along the bottom of the lake. The male ruddies were in fine plumage, with deep chestnut backs, white throat-patches, and bright-blue bills; they seldom tried to fly, and then skimmed the water for a few yards only; the ones we shot were so fat that it is hard to understand how they could fly at all. Occasionally we saw a giant grebe. From a distance it resembled a loon; they are fast swimmers and expert divers. Our boatman always begged us to shoot these birds, as the natives are very fond of the flesh and, also, the skin of the breast with its beautiful white, silky feathers, brings a good price in the feather markets. Needless to say, none was shot for this purpose.

Figure 77. Rice-fields at the foot of the Andes Mountains, Sarmientos.

Among the reeds flitted a wonderful little bird, known as the military flycatcher, or "bird of seven colors." It is little larger than a wren, yellow underneath and green above, with the crest and under tail-coverts bright red; there are yellow stripes on the sides of the head and the cheeks are blue; the wings and tail are black. The bird is a sprightly little fellow, flitting and jumping about among the reeds in pursuit of small insects, and uttering its cheerful "*cheeps*" at frequent intervals; it gives a touch of color and dainty life to the sombre green of the vegetation, and to the reflections in the murky water below.

Presently we left the region of the *totoras* and emerged into the open lake. The surface was dotted with ducks, coots, and grebes—a squawking, diving, racing mass that defies description. We made right for the centre of action. The coots always waited until the boat was but a few yards away and then, after giving a few clucks, started to run and flop across the water, leaving a myriad of silvery, rippling paths in their wake, and making the marsh reverberate with the noise. Often this would frighten the ducks, and flocks would jump up all around in such vast numbers that we were lost in admiration watching the wonderful sight of the thousands of swishing, black forms hurtling into the wintry sky.

Our method of hunting was to paddle along slowly, squatting low in the boat until within range of a flock of ducks; then, by standing up suddenly, the flock would be frightened into taking wing, and the individuals we had selected could be picked off. We wanted birds in good plumage only, and this manner of hunting gave us the opportunity of selecting each individual separately. There were shovellers and cinnamon teal without number; the handsome males, in brightest plumage, were dashing around the inconspicuously colored females, swimming low and with bills flat on the water; usually there were not more than a dozen or fifteen in a party. Then there were scaups, tree-ducks, pintails, blackheads, and rosy-bills. The latter were wary; they always passed high above, in large flocks, and the rushing sound made by their wings could be heard a long distance away.

Dabbling in the mud-banks along the edge of the marsh were flocks of from four to thirty large white geese (*Casa-roba*). Black-necked swans, singly or in small groups, sailed about majestically. Of the two birds the geese were the more graceful, and by far the more beautiful. The swans were not very wild, but when the boat approached they began to utter shrill "*kee-wee's*"; finally they would launch into the air with a great deal of flapping, beating the water with powerful strokes of the wings, and keeping up their cry all the while. When we neared a flock of geese, they began to patrol the water ahead, swimming back and forth, and eying us with suspicion; they swam well out of the water, with a graceful carriage of the head and neck, and uttered constant loud, penetrating cries that sounded like "*honk-honk-queenk.*" What is more thrilling than the clear, piercing challenge of this spirit of the wild? Wafted across the watery waste on the wings of a crisp autumn wind, it comes as a message from the regions of snow and ice—a foreboding of the bleak, dark days to follow. I never tired of hearing it, and lost more than one shot at a flock coming over from another direction because I was so interested in listening to the fascinating notes of other birds ahead of us. When they finally decided to take wing, they rose from the water quickly and gracefully, and flew at great speed, stringing out in various formations. They always went far away before again dropping down into the water.

We continued paddling through the centre of the open water to a large mud-flat in search of flamingoes. The natives called them *choflos*, and said that a great many came to this spot each day to feed on the small snails and other mollusks which abound in the shallow places. When still a good distance away we could make out what seemed to be a long row of old piles driven into the centre of the mud-flat. The water had become so shallow that the boat could not proceed, so there was nothing to do but wade, not an altogether pleasant experience, as it was bitter cold and sheets of thin ice floated about everywhere. When we moved, the flamingoes stood stock-still and looked at us; when we stood motionless they lowered their heads, dabbled in the mud, and walked about. From a distance they seemed to be of enormous size, and until we were nearby they appeared coal-black. Finally they became restive, ran back and forth a few steps and then, beating the air with laborious strokes of the wings, flew away. Frequently, on other occasions, they circled around a few times before departing from the locality.

We returned to camp by way of the sedge marshes, although, on account of the bushes and shallow water, poling the boat through the tangle was hard work. In the tops of many of the bushes were immense nests, built of sticks and reed-stems; they apparently belonged to the giant coots, as many of these birds still used them for resting-places; also, nearly all of the platforms were piled with dead frogs which the coots had disembowelled. Our man said that during the months of December and January all the people living near the lagoons camp on the edge of the water and collect eggs; they gather immense numbers and take them to the markets of the neighboring towns to sell.

There were ducks everywhere, feeding or playing among the sedges, and flocks coming from the surrounding sloughs whistled past constantly and plumped down with a splash. Black-headed gulls flew back and forth overhead, and cormorants stood on snags, drying their outstretched wings. To shoot birds under such circumstances would be mere slaughter, and the number one could kill is limited only by the amount of ammunition at hand. The natives kill four or five hundred ducks each day during this season, and have done so for years, but the number of birds does not seem to diminish.

There were also numbers of noisy stilt-sandpipers, storks, and screamers, and occasionally we ran across a pectoral sandpiper which, as at Tafí, was so fat that it did not attempt to fly and could be caught by throwing a hat over it. Lapwings, too, passed over in small bunches, screaming and quarrelling as they went.

Nearly all the ducks were feeding on the small seeds of the water-plants, and were rolling in fat; but on several occasions we ran into small flocks of shovellers and teals which were near the bank and refused to fly; an examination of several of them showed that they were very light and probably diseased.

As we neared the landing, dusk was just enveloping the landscape. Red-breasted meadow-larks sang in the desert, yellow-shouldered blackbirds babbled in the thick reeds, and black ibises in flocks of many thousands were returning from their feeding-grounds miles away, to spend the night in the marshes.

We desired our birds principally for scientific purposes; that is, to prepare the skins for museum specimens, and had shot only a limited number of the best-plumaged individuals of each species; but even then our bag amounted to over half a hundred ducks, a number of geese and swans, and a fairly good collection of coots, grebes, herons, and other birds typical of the vast Argentine lake region.

The preparation of all this material presented a stupendous task. First they were cleaned thoroughly of all spots, then hung up in a safe place, where they remained in good condition on account of the cold. The days that followed were so stormy that outdoor work was impossible, so we were glad to remain in the tent disposing of the work in hand.

When the weather cleared we took other boat-trips through the marshes and out into the lake, but our bag was always limited to things we did not possess or needed for food. The geese were leaving in small flocks to breed in the high Andes, the natives said. Swans also started to drift southward; but still the number of remaining water-fowl, mostly ducks and coots that did not migrate, was incalculable. The water was constantly ruffled by the myriad of moving forms and, at times, the roar of rapidly beating wings reminded us of distant thunder.

The few people living in widely separated hovels around the borders of the lake lead miserable lives. They cultivate small areas in grain, but live mostly on fish, water-birds, and goat's milk. The winter season is most trying. Snow falls infrequently and in small quantities, but the cold is intense. The dust-storms, however, are the real tribulations which render life well-nigh unbearable. They frequently last many days at a time; the fine sand sifts through and into everything and is almost suffocating. One breathes it, eats it, wakes up in the morning covered with a layer, and lives in it continuously as in a thick, brown haze that is most exasperating and invites almost constant profanity, at least in thought. We were glad we visited Media Agua; but we were glad indeed when we found ourselves back in San Juan.

It requires but four hours to reach Mendoza from San Juan by train. This attractive city is really in the heart of the wine country, but the vineyards were almost depleted from the inroads of an insect called the *bicho de cesto*. The vegetation all about was covered with small, ragged cocoons from which the hungry hordes of destructive creatures would emerge in the spring. In places wide areas of weeds had been burned over to destroy the pest while still in the incipient stage; but enough always escaped to undo the work of the few careful growers who attempted to stamp out their enemy of the grape-vines. The slaughter of birds on a vast scale may account for the increase of the *bicho de cesto*. We saw vendors on the streets carrying baskets full of small birds of several species—mostly sparrows—which they sold by the dozen. The number killed weekly must run into the thousands. As a natural result of this wholesale killing, birds are not plentiful in the environs of Mendoza.

From the outskirts of the city one has a superb view of the Andean Range. The lofty mountains extend in an unbroken, snow-capped line as far as the eye can see. Aconcagua, the peer of the Argentine Andes, may be seen from a point several miles south of Mendoza, lording over his lesser satellites in a majestic, awe-inspiring way. The shifting mists, cloud-banks, and intermittent sunlight playing on the white peaks present an ever-varying series of pictures that are unexcelled for beauty and grandeur.

At Mendoza we met an Italian who claimed to be the champion condor-hunter of all South America. During his ten years of collecting he had killed more than sixteen thousand of the magnificent birds. His record for one day was one hundred and fourteen. Naturally, they had become greatly reduced in numbers, for the condor lays but a single egg and it takes many months to rear the young. His method was to drive a burro to some lonely gorge among the bleak mountain-tops favored by the birds, and then to kill the animal. He was very particular in stating that the burro had to be fat—a poor one would not do for bait. He then spread nets about the carcass, and when the condors gathered about to feast he pulled a rope and ensnared them; on one occasion he trapped sixty-seven at one throw of the net. The prisoners were despatched with a club and the long wing-feathers extracted to be exported to France to decorate women's hats. Formerly he had received about twenty *pesos* per bird. With his accumulated wealth he built a powder-mill; this promptly blew up, so he was again practically penniless. Of course there were still condors in the mountains—in fact, he knew of a ledge where upward of eight hundred congregated to spend the nights, but the price of feathers had gone down fifty per cent on account of the war. He ended his speech in a very dramatic manner: "What," he said, "me go out and slaughter such a wonderful, magnificent, and rare bird as the condor *for ten pesos each*? No, señor! Not me."

About the only animal that was abundant near Mendoza was the jack-rabbit, introduced into the Argentine some forty years ago. It has increased to such an extent as to be harmful, and has spread over the entire southern part of the plains country. Many are killed and sold in the markets under the name *liebre*.

We met Doctor Chapman at Mendoza. He had come from Chile over the Trans-Andean Railroad. A wire had been sent us to join him at Santiago, but it arrived three weeks too late to be of any service. After a few days spent in taking photographs of the country and collecting accessories for a habitat group of the rhea, we started eastward to Buenos Aires.

We left Mendoza at one o'clock P. M., September 3. At first there was a seemingly endless succession of vineyards; then a vast expanse of arid country more barren even than the desert of Santiago del Estero. At midnight we left the parched plains and entered the fertile wheat and grazing lands which constitute one of Argentina's chief sources of wealth and justly entitle that country to rank among the producing and great nations of the New World. Commodious ranch-houses standing in fields where thousands of head of live stock grazed were passed in steady succession. In some of the pastures hundreds of half-tamed rheas fed unconcernedly among the horses and cattle. Frequently we saw flocks of snowy gulls following a plough or resting in a bunch on the ground; lapwings circled about with angry screams, and ducks swam unconcernedly in the little sloughs beside the railroad. There were also rows of solemn, sedate storks, gravely contemplating the train as it passed, and flamingoes dabbling for mollusks in shallow pools.

After a continuous ride of twenty-five hours we reached Buenos Aires, and two weeks later the *Amazon* of the Royal Mail Line was speeding us homeward.

I am writing these last few pages in an aviation concentration-camp awaiting orders to go to new lands, and new and possibly far more exciting experiences; but almost daily my thoughts go back to the great wonderland that lies south of us, and which I have learned to love. Speed the day when I may again eagerly scan the horizon for a first, faint tinge of its palm-fringed shore-line!

Figure 78. Routes taken by the author in his South American explorations.

INDEX

A

B

C

D

E

F

G

H

I

J

K

L

M

N

O

P

Q

R

S

T

U

V

W

Y

Z